Canadian Mathematical Society
Société mathématique du Canada

Editors-in-Chief
Rédacteurs-en-chef
Jonathan Borwein
Peter Borwein

Springer
New York
Berlin
Heidelberg
Barcelona
Hong Kong
London
Milan
Paris
Singapore
Tokyo

CMS Books in Mathematics
Ouvrages de mathématiques de la SMC

Frank Deutsch

Best Approximation in Inner Product Spaces

Springer

Frank Deutsch
Department of Mathematics
Pennsylvania State University
417 McAllister Bldg.
University Park, PA 16802-6401
USA
deutsch@math.psu.edu

Editors-in-Chief
Rédacteurs-en-chef
Jonathan Borwein
Peter Borwein
Centre for Experimental and Constructive Mathematics
Department of Mathematics and Statistics
Simon Fraser University
Burnaby, British Columbia V5A 1S6
Canada

Mathematics Subject Classification (2000): 41A50, 41A65

Library of Congress Cataloging-in-Publication Data
Deutsch, F. (Frank), 1936–
    Best approximation in inner product spaces / Frank Deutsch.
        p. cm. — (CMS books in mathematics ; 7)
    Includes bibliographical references and index.

    1. Inner product spaces.   2. Approximation theory.   I. Title.   II. Series.
QA322.4 .D48 2001
515'.733—dc21   00-047092

Printed on acid-free paper.

Photocomposed copy prepared using the author's $\mathcal{A}_{\mathcal{M}}\mathcal{S}$-TEX files.

Printed in the United States of America.

9 8 7 6 5 4 3 2 1

ISBN 978-1-4419-2890-0

Springer-Verlag   New York Berlin Heidelberg
A member of BertelsmannSpringer Science+Business Media GmbH

To Mary: a wonderful wife, mother, and mema

An approximate answer to the right problem is worth a good deal more than an exact answer to an approximate problem.

—John Tukey

What is written without effort is in general read without pleasure.

—Samuel Johnson

PREFACE

PREFACE

This book evolved from notes originally developed for a graduate course, "Best Approximation in Normed Linear Spaces," that I began giving at Penn State University more than 25 years ago. It soon became evident that many of the students who wanted to take the course (including engineers, computer scientists, and statisticians, as well as mathematicians) did not have the necessary prerequisites such as a working knowledge of L_p-spaces and some basic functional analysis. (Today such material is typically contained in the first-year graduate course in analysis.) To accommodate these students, I usually ended up spending nearly half the course on these prerequisites, and the last half was devoted to the "best approximation" part. I did this a few times and determined that it was not satisfactory: Too much time was being spent on the presumed prerequisites. To be able to devote most of the course to "best approximation," I decided to concentrate on the simplest of the normed linear spaces—the inner product spaces—since the theory in inner product spaces can be taught from first principles in much less time, and also since one can give a convincing argument that inner product spaces are the *most important* of all the normed linear spaces anyway. The success of this approach turned out to be even better than I had originally anticipated: *One can develop a fairly complete theory of best approximation in inner product spaces from first principles*, and such was my purpose in writing this book.

Because of the rich geometry that is inherent in inner product spaces, most of the fundamental results have simple *geometric* interpretations. That is, one can "draw pictures," and this makes the theorems easier to understand and recall. Several of these pictures are scattered throughout the book. This geometry also suggests the important role played by "duality" in the theory of best approximation. For example, in the Euclidean plane, it is very easy to convince yourself (draw a picture!) that the distance from a point to a convex set is the maximum of the distances from the point to all lines that separate the point from the convex set. This suggests the conjecture (by extrapolating to *any* inner product space) that the distance from a point to a convex set is the maximum of the distances from the point to "hyperplanes" that separate the point from the set. In fact, this conjecture is true (see Theorem 6.25)! Moreover, when this result is formulated analytically, it states that a certain minimization problem in an inner product space has an equivalent formulation as a maximization problem in the dual space. This is a classic example of the role played by duality in approximation theory.

Briefly, this is a book about the "theory and application of best approximation in inner product spaces" and, in particular, in "Hilbert space" (i.e., a complete inner product space). In this book geometric considerations play a prominent role in developing and understanding the theory. The only prerequisites for reading it are some "advanced calculus" and a little "linear algebra," where, for the latter subject, a first course is more than sufficient. Every author knows that it is impossible to write a book that proceeds at the correct pace for every reader. It will invariably be too slow for some and too fast for others. In writing this book I have tried to err on the side of including too much detail, rather than too little, especially in the early chapters, so that the book might prove valuable for self-study as well. It is my hope that the book proves to be useful to mathematicians, statisticians, engineers, computer scientists, and to any others who need to use best approximation principles in their work.

What do we mean by "best approximation in inner product spaces"? To explain this, let X be an inner product space (the simplest model is Euclidean n-space) and let K be a nonempty subset of X. An element x_0 in K is called a *best approximation* to x from K if x_0 is closest to x from among all the elements of K. That is, $\|x - x_0\| = \inf\{\|x - y\| \mid y \in K\}$. The theory of best approximation is mainly concerned with the following fundamental questions: (1) *Existence of best approximations*: When is it true that each x in X has a best approximation in K? (2) *Uniqueness of best approximations*: When is it true that each x in X has a unique best approximation in K? (3) *Characterization of best approximations*: How does one recognize which elements in K are best approximations to x? (4) *Continuity of best approximations*: How do best approximations in K to x vary as a function of x? (5) *Computation of best approximations*: What algorithms are available for the actual computation of best approximations? (6) *Error of approximation*: Can one compute the *distance* from x to K, i.e., $\inf\{\|x - y\| \mid y \in K\}$, or at least get good upper and/or lower bounds on this distance?

These are just a few of the more basic questions that one can ask concerning best approximation. In this book we have attempted to answer these questions, among others, in a systematic way. Typically, one or more general theorems valid in any inner product space are first proved, and this is then followed by deducing specific applications or examples of these theorems. The theory is the richest and most complete when the set K is a closed and *convex* set, and this is the situation that the bulk of the book is concerned with. It is well known, for example, that if K is a (nonempty) closed convex subset of a Hilbert space X, then each point x in X has a unique best approximation $P_K(x)$ in K. Sets that always have unique best approximations to each point in the space are called *Chebyshev sets*. Perhaps the major unsolved problem in (abstract) best approximation theory today is whether or not the converse is true. That is, must every Chebyshev set in a Hilbert space be convex? If X is finite-dimensional, the answer is yes. Much of what is known about this question is assembled in Chapter 12.

The book is organized as follows. In Chapter 1, the motivation for studying best approximation in inner product spaces is provided by listing five basic problems. These problems all appear quite different on the surface. But after defining inner product spaces and Hilbert spaces, and giving several examples of such spaces, we observe that each of these five problems is a *special case of the problem of best approximation from a certain convex subset of a particular inner product space*. The general problem of best approximation is discussed in Chapter 2. Existence and uniqueness theorems for best approximations are given in Chapter 3. In Chapter 4 a characterization of best approximations from convex sets is given, along with several improvements and refinements when the convex set K has a special form (e.g., a convex cone or a linear subspace). When K is a Chebyshev set, the mapping $x \mapsto P_K(x)$ that associates to each x in X its unique best approximation in K is called the *metric projection* onto K. In Chapter 5 a thorough study is made of the metric projection. In particular, P_K is always nonexpansive and, if K is a linear subspace, P_K is just the so-called orthogonal projection onto K. In Chapter 6 the bounded linear functionals on an inner product space are studied. Representation theorems for such functionals are given, and these are applied to deduce detailed results in approximation by hyperplanes or half-spaces. In Chapter 7 a general duality theorem for the error of approximation is given. A new elementary proof of the Weierstrass approximation theorem is established, and explicit formulas are obtained for the distance from any monomial to a polynomial subspace. These are then used to establish an elegant approximation theorem of Müntz.

In Chapter 8 we seek solutions to the operator equation $Ax = b$, where A is a bounded linear operator from the inner product space X to the inner product space Y, and $b \in Y$. If X and Y are Euclidean n-space and Euclidean m-space respectively, then this operator equation reduces to m *linear equations in n unknowns*. To include the possible situation when no solution x exists, we reformulate the problem as follows: Minimize $\|Ax - b\|$ over all x in X. This study gives rise to the notions of "generalized solutions" of equations, "generalized inverses" of operators, etc.

The theory of Dykstra's cyclic projections algorithm is developed in Chapter 9. This algorithm is an iterative scheme for computing best approximations from the intersection $K = \cap_1^m K_i$, where each of the sets K_i is a closed convex set, in terms of computing best approximations from the *individual* sets K_i. When all the K_i are linear subspaces, this algorithm is called von Neumann's method of alternating projections. It is now known that there are at least fifteen different areas of mathematics for which this algorithm has proven useful. They include linear equations and linear inequalities, linear prediction theory, linear regression, computed tomography, and image restoration. The chapter concludes with several representative applications of the algorithm.

In Chapter 10 we consider the problem of best approximation from a set of the type $K = C \cap A^{-1}(b)$, where C is a closed convex set, A is a bounded linear mapping from X to Y, and $b \in Y$. This problem includes as a special case the general shape-preserving interpolation problem, which is another one of the more important applications of best approximation theory. It is shown that one can always replace the problem of determining best approximations to x from K by the (generally easier) problem of determining best approximations to a certain perturbation of x from the set C (or from a specified extremal subset of C).

In Chapter 11 the general problem of (finite) interpolation is considered. Also studied are the problems of simultaneous approximation and interpolation, simultaneous interpolation and norm-preservation, simultaneous approximation, interpolation, and norm-preservation, and the relationship among these properties. The last chapter (Chapter 12) examines the question of whether every Chebyshev set in a Hilbert space must be convex.

Throughout each chapter, examples and applications have been interspersed with the theoretical results. At the end of each chapter there are numerous exercises. They vary in difficulty from the almost trivial to the rather challenging. I have also occasionally given as exercises certain results that are proved in later chapters. My purpose in doing this is to allow the students to discover the proofs of some of these important results for themselves, since for each of these exercises all the necessary machinery is already at hand. Following each set of exercises there is a section called "Historical Notes," in which I have tried to put the results of that chapter into a historical perspective. The absence of a citation for a particular result should not, however, be interpreted as a claim that the result is new or my own. While I believe that some of the material included in this book is new, it was not always possible for me to determine just who proved what or when.

Much of the material of the book has not yet appeared in book form, for example, most of Chapters 9–12. Surprisingly, I have not seen even some of the more basic material on inner product spaces in any of the multitude of books on Hilbert space theory. As an example, the well known Fréchet–Riesz representation theorem states that *every* bounded linear functional on a Hilbert space has a "representer" in the space (Theorem 6.10), and this result *is* included in just about every book on Hilbert space theory. In contrast to this, not every bounded linear functional on an (incomplete) inner product space has such a representation. Nevertheless, one can specify *exactly* which functionals do have such representations (Theorem 6.12), and this condition is especially useful in approximation by hyperplanes (Theorem 6.17) or half-spaces (Theorem 6.31).

If we had worked only in *Hilbert* spaces, parts of the book could have been shortened and simplified. The lengthy Chapter 6, for example, could have been considerably abbreviated. Thus we feel an obligation to explain why we spent the extra time, effort, and pages to develop the theory in arbitrary inner product spaces. In many applications of the theory of best approximation the natural space to work in is a space of *continuous* or *piecewise continuous* functions defined on an interval, and endowed with the L_2-norm (i.e., the norm of a function is the square root of the integral of the square of the function). Such spaces are inner product spaces that are not Hilbert spaces. Moreover, for a variety of reasons, it is not always satisfactory to work in the Hilbert space *completion* of the given inner product space. For example, a certain physical problem might demand that a solution, if any exists, be a piecewise continuous function. Thus it seemed important to me to develop the theory of best approximation in *any inner product space* and not just in a Hilbert space.

In what way does this book differ from the collection of books available today on approximation theory? In most of these books, if any best approximation theory in a Hilbert space setting was included, the sum total rarely amounted to more than one or two chapters. Here we have attempted to produce the first systematic study of best approximation theory in inner product spaces, and without elaborate prerequisites. While the choice of topics that one includes in a book is always a subjective matter, we have tried to include a fairly complete study of the *fundamental* questions concerning best approximation in inner product spaces.

Each of the first six chapters of the book depends on the material of the chapter preceding it. However (with the exception of Chapters 9 and 10, each of which depends upon a few basic facts concerning adjoint operators given in Chapter 8, specifically, Theorems 8.25–8.33), each of the remaining six chapters (7–12) depends only on the first six. In particular, Chapters 7–12 are essentially independent of one another, and any one of these may be studied after one has digested the first six chapters. The reader already familiar with the basic fundamentals of Hilbert space theory can skim the first six chapters to obtain the relevant notation and terminology, and then delve immediately into any of the remaining chapters. My own experience with teaching courses based on this book has shown that in a one-semester course it is possible to cover essentially all of the first seven chapters and at least two or three of the remaining ones. One way of reducing the amount of time needed to cover the material would be to concentrate on just the Hilbert space case, and to omit the results that require more involved technicalities necessitated by the possible incompleteness of the space. Also, for those students whose background already includes an elementary course in Hilbert space theory, the instructor should be able to cover the whole book in one semester. Actually, one can even regard this book as *an introduction to the theory of inner product spaces* developed simultaneously with an important application in mind.

Finally, for the sake of completeness and in deference to the experts, we have included some examples and exercises that *do* require general measure and integration theory (which are *not* prerequisites for this book). Such examples and exercises are clearly marked with a ★ and can be skipped without any loss of continuity.

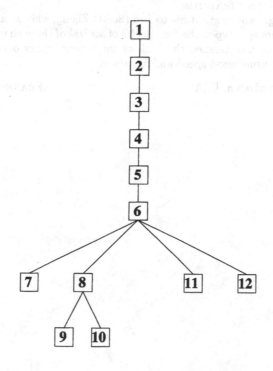

CHAPTER INTERDEPENDENCY*

***A chapter is dependent on every chapter to which it is
connected by a sequence of rising lines.**

Acknowledgments

I have been fortunate to have had the sound advice of many of my students over the years that I have taught the topics in this book. They have helped me to see how to present the material in a way that is most beneficial to a student's understanding, and they forced me to clarify some arguments that were originally somewhat opaque.

I am grateful to the following people, each of whom read certain parts of earlier drafts of the book and offered valuable constructive criticisms: Heinz Bauschke, Ward Cheney, Charles Groetsch, Hein Hundal, Vic Klee, Wu Li, Daniel Murphy, Adrian Ocneanu, Bob Phelps, Allan Pinkus, Ivan Singer, Don Solmon, Leonid Vlasov, Joe Ward, and Isao Yamada. Moreover, Heinz Bauschke read the entire manuscript, offered many useful critical comments, and corrected numerous typos. In short, he was an editor's dream.

I am also indebted to Bob Seeds and Frances Weller of the Penn State Mathematics Library for their assistance in so many ways related to reference materials. In addition, I am grateful to Doug Arnold who bailed this TeX novice out of some serious difficulties with his "TeXpertise."

Finally, I owe a large debt of gratitude to Jun (Scott) Zhong, who, as a graduate student about ten years ago, typed the first drafts of several of these chapters, and to Kathy Wyland, who has handled the bulk of the typing chores over the last several years with her usual good speed and accuracy.

University Park, Pennsylvania, USA   FRANK DEUTSCH

CONTENTS

CHAPTER 1

INNER PRODUCT SPACES

Five Basic Problems

To motivate the subject matter of this book, we begin this chapter by listing five basic problems that arise in various applications of "least-squares" approximation. While these problems seem to be quite different on the surface, we will later see that the first three (respectively the fourth and fifth) are special cases of the general problem of *best approximation in an inner product space by elements of a finite-dimensional subspace* (respectively *convex set*). In this latter formulation, the problem has a rather simple geometric interpretation: A certain vector must be orthogonal to the linear subspace.

The remainder of the chapter is devoted to defining the main spaces of interest: the inner product spaces and Hilbert spaces, giving some examples of these spaces, and recording a few of their elementary, but useful, properties.

Problem 1. (Best least-squares polynomial approximation to data) Let $\{(t_j, x(t_j)) \mid j = 1, 2, \ldots, m\}$ be a table of data (i.e., the graph of a real function x defined on the t_j's). For any fixed integer $n < m$, find a polynomial $p(t) = \sum_{i=0}^{n} \alpha_i t^i$, of degree at most n, so that the expression

$$\sum_{k=1}^{m} [x(t_k) - p(t_k)]^2$$

is minimized.

Problem 2. (Solution to an over-determined system of equations) Consider the linear system of m equations in the n unknowns $x_1, x_2, \ldots, x_n$:

$$a_{11}x_1 + a_{12}x_2 + \cdots + a_{1n}x_n = b_1$$
$$a_{21}x_1 + a_{22}x_2 + \cdots + a_{2n}x_n = b_2$$
$$\cdots$$
$$a_{m1}x_1 + a_{m2}x_2 + \cdots + a_{mn}x_n = b_m.$$

In matrix–vector notation, this can be written as $Ax = b$. In the absence of further restrictions on A or b, this system may fail to have a solution x. The next best thing we can ask for is to make the residual vector $r := Ax - b$ "small" in some sense. This suggests the following problem: Find a vector $x = (x_1, x_2, \ldots, x_n)$ that minimizes the expression

$$\sum_{1}^{m} r_i^2 = \sum_{i=1}^{m} \left(\sum_{j=1}^{n} a_{ij}x_j - b_i \right)^2.$$

Problem 3. (Best least-squares polynomial approximation to a function) Let x be a real continuous function on the interval $[a, b]$. Find a polynomial $p(t) = \sum_{i=0}^{n} \alpha_i t^i$, of degree at most n, such that the expression

$$\int_a^b [x(t) - p(t)]^2 dt$$

is minimized.

Problem 4. (A control problem) The position θ of the shaft of a dc motor driven by a variable current source u is governed by the differential equation

$$(4.1) \qquad \theta''(t) + \theta'(t) = u(t), \qquad \theta(0) = \theta'(0) = 0,$$

where $u(t)$ is the field current at time t. Suppose that the boundary conditions are given by

$$(4.2) \qquad \theta(1) = 1, \quad \theta'(1) = 0,$$

and the energy is proportional to $\int_0^1 u^2(t) dt$. Find the function u having minimum energy in the class of all real continuous functions on $[0, 1]$ for which the system (4.1) and (4.2) has a solution θ.

Problem 5. (Positive constrained interpolation) Let $\{x_1, x_2, \ldots, x_n\}$ be a finite set of real square-integrable functions on $[0, 1]$ and let $\{b_1, b_2, \ldots, b_n\}$ be real numbers. In the class of all square-integrable functions x on $[0, 1]$ such that

$$x(t) \geq 0 \quad \text{for} \quad t \in [0, 1]$$

and

$$\int_0^1 x(t) x_i(t) dt = b_i \quad (i = 1, 2, \ldots, n),$$

find the one for which

$$\int_0^1 x^2(t) dt$$

is minimized.

Inner Product Spaces

The natural setting for these and similar problems is a real inner product space. Let X be a real linear space, that is, a linear space over the field $\mathbb{R}$ of real numbers.

1.1 Definition. *A real linear space X is called an **inner product space** if for each pair of elements x, y in X there is defined a real scalar $\langle x, y \rangle$ having the following properties (for every x, y, z in X and $\alpha \in \mathbb{R}$):*
  (1) $\langle x, x \rangle \geq 0$,
  (2) $\langle x, x \rangle = 0$ *if and only if* $x = 0$,
  (3) $\langle x, y \rangle = \langle y, x \rangle$,
  (4) $\langle \alpha x, y \rangle = \alpha \langle x, y \rangle$,
  (5) $\langle x + y, z \rangle = \langle x, z \rangle + \langle y, z \rangle$.

The function $\langle \cdot, \cdot \rangle$ on $X \times X$ is called an **inner product** (or scalar product) on X. Properties (1) and (2) state that the inner product is "positive definite," and property (3) says that it is "symmetric."

From (4), (5), and induction, we obtain that

$$(1.1.1) \qquad \left\langle \sum_1^n \alpha_i x_i, y \right\rangle = \sum_1^n \alpha_i \langle x_i, y \rangle .$$

From (3) and (1.1.1) we deduce

$$(1.1.2) \qquad \left\langle x, \sum_1^n \alpha_i y_i \right\rangle = \sum_1^n \alpha_i \langle x, y_i \rangle .$$

Properties (1.1.1) and (1.1.2) state that the inner product is "bilinear;" i.e., linear in each of its arguments. An immediate consequence of (4) is

$$(1.1.3) \qquad \langle 0, y \rangle = 0 \quad \text{for every} \quad y \in X.$$

Also, from property (2), we immediately deduce

$$(1.1.4) \qquad \langle x, y \rangle = 0 \quad \text{for every} \quad x \in X \quad \text{implies that} \quad y = 0.$$

Inner product spaces are sometimes called **pre-Hilbert spaces**.

Unless explicitly stated otherwise, **we shall henceforth assume throughout the entire book that X is a real inner product space!**

Before giving examples of inner product spaces, it is convenient to state a fundamental inequality. For each $x \in X$, the **norm** of x is defined by

$$\|x\| := \sqrt{\langle x, x \rangle} .$$

Theorem 1.2 (Schwarz Inequality). *For any pair x, y in X,*

$$(1.2.1) \qquad |\langle x, y \rangle| \le \|x\|\, \|y\| .$$

Moreover, equality holds in this inequality if and only if x and y are linearly dependent.

Proof. If x and y are linearly dependent, say $x = \alpha y$ for some scalar α, then both sides of (1.2.1) equal $|\alpha|\, \|y\|^2$. If x and y are linearly independent, then for every scalar λ we see that $x - \lambda y \ne 0$, so that

$$0 < \langle x - \lambda y,\ x - \lambda y \rangle = \|x\|^2 - 2\lambda \langle x, y \rangle + \lambda^2 \|y\|^2.$$

In particular, setting $\lambda = \langle x, y \rangle \|y\|^{-2}$ yields

$$0 < \|x\|^2 - |\langle x, y \rangle|^2 \|y\|^{-2},$$

and hence strict inequality holds in (1.2.1). ∎

1.3 Examples of Inner Product Spaces. All of the examples of linear spaces that are considered below are spaces of real functions x defined on a certain domain. The operations of addition and scalar multiplication of functions are defined pointwise:

$$(x + y)(t) := x(t) + y(t), \quad (\alpha x)(t) := \alpha x(t).$$

Let T be any set and let $\mathcal{F}(T)$ denote the set of all real functions defined on T. It is easy to verify that $\mathcal{F}(T)$ is a linear space with zero element the function identically 0. Thus, in each example below, to verify that X is a linear space is equivalent to verifying that X is a (linear) *subspace* of $\mathcal{F}(T)$ (for some T). And for this it suffices to verify that $x + y$ and αx are in X whenever x and y are in X and $\alpha \in \mathbb{R}$.

(1) For any positive integer n, $l_2(n)$ will denote the functions $x : \{1, 2, \ldots, n\} \to \mathbb{R}$ with the inner product

$$\langle x, y \rangle = \sum_{1}^{n} x(i)y(i).$$

The inner product space $l_2(n)$ is called **Euclidean n-space**. The norm is given by

$$\|x\| = \left[\sum_{1}^{n} |x(i)|^2 \right]^{\frac{1}{2}}.$$

Observe that there is a one-to-one correspondence between each $x \in l_2(n)$ and its ordered array of values

$$(x(1), x(2), \ldots, x(n)) \in \mathbb{R}^n.$$

Thus we may regard the space $l_2(n)$ and $\mathbb{R}^n$ as one and the same. This identification of $l_2(n)$ with $\mathbb{R}^n$ will aid our geometric intuition. (We can "draw pictures" in $\mathbb{R}^n$ for $n = 1, 2,$ or 3!) More often than not, this geometric intuition in $\mathbb{R}^n$ will enable us to conjecture results that are true in *any* inner product space.

(2) l_2 will denote the space of all real-valued functions x defined on the *natural numbers* $\mathbb{N} := \{1, 2, \ldots, n, \ldots\}$ with the property that $\sum_{1}^{\infty} x^2(i) < \infty$. If we define

(1.3.1) $$\langle x, y \rangle = \sum_{1}^{\infty} x(i)y(i),$$

then l_2 is an inner product space. To see this, first note that by the Schwarz inequality in $l_2(n)$, we have, for each x, y in l_2,

$$\sum_{1}^{n} |x(i)y(i)| \leq \left[\sum_{1}^{n} x^2(i) \right]^{\frac{1}{2}} \left[\sum_{1}^{n} y^2(i) \right]^{\frac{1}{2}} \leq \left[\sum_{1}^{\infty} x^2(i) \right]^{\frac{1}{2}} \left[\sum_{1}^{\infty} y^2(i) \right]^{\frac{1}{2}} =: c$$

for all n. Hence, by passing to the limit as n tends to infinity,

$$\sum_{1}^{\infty} |x(i)y(i)| \leq c < \infty,$$

so $\sum_1^\infty x(i)y(i)$ converges absolutely, and (1.3.1) is well-defined. The remaining inner product properties are easily verified.

To verify that l_2 is actually a linear space, it suffices to show that $x + y$ and αx are in l_2 whenever x and y are in l_2 and $\alpha \in \mathbb{R}$. Clearly, $\alpha x \in l_2$ if $x \in l_2$. If $x, y \in l_2$, then for every $n \in \mathbb{N}$, again using Schwarz's inequality in $l_2(n)$,

$$\sum_1^n [x(i) + y(i)]^2 = \sum_1^n x^2(i) + 2\sum_1^n x(i)y(i) + \sum_1^n y^2(i)$$

$$\leq \sum_1^n x^2(i) + 2\left[\sum_1^n x^2(i)\right]^{\frac{1}{2}}\left[\sum_1^n y^2(i)\right]^{\frac{1}{2}} + \sum_1^n y^2(i)$$

$$\leq \sum_1^\infty x^2(i) + 2\left[\sum_1^\infty x^2(i)\right]^{\frac{1}{2}}\left[\sum_1^\infty y^2(i)\right]^{\frac{1}{2}} + \sum_1^\infty y^2(i) =: c.$$

Thus

$$\sum_1^\infty [x(i) + y(i)]^2 \leq c < \infty,$$

and so $x + y \in l_2$. Thus l_2 is an inner product space. The norm in l_2 is given by

$$\|x\| = \left[\sum_1^\infty x^2(i)\right]^{\frac{1}{2}}.$$

(3) Let $C_2[a, b]$ denote the space of all continuous real functions x on the finite interval $[a, b]$, and define

$$\langle x, y \rangle = \int_a^b x(t)y(t)dt.$$

It is easy to verify that $C_2[a, b]$ is an inner product space. The norm is given by

$$\|x\| = \left[\int_a^b x^2(t)dt\right]^{\frac{1}{2}}.$$

(4) We consider a more general inner product space than that of example (3). Let I be any closed interval of the real line $\mathbb{R}$. Thus I is one of the four intervals $[a, b]$, $(-\infty, b]$, $[a, \infty)$, or $\mathbb{R} = (-\infty, \infty)$. Let w be any function that is positive and continuous on the interior of I and is integrable:

$$\int_I w(t)dt < \infty.$$

Let $C_2(I; w)$ denote the set of all bounded continuous real functions on I with an inner product defined by

$$\langle x, y \rangle = \int_I w(t)x(t)y(t)\, dt.$$

In the same way one can show that $C_2(I; w)$ is an *inner product space*. The norm is thus given by

$$\|x\| = \left[\int_I w(t) x^2(t) \, dt \right]^{\frac{1}{2}}.$$

Note that in the special case where $w(t) \equiv 1$ and $I = [a, b]$, $C_2(I; w)$ is just the space we denoted by $C_2[a, b]$ in (3).

★ (5) A somewhat larger space than $C_2[a, b]$ is the space of all (Lebesgue) measurable real functions x on $[a, b]$ with the property that (the Lebesgue integral) $\int_a^b x^2(t) dt < \infty$. If we agree to identify two functions that are equal almost everywhere, then the expression

$$\langle x, y \rangle = \int_a^b x(t) y(t) \, dt$$

defines an inner product, and *the resulting inner product space is denoted by* $L_2[a, b]$. The norm is the same as in $C_2[a, b]$.

(6) The present example contains the first and second as special cases. Let I denote any index set and consider the space $l_2(I)$ of all functions $x : I \to \mathbb{R}$ with the property that $\{i \in I \mid x(i) \neq 0\}$ is a countable set $J = \{j_1, j_2, \dots\}$ and $\sum_k x^2(j_k) < \infty$. Since this sum converges absolutely, every rearrangement of the terms yields the same sum. Thus we can unambiguously define

$$\sum_{j \in I} x^2(j) := \sum_{j \in J} x^2(j) = \sum_k x^2(j_k).$$

The same argument that shows that l_2 is a linear space also shows that $l_2(I)$ is a linear space. Similarly, the expression

$$\langle x, y \rangle = \sum_{i \in I} x(i) y(i)$$

can be shown to be an (unambiguously defined) inner product on $l_2(I)$. Thus $l_2(I)$ is an *inner product space*. The norm in $l_2(I)$ is given by

$$\|x\| = \left[\sum_{i \in I} x^2(i) \right]^{\frac{1}{2}}.$$

Note that when $I = \{1, 2, \dots, n\}$ (respectively $I = \mathbb{N} = \{1, 2, \dots\}$), then $l_2(I)$ is just the space we earlier denoted by $l_2(n)$ (respectively l_2). (An alternative approach to defining the space $l_2(I)$ is given in Exercise 7 at the end of this chapter.)

★ (7) This example contains the first, second, fifth, and sixth as special cases. Let $(T, \mathcal{S}, \mu)$ be a measure space. That is, T is a set, $\mathcal{S}$ is a σ-algebra of subsets of T, and μ is a measure on $\mathcal{S}$. Let $L_2(\mu) = L_2(T, \mathcal{S}, \mu)$ denote the set of all measurable real functions x on T such that

$$\int_T |x|^2 \, d\mu < \infty.$$

(Here, as in example 5, we identify functions that are equal almost everywhere.) Then $L_2(\mu)$ is a linear space. If we define

$$\langle x, y \rangle := \int_T xy \, d\mu \text{ for } x, y \in L_2(\mu),$$

then $L_2(\mu)$ is an inner product space.

If $T = I$ is a closed interval in $\mathbb{R}$ and μ is Lebesgue measure on T, we often denote $L_2(\mu)$ by $L_2(I)$. Note that if T is any set, S consists of all subsets of T, and μ is "counting measure" on T (i.e., $\mu(E)$ is the number of elements in E if E is finite, and $\mu(E) = \infty$ otherwise), then $L_2(T, S, \mu) = l_2(T)$.

1.4 Properties of the Norm. *The norm function* $x \mapsto \|x\| := \sqrt{\langle x, x \rangle}$ *has the following properties:*

(1) $\|x\| \geq 0$,
(2) $\|x\| = 0$ *if and only if* $x = 0$,
(3) $\|\alpha x\| = |\alpha| \, \|x\|$,
(4) $\|x + y\| \leq \|x\| + \|y\|$ *("triangle inequality")*.

Moreover, equality holds in the triangle inequality if and only if x *or* y *is a nonnegative multiple of the other.*

Proof. Properties (1), (2), and (3) follow immediately from the properties of the inner product. To prove (4), we use the Schwarz inequality to obtain

$$\|x + y\|^2 = \langle x + y, x + y \rangle = \|x\|^2 + 2\langle x, y \rangle + \|y\|^2$$
$$\leq \|x\|^2 + 2|\langle x, y \rangle| + \|y\|^2$$
$$\leq \|x\|^2 + \|x\| \, \|y\| + \|y\|^2 = (\|x\| + \|y\|)^2,$$

which implies (4).

From the proof of (4), we see that equality holds in the triangle inequality if and only if

(1.4.1) $$\langle x, y \rangle = |\langle x, y \rangle| = \|x\| \, \|y\|.$$

Using the condition of equality in Schwarz's inequality, it follows that the second equality in (1.4.1) holds if and only if one of x or y is a scalar multiple of the other. Then the first equality in (1.4.1) shows that this scalar must be nonnegative. ∎

We note that the norm function is a generalization of the "absolute value" function on the space of real numbers $\mathbb{R}$. More generally, the norm of the vector $x - y$ in Euclidean 2- or 3-space is just the usual "Euclidean distance" between x and y. This suggests that we define the **distance** between any two vectors x and y in an inner product space X by $\|x - y\|$, and the **length** of x by $\|x\|$.

Any linear space that has a function $x \mapsto \|x\|$ defined on it satisfying the properties (1)–(4) of Proposition 1.4 is called a **normed linear space**. Thus an inner product space is a particular kind of normed linear space. What differentiates an inner product space from an arbitrary normed linear space is the following "parallelogram law." In fact, a normed linear space is an inner product space if and only if the parallelogram law holds (see Exercise 13 at the end of the chapter).

1.5 Parallelogram Law. *For every x, y in an inner product space,*

$$\|x + y\|^2 + \|x - y\|^2 = 2(\|x\|^2 + \|y\|^2) .$$

Proof. We have

$$\|x \pm y\|^2 = \langle x \pm y, x \pm y \rangle = \|x\|^2 \pm 2 \langle x, y \rangle + \|y\|^2 .$$

Adding these two equations yields the result. ∎

The geometric interpretation of the parallelogram law is that in a parallelogram, the sum of the squares of the lengths of the diagonals equals the sum of the squares of the lengths of all four sides (see Figure 1.5.1).

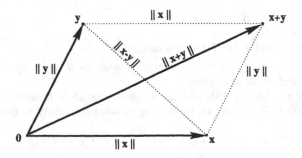

Figure 1.5.1. Parallelogram law

Orthogonality

It is a well-known elementary geometric fact in $\mathbb{R}^2$ (or $\mathbb{R}^3$) that two vectors x and y are perpendicular if and only if $\sum_1^2 x(i)y(i) = 0$ (or $\sum_1^3 x(i)y(i) = 0$). That is, in $\mathbb{R}^2$ or $\mathbb{R}^3$, two vectors x and y are perpendicular if and only if $\langle x, y \rangle = 0$. Motivated by this, we make the following definition in any inner product space.

1.6 Definition. *Two vectors x and y are said to be **orthogonal**, written $x \perp y$, if $\langle x, y \rangle = 0$. The vector x is said to be **orthogonal to the set** Y, written $x \perp Y$, if $x \perp y$ for every $y \in Y$. A set A is called an **orthogonal set** if each pair of distinct vectors in A are orthogonal. An **orthonormal set** is an orthogonal set in which each vector has norm one. Thus $\{x_i \mid i \in I\}$ is an orthonormal set if and only if*

$$\langle x_i, x_j \rangle = \delta_{ij} := \begin{cases} 0 & \text{if } i \neq j, \\ 1 & \text{if } i = j. \end{cases}$$

1.7 Orthogonal Sets are Linearly Independent. *Let A be an orthogonal set of nonzero vectors. Then A is a linearly independent set.*

Proof. Let $\{x_1, x_2, \ldots, x_n\} \subset A$ and suppose $\sum_1^n \alpha_i x_i = 0$. Then for each k,

$$0 = \left\langle \sum_{i=1}^n \alpha_i x_i, \; x_k \right\rangle = \sum_{i=1}^n \alpha_i \langle x_i, x_k \rangle = \alpha_k \langle x_k, x_k \rangle = \alpha_k \|x_k\|^2$$

implies $\alpha_k = 0$, since $x_k \neq 0$. This proves that A is linearly independent. ∎

The converse of Proposition 1.7 is false: There are linearly independent sets A that are not orthogonal (e.g., $A = \{(1,0),(1,1)\}$ in $l_2(2)$). However, in Proposition 4.11 a constructive procedure is presented (the Gram-Schmidt process) for obtaining an orthonormal set B from A with the property that span B = span A. (Here span A denotes the subspace spanned by A.)

1.8 Examples of Orthonormal Sets.

(1) In $l_2(I)$, the set $\{e_i \mid i \in I\}$, where $e_i(j) = \delta_{ij}$ for all i, j in I, is orthonormal. In particular, in $l_2(n)$, the vectors $e_1 = (1, 0, \ldots, 0)$, $e_2 = (0, 1, 0, \ldots, 0), \ldots, e_n = (0, \ldots, 0, 1)$ form an orthonormal set.

(2) In $C_2[-\pi, \pi]$ (or $L_2[-\pi, \pi]$), the set of functions

$$\left\{ \frac{1}{\sqrt{2\pi}}, \frac{1}{\sqrt{\pi}} \cos t, \frac{1}{\sqrt{\pi}} \sin t, \ldots, \frac{1}{\sqrt{\pi}} \cos nt, \frac{1}{\sqrt{\pi}} \sin nt, \ldots \right\}$$

is orthonormal. This follows from the relations

$$\int\limits_{-\pi}^{\pi} \cos nt \cos mt \, dt = \int\limits_{-\pi}^{\pi} \sin nt \sin mt \, dt = \pi \delta_{nm}$$

and

$$\int\limits_{-\pi}^{\pi} \cos nt \, dt = \int\limits_{-\pi}^{\pi} \sin nt \, dt = \int\limits_{-\pi}^{\pi} \sin mt \cos nt \, dt = 0.$$

Just as in Euclidean 2-space, the Pythagorean theorem has a valid analogue in any inner product space (see Figure 1.9.1).

1.9 Pythagorean Theorem. Let $\{x_1, x_2, \ldots, x_n\}$ be an orthogonal set. Then

$$\left\| \sum_1^n x_i \right\|^2 = \sum_1^n \|x_i\|^2.$$

In particular, if $x \perp y$, then

$$\|x + y\|^2 = \|x\|^2 + \|y\|^2.$$

Proof.

$$\left\| \sum_1^n x_i \right\|^2 = \left\langle \sum_1^n x_i, \sum_1^n x_j \right\rangle = \sum_i \sum_j \langle x_i, x_j \rangle = \sum_i \|x_i\|^2. \quad ∎$$

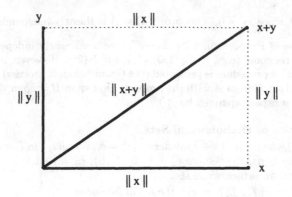

Figure 1.9.1. Pythagorean theorem

Again motivated by the situation in Euclidean 2 and 3-space, we can define the angle between two vectors in an inner product space as follows.

1.10 Definition. *For any two nonzero vectors x and y, the* **angle** $\theta = \theta(x, y)$ *between them is defined by*

$$\theta = \arccos \frac{\langle x, y \rangle}{\|x\| \, \|y\|} \, .$$

This expression is well-defined, since the Schwarz inequality implies

$$\frac{|\langle x, y \rangle|}{\|x\| \, \|y\|} \leq 1,$$

and moreover, $0 \leq \theta \leq \pi$. Observe that x and y make an angle of at least (respectively at most) $\pi/2$ if and only if $\langle x, y \rangle \leq 0$ (respectively $\langle x, y \rangle \geq 0$). Also $x \perp y$ if and only if $\theta(x, y) = \frac{\pi}{2}$. Further, $\theta(x, y) = 0$ (respectively $\theta(x, y) = \pi$) if and only if one of x or y is a positive (respectively negative) multiple of the other. (See Exercise 11 at the end of the chapter.)

Topological Notions

1.11 Definition. *A sequence of vectors $\{x_n\}$ in X is said to* **converge** *to $x \in X$ and x is called the* **limit** *of $\{x_n\}$, denoted by $x_n \to x$ or $x = \lim x_n$, provided that $\lim_{n \to \infty} \|x_n - x\| = 0$. That is, for each $\epsilon > 0$, there exists an integer N such that*

$$\|x_n - x\| < \epsilon \quad \text{for all} \quad n \geq N.$$

The next result shows, essentially, that the norm, inner product, addition, and scalar multiplication are all continuous functions.

1.12 Theorem. (1) *For all x, y,*

$$|\, \|x\| - \|y\| \,| \leq \|x - y\| \, .$$

(2) If $x_n \to x$, $y_n \to y$, and $\alpha_n \to \alpha$, then:

(a) $\|x_n\| \to \|x\|$,
(b) $\langle x_n, y_n \rangle \to \langle x, y \rangle$,
(c) $x_n + y_n \to x + y$,
(d) $\alpha_n x_n \to \alpha x$.

Proof. (1) Now,

$$\|x\| = \|(x - y) + y\| \le \|x - y\| + \|y\|$$

implies

$$\|x\| - \|y\| \le \|x - y\|.$$

Interchanging the roles of x and y yields

$$\|y\| - \|x\| \le \|y - x\| = \|x - y\|.$$

These two inequalities combine to yield (1).

(2) (a) This follows immediately from (1).

(b) Using Schwarz's inequality, we obtain

$$
\begin{aligned}
|\langle x_n, y_n \rangle - \langle x, y \rangle| &= |\langle x_n - x, y_n \rangle + \langle x, y_n - y \rangle| \\
&\le |\langle x_n - x, y_n \rangle| + |\langle x, y_n - y \rangle| \\
&\le \|x_n - x\| \, \|y_n\| + \|x\| \, \|y_n - y\| \to 0.
\end{aligned}
$$

(c) $\|x_n + y_n - (x + y)\| \le \|x_n - x\| + \|y_n - y\| \to 0.$

(d)

$$
\begin{aligned}
\|\alpha_n x_n - \alpha x\| &= \|\alpha_n (x_n - x) + (\alpha_n - \alpha)x\| \\
&\le |\alpha_n| \, \|x_n - x\| + |\alpha_n - \alpha| \, \|x\| \to 0. \quad \blacksquare
\end{aligned}
$$

A set K is said to be **closed** if $x \in K$ whenever $\{x_n\} \subset K$ and $x_n \to x$. That is, K contains the limits of all its convergent sequences. K is called **open** if its complement $X \setminus K := \{x \in X \mid x \notin K\}$ is closed. There are two sets that arise so frequently that they warrant a special notation. The **closed ball** and **open ball** centered at x with radius r are defined by

$$B[x, r] := \{y \in X \mid \|x - y\| \le r\} \text{ and}$$

$$B(x, r) := \{y \in X \mid \|x - y\| < r\}), \text{ respectively.}$$

The justification for these names is included in the next lemma, which also records some elementary topological properties of open and closed sets.

1.13 Lemma. (1) The empty set $\emptyset$ and full space X are both open and closed.

(2) The intersection of any collection of closed sets is closed.

(3) The union of a finite number of closed sets is closed.

(4) Any finite set is closed.

(5) The union of any collection of open sets is open.

(6) The intersection of a finite number of open sets is open.

(7) The set U is open if and only if for each $x \in U$, there exists an $\epsilon > 0$ such that $B(x, \epsilon) \subset U$.

(8) Any open ball $B(x, r)$ is open.

(9) Any closed ball $B[x, r]$ is closed.

Proof. Statements (1)–(6) are easy consequences of the definitions and DeMorgan's laws. They are left as an exercise.

(7) Let U be open and $x \in U$. If $B(x, \epsilon) \setminus U \neq \emptyset$ for each $\epsilon > 0$, then for each integer n, we can choose $x_n \in B(x, 1/n) \setminus U$. Then $x_n \to x$ and $x \in X \setminus U$, since $X \setminus U$ is closed. This contradiction shows that $B(x, \epsilon) \subset U$ for some $\epsilon > 0$.

Conversely, if U is not open, then $K = X \setminus U$ is not closed. Thus there exist a sequence $\{x_n\}$ in K and an $x \notin K$ such that $x_n \to x$. Then $x_n \notin U$ for all n, so that for each $\epsilon > 0$, $x_n \in B(x, \epsilon) \setminus U$ for n sufficiently large. Hence $B(x, \epsilon) \setminus U \neq \emptyset$, i.e., $B(x, \epsilon) \not\subset U$, for every $\epsilon > 0$. This proves (7).

(8) Let $C = X \setminus B(x, r)$. Let $x_n \in C$ and $x_n \to x_0$. Then $\|x_n - x\| \geq r$ for all n, and by Theorem 1.12 (2), $r \leq \lim \|x_n - x\| = \|x_0 - x\|$, so $x_0 \in C$. Thus C is closed and $B(x, r)$ is open.

The proof of (9) is similar to that of (8). ∎

Next we give a procedure for obtaining special closed and open sets generated by a given set.

1.14 Definition. *The* **closure** *of a set A, denoted by $\overline{A}$, is the intersection of all closed sets that contain A. Similarly, the* **interior** *of A, denoted by int A, is the union of all open sets that are contained in A.*

Some elementary properties of the closure and interior operations, which follow easily from Lemma 1.13, are listed next.

1.15 Lemma. (1) *$\overline{A}$ is the smallest closed set that contains A.*

(2) *A is closed if and only if $A = \overline{A}$.*

(3) *$x \in \overline{A}$ if and only if $B(x, \epsilon) \cap A \neq \emptyset$ for each $\epsilon > 0$.*

(4) *$x \in \overline{A}$ if and only if there exists a sequence $\{x_n\}$ in A such that $x_n \to x$.*

(5) *int A is the largest open set contained in A.*

(6) *A is open if and only if $A = $ int A.*

(7) *$x \in$ int A if and only if there exists $\epsilon > 0$ such that $B(x, \epsilon) \subset A$.*

(8) *$B[x, r] = \overline{B(x, r)}$.*

Proof. We prove only statement (3) and leave the remaining statements as an easy exercise.

Let $x \in \overline{A}$. If $B(x, \epsilon) \cap A = \emptyset$ for some $\epsilon > 0$, set $F = X \setminus B(x, \epsilon)$. Then F is closed and $F \supset A$. Hence $F \supset \overline{A}$ by (1), which implies that $\overline{A} \cap B(x, \epsilon) = \emptyset$. But this contradicts $x \in \overline{A} \cap B(x, \epsilon)$. Hence $B(x, \epsilon) \cap A \neq \emptyset$ for every $\epsilon > 0$.

Conversely, if $x \notin \overline{A}$, then x is in the open set $U := X \setminus \overline{A}$, so there exists an $\epsilon > 0$ such that $B(x, \epsilon) \subset U$. Therefore $B(x, \epsilon) \cap \overline{A} = \emptyset$, and so $B(x, \epsilon) \cap A = \emptyset$. ∎

1.16 Definition. *A sequence $\{x_n\}$ is called a* **Cauchy sequence** *if for each $\epsilon > 0$, there exists an integer N such that*

$$\|x_n - x_m\| < \epsilon \quad \text{for all} \quad n, m \geq N.$$

A nonempty set B is called **bounded** *if $\sup\{\|x\| \mid x \in B\} < \infty$.*

1.17 Lemma. (1) *Every convergent sequence is a Cauchy sequence.*

(2) *Every Cauchy sequence is bounded.*

(3) *The space $C_2[a, b]$ contains a Cauchy sequence that does not converge.*

Proof. (1) If $x_n \to x$, then

$$\|x_n - x_m\| \leq \|x_n - x\| + \|x - x_m\| \to 0$$

as $n, m \to \infty$. Thus $\{x_n\}$ is Cauchy.

(2) Let $\{x_n\}$ be Cauchy. Then there exists N such that $\|x_n - x_m\| \leq 1$ for all $n, m \geq N$. In particular,

$$\|x_n\| \leq \|x_n - x_N\| + \|x_N\| \leq 1 + \|x_N\|$$

for all $n \geq N$. Hence $\sup_n \|x_n\| \leq \max\{\|x_1\|, \|x_2\|, \dots, \|x_{N-1}\|, 1 + \|x_N\|\} < \infty$.

(3) Without loss of generality, we may assume $[a, b] = [-1, 1]$. Consider the functions $x_n \in C_2[-1, 1]$ defined by

$$x_n(t) = \begin{cases} 0 & \text{if } -1 \leq t \leq 0, \\ nt & \text{if } 0 \leq t \leq \dfrac{1}{n}, \\ 1 & \text{if } \dfrac{1}{n} \leq t \leq 1. \end{cases}$$

(See Figure 1.17.1.)

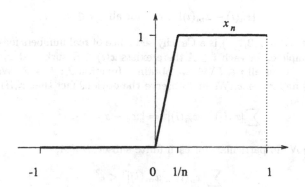

Figure 1.17.1

It is easy to verify that $\{x_n\}$ is a Cauchy sequence. Suppose there *did* exist some $x \in C_2[-1, 1]$ such that $x_n \to x$. Then, since

$$\|x_n - x\|^2 = \int_{-1}^{1} [x_n(t) - x(t)]^2 dt$$

$$= \int_{-1}^{0} x^2(t) dt + \int_{0}^{1/n} [nt - x(t)]^2 dt + \int_{1/n}^{1} [1 - x(t)]^2 dt$$

and $\|x_n - x\| \to 0$, it follows that $x(t) = 0$ for $-1 \leq t \leq 0$ and $x(t) = 1$ for $0 < t \leq 1$. But this contradicts the continuity of x. Hence $\{x_n\}$ does not converge. ∎

Hilbert Space

Those inner product spaces in which every Cauchy sequence converges have many pleasant properties. Some important theorems can be proved in such spaces that are not true in general inner product spaces. They therefore warrant a special name.

1.18 Definition. *An inner product space X is called* **complete**, *or a* **Hilbert space**, *if each Cauchy sequence in X converges to a point in X.*

1.19 Examples. (1) *The space $\mathbb{R} = l_2(1)$ is complete.* This is just the statement that every Cauchy sequence of real numbers converges to a real number. This, in turn, is equivalent to the completeness axiom of the real numbers.

(2) *The space $l_2(I)$ is a Hilbert space for any index set I.* In particular, $l_2(n)$ and l_2 are both Hilbert spaces.

To see this, let $\{x_n\}$ be a Cauchy sequence in $l_2(I)$. Then each x_n vanishes except on the countable set

$$A_n := \{i \in I \mid x_n(i) \neq 0\}.$$

Let $A = \cup_1^\infty A_n$, so A is also countable. Given $\epsilon > 0$, choose an integer N such that

$$\|x_n - x_m\| < \epsilon \quad \text{for all} \quad n, m \geq N.$$

In particular, for all $n, m \geq N$,

$$|x_n(i) - x_m(i)| < \epsilon \quad \text{for all} \quad i \in A.$$

That is, $\{x_n(i) \mid n = 1, 2, \dots\}$ is a Cauchy sequence of real numbers for each $i \in A$. Since $\mathbb{R}$ is complete, for each $i \in A$ there exists $x(i) \in \mathbb{R}$ such that $x_n(i) \to x(i)$. Defining $x(i) = 0$ for all $i \in I \setminus A$, we obtain a function $x : I \to \mathbb{R}$. We will show that $x \in l_2(I)$ and $x_n \to x$. We first observe the obvious fact that $x_n(i) \to x(i)$ for each $i \in I$. Also,

$$\sum_{i \in A} |x_n(i) - x_m(i)|^2 = \|x_n - x_m\|^2 < \epsilon^2$$

for all $n, m \geq N$. In particular, for each finite subset J of A,

$$\sum_{i \in J} |x_n(i) - x_m(i)|^2 < \epsilon^2$$

for all $n, m \geq N$. Letting $m \to \infty$ in this inequality, we obtain

$$\sum_{i \in J} |x_n(i) - x(i)|^2 \leq \epsilon^2 \text{ for all } n \geq N.$$

Since this is true for every finite subset $J \subset A$, it follows that

$$(1.19.1) \qquad \|x_n - x\|^2 = \sum_{i \in A} |x_n(i) - x(i)|^2 \leq \epsilon^2$$

for all $n \geq N$. This shows that for $n \geq N$, $x - x_n \in l_2(I)$. Since $l_2(I)$ is a linear space and $x_n \in l_2(I)$, we have that $x = (x - x_n) + x_n \in l_2(I)$. Finally, from (1.19.1) we see that $x_n \to x$.

It can be shown that *every* Hilbert space can be identified with a space of type $l_2(I)$ for some index set I (see Appendix 2). *Thus $l_2(I)$ represents the most general Hilbert space.*

(3) *Every finite-dimensional inner product space is complete.* (This will follow as a consequence of stronger results proved in Chapter 3; see Theorem 3.7.)

(4) $C_2[a, b]$ *is not a Hilbert space.* (This follows by Lemma 1.17(3).)

(5) More generally, *the weighted space $C_2(I; w)$* (see Example 1.3(4)) *is not a Hilbert space.* (The proof is similar to that of (4).)

★ (6) $L_2[a, b]$ *is a Hilbert space.* This is a special case of example (7) below.

★ (7) $L_2(\mu)$ *is a Hilbert space.* The proof of this fact can be found in almost any book on real analysis.

Exercises

1. Limits are unique. That is, if $x_n \to x$ and $x_n \to y$, then $x = y$.
2. A sequence $\{x_n\}$ converges to x if and only if for each $\epsilon > 0$, $x_n \in B(x, \epsilon)$ for all n sufficiently large.
3. A Cauchy sequence converges if and only if some subsequence converges.
4. Verify statements (1)–(6) of Lemma 1.13.
5. Let $\{x_n\} \subset X$, $x \in X$. Show that $x_n \to x$ if and only if $\langle x_n, y \rangle \to \langle x, y \rangle$ for every $y \in X$ and $\|x_n\| \to \|x\|$.
6. Call a subset A of X **complete** if each Cauchy sequence in A converges to a point in A. Verify that:
   (a) every complete set is closed;
   (b) a subset of a Hilbert space is complete if and only if it is closed.
7. An alternative, but equivalent, way of defining the space $l_2(I)$ is as follows. Let $l_2(I)$ denote the set of all functions $x : I \to \mathbb{R}$ such that $\sum_{i \in I} x^2(i) < \infty$, where

$$\sum_{i \in I} x^2(i) := \sup \left\{ \sum_{j \in J} x^2(j) \mid J \subset I, \ J \text{ finite} \right\}.$$

(a) Show that if $x \in l_2(I)$, then the **support** of x,

$$\operatorname{supp} x := \{i \in I \mid x(i) \neq 0\},$$

is countable.

(b) Show that for each $x \in l_2(I)$, if $J = \{j_1, j_2, \dots\} = \operatorname{supp} x$, then the sum

$$\sum_k x^2(j_k)$$

does not depend on the order of the indices j_k in this sum and

$$\sum_k x^2(j_k) = \sum_{i \in I} x^2(i).$$

(c) Show that for each $x, y \in l_2(I)$, the expression

$$\sum_{i \in I} x(i)y(i) := \sum_k x(j_k)y(j_k)$$

is a well-defined real number that is independent of the ordering of the indices $j_k \in J := (\text{supp } x) \cup (\text{supp } y)$.

(d) Show that the expression

$$\langle x, y \rangle := \sum_{i \in I} x(i) y(i)$$

defines an inner product on $l_2(I)$.

8. (a) If $\{x_n\}$ is a Cauchy sequence, then $\{\|x_n\|\}$ converges.

(b) If $\{x_n\}$ is a Cauchy sequence, then $\{\langle x_n, x \rangle\}$ converges for each $x \in X$.

(c) Give an example in l_2 of a sequence $\{x_n\}$ such that $\{\langle x_n, x \rangle\}$ converges for each $x \in l_2$, but $\{x_n\}$ is not a Cauchy sequence.

9. Let $x \in X$ and $r > 0$. (a) Show that the set

$$A = \{y \in X \mid \|x - y\| \le r\} \cap \{y \in X \mid \|y\| \ge r\}$$

is closed.

(b) Show that the set

$$B = \{y \in X \mid \|x - y\| > r\}$$

is open.

10. Prove the "cosine law":

$$\|x - y\|^2 = \|x\|^2 + \|y\|^2 - 2\|x\| \, \|y\| \cos \theta(x, y).$$

Interpret geometrically.

11. Verify that if x and y are nonzero vectors, then $\theta(x, y) = 0$ (respectively $\theta(x, y) = \pi$) if and only if one of them is a positive (respectively negative) multiple of the other.

12. Estimate

$$\int_0^1 \frac{t^n}{1 + t} dt$$

from above by the Schwarz inequality. Compare with the exact answer for $n = 6$.

13. Let X be a real normed linear space whose norm satisfies the parallelogram law:

$$\|x + y\|^2 + \|x - y\|^2 = 2(\|x\|^2 + \|y\|^2).$$

Define

$$\langle x, y \rangle := \frac{1}{4} \left[\|x + y\|^2 - \|x - y\|^2 \right] \quad \text{for} \quad x, y \in X.$$

Show that this expression defines an inner product on X such that $\sqrt{\langle x, x \rangle} = \|x\|$. Thus verify that a normed linear space "is" an inner product space if and only if its norm satisfies the parallelogram law. [Hint: Verifying additivity $\langle x + y, z \rangle = \langle x, z \rangle + \langle y, z \rangle$ requires the parallelogram law. From this one gets $\langle nx, y \rangle = n \langle x, y \rangle$ for all $n \in \mathbb{N}$ by induction. Then obtain

$\langle rx, y \rangle = r \langle x, y \rangle$ for all rational numbers r. Finally, deduce $\langle \alpha x, y \rangle = \alpha \langle x, y \rangle$ for any real number α by taking a sequence of rationals $r_n \to \alpha$.]

14. (**Weighted** l_2) This example generalizes the space $l_2(I)$. Let I be an index set and w a positive function on I. Let $l_2(I; w)$ denote the set of all functions $x : I \to \mathbb{R}$ such that $\sum_{i \in I} w(i) x^2(i) < \infty$. Then

$$\langle x, y \rangle = \sum_{i \in I} w(i) x(i) y(i)$$

is a well-defined inner product on $l_2(I; w)$, and $l_2(I; w)$ is complete.

15. (**Weighted** $C_2[a, b]$) Show that the space $C_2(I; w)$ defined in Example 1.3(4) is an inner product space that is not complete.

16. (**Discrete spaces**) Let $\mathcal{P}_n$ denote the set of all polynomials of degree at most n:

$$\mathcal{P}_n = \left\{ x \;\middle|\; x(t) = \sum_0^n \alpha_i t^i, \alpha_i \in \mathbb{R} \right\}.$$

Let $T = \{t_1, t_2, \ldots, t_m\}$ be a set of m distinct real numbers and define

(16.1) $$\langle x, y \rangle = \sum_{t \in T} x(t) y(t).$$

(a) If $m \geq n + 1$, then (16.1) defines an inner product on $\mathcal{P}_n$. Moreover, $\mathcal{P}_n$ is complete. [Hint: Induction on n.]

(b) If $m < n + 1$, what property of the inner product does (16.1) fail to possess?

(c) More generally, if w is a positive function on T and $m \geq n + 1$, then

(16.2) $$\langle x, y \rangle = \sum_{t \in T} w(t) x(t) y(t)$$

defines an inner product on $\mathcal{P}_n$, and $\mathcal{P}_n$ is complete.

17. Complete the proof of Lemma 1.15.

18. Let X be a *complex* linear space, i.e., a linear space over the field $\mathbb{C}$ of complex numbers. X is called a **complex inner product space** if there is a function $\langle \cdot, \cdot \rangle$ from $X \times X$ into $\mathbb{C}$ having the properties

(1) $\langle x, x \rangle \geq 0$,

(2) $\langle x, x \rangle = 0$ if and only if $x = 0$,

(3) $\langle x, y \rangle = \overline{\langle y, x \rangle}$ (where the bar denotes complex conjugation),

(4) $\langle \alpha x, y \rangle = \alpha \langle x, y \rangle$,

(5) $\langle x + y, z \rangle = \langle x, z \rangle + \langle y, z \rangle$.

The function $\langle \cdot, \cdot \rangle$ is called an **inner product** on X. Note that the only difference between inner products on real spaces and on complex spaces is that in the latter case, $\langle x, y \rangle = \overline{\langle y, x \rangle}$ rather than $\langle x, y \rangle = \langle y, x \rangle$. (Of course, if X is a real space, then since the inner product is real-valued, $\langle x, y \rangle = \langle y, x \rangle = \overline{\langle y, x \rangle}$ and all the conditions (1)–(5) hold.)

(a) Verify that the inner product is a linear function in the first argument and conjugate linear in the second; i.e.,

$$\langle \alpha x + \beta y, z \rangle = \alpha \langle x, z \rangle + \beta \langle y, z \rangle$$

and

$$\langle z, \alpha x + \beta y \rangle = \overline{\alpha} \langle z, x \rangle + \overline{\beta} \langle z, y \rangle.$$

(b) Show that Schwarz's inequality is valid for complex inner product spaces.

(c) Show that the parallelogram law and the Pythagorean theorem are valid in complex inner product spaces.

(d) Define the angle $\theta(x, y)$ between nonzero vectors x and y by

$$\theta(x, y) = \arccos\left(\frac{\operatorname{Re}\langle x, y\rangle}{\|x\|\,\|y\|}\right).$$

Show that the cosine law (Exercise 10) is valid.

(e) Show that the norm and inner product in a complex inner product space are related by

$$\langle x, y\rangle = \frac{1}{4}[\|x + y\|^2 - \|x - y\|^2 + i(\|x + iy\|^2 - \|x - iy\|^2)]$$

(the "polarization identity").

(f) If X is any complex normed linear space whose norm satisfies the parallelogram law, then X is an inner product space (and conversely). [Hint: Define an inner product on X suggested by the polarization identity.]

19. If I is any index set, consider the set of all functions $x : I \to \mathbb{C}$ such that $\sum_{i \in I} |x(i)|^2 < \infty$. Then

$$\langle x, y\rangle = \sum_{i \in I} x(i)\overline{y(i)}$$

defines an inner product on this space. We call this space "complex $l_2(I)$." Show that it is a Hilbert space. [Hint: Proof just as for "real" $l_2(I)$.]

20. If in Exercise 15 we consider bounded continuous *complex* functions x on I with the inner product

$$\langle x, y\rangle = \int_I w(t)x(t)\overline{y(t)}\, dt,$$

we obtain the inner product space called "complex $C_2(I; w)$."

★ 21. Let (T, S, μ) be a measure space and consider the linear space of all measurable complex functions x on T with

$$\int_T |x|^2\, d\mu < \infty.$$

If we also define

$$\langle x, y\rangle = \int_T x\overline{y}\, d\mu$$

(where the bar denotes complex conjugation), the resulting inner product space obtained is called "complex $L_2(\mu)$". Show that this space is a Hilbert space. [Hint: Proof as for "real" $L_2(\mu)$.]

Historical Notes

The Schwarz Inequality (Theorem 1.2) has a long history. In the specific case of a double integral over a domain in the complex plane, it was established by Schwarz

(1885; Vol. 1, pp. 223–269) in a highly original paper where he studied the problem of a vibrating membrane. The same inequality had actually been discovered earlier by Buniakowsky (1859), but went unnoticed by most mathematicians until 1885. The corresponding inequality for finite sums (i.e., in $\ell_2(n)$) goes back at least to Cauchy (1821; p. 373) and, in the special case when $n = 3$, even to Lagrange (1773; pp. 662-3). To appreciate the power and originality of Schwarz's (1885) paper, Dieudonné (1981; pp. 53–54) showed how Schwarz's method could be easily translated, with little effort, to the theory of compact self-adjoint operators on a separable Hilbert space.

Jordan and von Neumann (1935) showed that the parallelogram law actually *characterizes* those normed linear spaces that are inner product spaces (see also Exercise 13). The reader can find 350 characterizations of inner product or Hilbert spaces among all normed linear spaces in the unusual book of Amir (1986).

Hilbert (1906), inspired by the work of Fredholm in (1900) and (1903) on integral equations, recognized that the theory of integral equations could be subsumed in the theory of infinite systems of linear equations. This led him to a study of the space that we now call ℓ_2, and to the symmetric bilinear forms on this space. While he did not use the geometric language that later workers did, he seemed motivated by the analogy with n-dimensional Euclidean space. In this regard, we should mention that the axioms for finite-dimensional Euclidean space $\mathbb{R}^n$ were stated explicitly by Weyl (1923), who regarded $\mathbb{R}^n$ as the natural setting for problems in geometry and physics.

Hilbert's work inspired a transfer of Euclidean geometry to infinite dimensions. Indeed, Fréchet (1906) introduced the notion of *distance* between points in an arbitrary set, and the idea of a *metric space*. Schmidt (1908) defined what we now call (complex) ℓ_2, along with the notions of inner product and norm, orthogonality, closed sets, and linear subspaces. He even proved the existence of the orthogonal projection (in our terminology, the *best approximation*) of any point in ℓ_2 onto any given closed subspace. (See Chapter 2 for the definition of best approximation.)

It was von Neumann (1929a) who first defined and studied what we now call (an abstract) Hilbert space. (Actually, he defined a Hilbert space to have the additional properties of being infinite-dimensional and separable. It was Löwig (1934) and Rellich (1935) who studied abstract Hilbert spaces in general.) Von Neumann (1929a) proved the Schwarz inequality in its present form (Theorem 1.2), that Hilbert spaces are normed linear spaces, studied orthonormal and complete orthonormal sets (or "abstract" Fourier analysis), orthogonal projections onto closed subspaces, and general self-adjoint operators on Hilbert spaces. In two subsequent papers, von Neumann (1929b), (1932) continued to develop the theory of most of the important linear operators on Hilbert spaces. In short, much of what we know today about Hilbert space and the linear operators on such a space can be traced to these three fundamental papers of von Neumann.

CHAPTER 2

BEST APPROXIMATION

Best Approximation

We will describe the general problem of best approximation in an inner product space. A uniqueness theorem for best approximations from convex sets is also provided. The five problems posed in Chapter 1 are all shown to be special cases of best approximation from a convex subset of an appropriate inner product space.

2.1 Definition. *Let K be a nonempty subset of the inner product space X and let $x \in X$. An element $y_0 \in K$ is called a* **best approximation,** *or nearest point, to x from K if*

$$\|x - y_0\| = d(x, K),$$

where $d(x, K) := \inf_{y \in K} \|x - y\|$. The number $d(x, K)$ is called the **distance** *from x to K, or the error in approximating x by K.*

The (possibly empty) set of all best approximations from x to K is denoted by $P_K(x)$. Thus

$$P_K(x) := \{ y \in K \mid \|x - y\| = d(x, K) \}.$$

This defines a mapping P_K from X into the subsets of K called the **metric projection** onto K. (Other names for metric projection include projector, nearest point mapping, proximity map, and best approximation operator.)

If each $x \in X$ has at least (respectively exactly) one best approximation in K, then K is called a **proximinal** (respectively **Chebyshev**) set. Thus K is proximinal (respectively Chebyshev) if and only if $P_K(x) \neq \emptyset$ (respectively $P_K(x)$ is a singleton) for each $x \in X$.

The general theory of best approximation may be briefly outlined as follows: It is the mathematical study that is motivated by the desire to seek answers to the following basic questions, among others.

(1) (Existence of best approximations) Which subsets are proximinal?
(2) (Uniqueness of best approximations) Which subsets are Chebyshev?
(3) (Characterization of best approximations) How does one recognize when a given element $y \in K$ is a best approximation to x?
(4) (Error of approximation) How does one compute the error of approximation $d(x, K)$, or at least get sharp upper or lower bounds for it?
(5) (Computation of best approximations) Can one describe some useful algorithms for actually computing best approximations?
(6) (Continuity of best approximations) How does $P_K(x)$ vary as a function of x (or K)?

In this book we will answer each of these questions by proving one or more general theorems. It will turn out that most of the theorems are suggested by rather elementary *geometric* considerations.

A number of applications to specific inner product spaces will also be made. In addition, some problems will be studied that are not immediately suggested by the above six questions. For example, it is known (see Chapter 3) that every closed convex subset of a Hilbert space is a Chebyshev set. The question whether the converse holds (i.e., must every Chebyshev subset of a Hilbert space be convex?) is still unanswered. However, if the Hilbert space is finite-dimensional, the answer is affirmative (see Chapter 12).

Convex Sets

Since the theory of best approximation is the most well developed when the approximating set K is a linear subspace, or more generally, a convex set, we will first describe such sets.

2.2 Definition. *A subset K of X is called* **convex** *if $\lambda x + (1-\lambda)y \in K$ whenever $x, y \in K$ and $0 \leq \lambda \leq 1$.*

Geometrically, a set is convex if and only if it contains the line segment

$$\{\, \lambda x + (1 - \lambda)y \mid 0 \leq \lambda \leq 1 \,\}$$

joining each pair of its points x, y (see Figure 2.2.1). By a simple induction argument one can show that K is convex if and only if $\sum_1^n \lambda_i x_i \in K$ whenever $x_i \in K, \lambda_i \geq 0$, and $\sum_1^n \lambda_i = 1$. (See also Exercise 10 at the end of the chapter.)

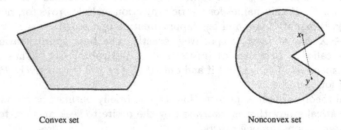

Convex set   Nonconvex set

Figure 2.2.1

2.3 Examples of Convex Sets.

(1) *The empty set, the whole space X, and any singleton set are convex.*
(2) *The intersection of any collection of convex sets is convex.*
(3) *The "closed ball centered at x with radius r," which is defined by*

$$B[x, r] := \{y \in X \mid \|x - y\| \leq r\},$$

is convex.

(4) *The "open ball centered at x with radius r," which is defined by*

$$B(x, r) := \{y \in X \mid \|x - y\| < r\},$$

is convex.

(5) *If A and B are convex, then their sum*

$$A + B := \{a + b \mid a \in A,\ b \in B\}$$

is also convex. In particular, any translate $K + x$ of a convex set K is convex.

Proof. (1) This is obvious.

(2) If $\{K_i \mid i \in I\}$ is a collection of convex sets, $x, y \in \cap_i K_i$, and $0 \le \lambda \le 1$, then $x, y \in K_i$ for all i implies $\lambda x + (1 - \lambda)y \in K_i$ for all i. Hence $\lambda x + (1 - \lambda)y \in \cap_i K_i$ and $\cap_i K_i$ is convex.

(3) Let $y_1,\ y_2 \in B[x, r]$ and $0 \le \lambda \le 1$. Then

$$\|x - [\lambda y_1 + (1 - \lambda)y_2]\| = \|\lambda(x - y_1) + (1 - \lambda)(x - y_2)\|$$
$$\le \lambda\|x - y_1\| + (1 - \lambda)\|x - y_2\|$$
$$\le \lambda r + (1 - \lambda)r = r$$

implies that $\lambda y_1 + (1 - \lambda)y_2 \in B[x, r]$. Thus $B[x, r]$ is convex.

(4) The proof is similar to (3).

(5) Let $y_1, y_2 \in A + B$ and $0 \le \lambda \le 1$. Then $y_1 = a_1 + b_1$ and $y_2 = a_2 + b_2$ for some $a_i \in A$ and $b_i \in B$. Then since A and B are convex,

$$\lambda y_1 + (1 - \lambda)y_2 = [\lambda a_1 + (1 - \lambda)a_2] + [\lambda b_1 + (1 - \lambda)b_2] \in A + B.$$

Thus $A + B$ is convex. ∎

2.4 Uniqueness of Best Approximations. *Let K be a convex subset of X. Then each $x \in X$ has at most one best approximation in K.*

In particular, every convex proximinal set is Chebyshev.

Proof. Let $x \in X$ and suppose y_1 and y_2 are in $P_K(x)$. Then $(y_1 + y_2)/2 \in K$ by convexity and

$$d(x, K) \le \left\| x - \frac{1}{2}(y_1 + y_2) \right\| = \left\| \frac{1}{2}(x - y_1) + \frac{1}{2}(x - y_2) \right\|$$
$$\le \frac{1}{2}\|x - y_1\| + \frac{1}{2}\|x - y_2\| = d(x, K).$$

Hence equality must hold throughout these inequalities. By the condition of equality in the triangle inequality, $x - y_1 = \rho(x - y_2)$ for some $\rho \ge 0$. But $\|x - y_1\| = d(x, K) = \|x - y_2\|$ implies $\rho = 1$ and hence $y_1 = y_2$. ∎

Alternative proofs of Theorem 2.4 are described in Exercises 8 and 9 at the end of the chapter. One proof is based on the fact that $P_K(x) = K \cap B[x, d(x, K)]$ is a convex set that is contained in a "sphere" (i.e., the boundary of a ball), and the only convex subsets of spheres are singleton sets. Another proof is based on the parallelogram law.

Theorem 2.4 is important for our purposes. Since we will be approximating almost exclusively by convex sets, uniqueness of best approximations is always guaranteed whenever best approximations exist. Hence every convex proximinal set is automatically Chebyshev, and its metric projection is a point-valued mapping. We will use this fact frequently in the subsequent material without making explicit reference to Theorem 2.4.

There are certain convex sets that often arise in the various applications.

2.5 Definitions. (1) *A subset C of X is called a* **convex cone** *if $\alpha x + \beta y \in C$ whenever x, $y \in C$ and α, $\beta \geq 0$.*

(2) *A nonempty subset M of X is called a (linear)* **subspace** *if $\alpha x + \beta y \in M$ whenever x, $y \in M$ and α, $\beta \in \mathbb{R}$.*

(3) *If A is any nonempty subset of X, the* **subspace spanned** *by A, written* span(A), *is the set of all finite linear combinations of elements of A. That is,*

$$\text{span}(A) = \left\{ \sum_{1}^{n} \alpha_i x_i \mid x_i \in A,\ \alpha_i \in \mathbb{R},\ n \in \mathbb{N} \right\}.$$

Equivalently, span(A) *is the smallest subspace of X that contains A (see Exercise 12 at the end of the chapter).*

(4) *The* **dimension** *of a subspace M, denoted by* dim M, *is the number of vectors in a basis for M. In particular, if $\{x_1, x_2, \ldots, x_n\}$ is a linearly independent set and $M = \text{span}\{x_1, x_2, \ldots, x_n\}$, then* dim $M = n$.

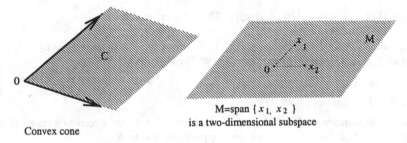

M=span $\{x_1,\ x_2\}$
is a two-dimensional subspace

Convex cone

Figure 2.5.1

2.6 Examples of Cones and Subspaces.

(1) *The empty set, the whole space X, and the set $\{0\}$ consisting of just the zero vector are all subspaces.*

(2) *Every subspace is a convex cone, but not conversely.*

(3) *The intersection of any collection of convex cones (respectively subspaces) is a convex cone (respectively subspace).*

(4) *In $C_2[a, b]$, the set of all nonnegative functions*

$$C = \{x \in C_2[a, b] \mid x(t) \geq 0 \ \text{for all } t \in [a, b]\}$$

is a convex cone that is not a subspace.

(5) *The set of all* **polynomials** *of degree at most n,*

$$\mathcal{P}_n = \left\{ x \;\middle|\; x(t) = \sum_{0}^{n} \alpha_i t^i, \alpha_i \in \mathbb{R} \right\}$$

is an $(n + 1)$-dimensional subspace of $C_2[a, b]$.

Proof. (3) Let $\{C_i \mid i \in I\}$ be a collection of convex cones and $C = \cap_i C_i$. If $x, y \in C$ and $\alpha, \beta \geq 0$, then $x, y \in C_i$ for all i implies $\alpha x + \beta y \in C_i$ for all i. Hence $\alpha x + \beta y \in C$, so C is a convex cone. The proof that the intersection of subspaces is a subspace is similar.

The proofs of the remaining statements are left as exercises (see Exercises 6(b), 7, and 17 at the end of the chapter). ∎

Certain elementary properties of the distance function allow us to replace the problem of approximating from a set K with that of approximating from a translate of K. Moreover, they also allow us to deduce that translates or scalar multiples of Chebyshev sets or proximinal sets are again sets of the same kind. We record a few of these properties next.

2.7 Invariance by Translation and Scalar Multiplication.

(1) *Let K be a nonempty subset of X. Then:*

(i) $d(x + y, K + y) = d(x, K)$ *for every* $x, y \in X$.

(ii) $P_{K+y}(x + y) = P_K(x) + y$ *for every* $x, y \in X$.

(iii) $d(\alpha x, \alpha K) = |\alpha| d(x, K)$ *for every* $x \in X$ *and* $\alpha \in \mathbb{R}$.

(iv) $P_{\alpha K}(\alpha x) = \alpha P_K(x)$ *for every* $x \in X$ *and* $\alpha \in \mathbb{R}$.

(v) *K is proximinal (respectively Chebyshev) if and only if $K + y$ is proximinal (respectively Chebyshev) for any given $y \in X$.*

(vi) *K is proximinal (respectively Chebyshev) if and only if αK is proximinal (respectively Chebyshev) for any given $\alpha \in \mathbb{R} \backslash \{0\}$.*

(2) *Let M be a nonempty subspace of X. Then:*

(i) $d(x + y, M) = d(x, M)$ *for every* $x \in X$ *and* $y \in M$.

(ii) $P_M(x + y) = P_M(x) + y$ *for every* $x \in X$ *and* $y \in M$.

(iii) $d(\alpha x, M) = |\alpha| d(x, M)$ *for every* $x \in X$ *and* $\alpha \in \mathbb{R}$.

(iv) $P_M(\alpha x) = \alpha P_M(x)$ *for every* $x \in X$ *and* $\alpha \in \mathbb{R}$.

(v) $d(x + y, M) \leq d(x, M) + d(y, M)$ *for every* x *and* y *in* X.

Proof. (1) (i) For any $x, y \in X$,

$$d(x + y, K + y) = \inf_{z \in K} \|x + y - (z + y)\| = \inf_{z \in K} \|x - z\| = d(x, K).$$

(ii) Using (i), $y_0 \in P_{K+y}(x + y)$ if and only if $y_0 \in K + y$ and $\|x + y - y_0\| = d(x + y, K + y)$ if and only if $y_0 - y \in K$ and $\|x - (y_0 - y)\| = d(x, K)$ if and only if $y_0 - y \in P_K(x)$; i.e., $y_0 \in P_K(x) + y$.

(iii) We have

$$d(\alpha x, \alpha K) = \inf_{y \in K} \|\alpha x - \alpha y\| = |\alpha| \inf_{y \in K} \|x - y\| = |\alpha| d(x, K).$$

(iv) If $\alpha = 0$, the result is trivially true. Thus assume $\alpha \neq 0$. Using (iii) we see that $y_0 \in P_{\alpha K}(\alpha x)$ if and only if $y_0 \in \alpha K$ and $\|\alpha x - y_0\| = d(\alpha x, \alpha K)$ if and only if $\frac{1}{\alpha} y_0 \in K$ and $|\alpha| \|x - \frac{1}{\alpha} y_0\| = |\alpha| d(x, K)$. However, this is equivalent to $\frac{1}{\alpha} y_0 \in P_K(x)$; i.e., $y_0 \in \alpha P_K(x)$.

(v) This is an immediate consequence of (ii), and (vi) follows from (iv).

(2) (i) and (ii) follow from (1) (i) and (ii), and the fact that if M is a subspace and $y \in M$, then $M + y = M$.

(iii) and (iv) follow from (1) (iii) and (iv), and the fact that if M is a subspace and $\alpha \neq 0$, then $\alpha M = M$.

The proof of (v) follows from the relations

$$d(x + y, M) = \inf_{m \in M} \|x + y - m\| = \inf_{m, m' \in M} \|x + y - (m + m')\|$$

$$\leq \inf_{m, m' \in M} [\|x - m\| + \|y - m'\|]$$

$$= \inf_{m \in M} \|x - m\| + \inf_{m' \in M} \|y - m'\| = d(x, M) + d(y, M). \quad \blacksquare$$

We next show that the interior and closure of a convex set are convex.

2.8 Theorem. *Let K be a convex set in X. Then both $\overline{K}$ and int K are convex. Moreover, if int $K \neq \emptyset$, then for every $x \in$ int K, $y \in \overline{K}$, and $0 < \lambda \leq 1$, there follows*

$$(2.8.1) \qquad\qquad \lambda x + (1 - \lambda)y \in \text{int } K,$$

and hence

$$(2.8.2) \qquad\qquad \overline{\text{int } K} = \overline{K} \quad \text{and} \quad \text{int } \overline{K} = \text{int } K.$$

Proof. Let $x, y \in \overline{K}$ and $0 \leq \lambda \leq 1$. By Lemma 1.15(4), we can choose sequences $\{x_n\}, \{y_n\}$ in K such that $x_n \to x$ and $y_n \to y$. Then $\lambda x_n + (1 - \lambda)y_n \in K$ by convexity, and

$$\lambda x + (1 - \lambda)y = \lim[\lambda x_n + (1 - \lambda)y_n] \in \overline{K}.$$

Thus $\overline{K}$ is convex.

Now fix any $x \in$ int K, $y \in \overline{K}$, $0 < \lambda < 1$, and set $z = \lambda x + (1 - \lambda)y$. Since $x \in$ int K, choose $\epsilon > 0$ such that $B(x, \epsilon) \subset K$.

Claim: $B(z, \lambda\epsilon) \subset K$ *and hence* $z \in$ int K.

To see this, let $u \in B(z, \lambda\epsilon)$, and hence $\|u - z\| < \lambda\epsilon$. Since $y \in \overline{K}$, it follows by Lemma 1.15(3) that

$$B(y, \frac{\lambda\epsilon - \|u - z\|}{1 - \lambda}) \cap K \neq \emptyset.$$

In particular, choose $u_1 \in K$ such that

$$\|u_1 - y\| < \frac{1}{1 - \lambda}[\lambda\epsilon - \|u - z\|].$$

Now set $u_0 := \frac{1}{\lambda}[u - (1 - \lambda)u_1]$. Then

$$\|u_0 - x\| = \|\frac{1}{\lambda}[u - (1 - \lambda)u_1] - x\| = \|\frac{1}{\lambda}[u - (1 - \lambda)u_1 - \lambda x]\|$$

$$= \frac{1}{\lambda}\|u - (1 - \lambda)u_1 - [z - (1 - \lambda)y]\| \leq \frac{1}{\lambda}\|u - z\| + \frac{(1 - \lambda)}{\lambda}\|y - u_1\|$$

$$< \frac{1}{\lambda}\|u - z\| + \frac{1}{\lambda}[\lambda\epsilon - \|u - z\|] = \epsilon.$$

Thus $u_0 \in B(x, \epsilon) \subset K$. By definition of u_0 and the convexity of K, $u = \lambda u_0 + (1 - \lambda)u_1 \in K$. This proves the claim, and hence verifies (2.8.1).

It is clear that int $K \subset K \subset \overline{K}$. Hence $\overline{\text{int } K} \subset \overline{K}$. For the reverse inclusion, let $y \in \overline{K}$ and choose any $x \in$ int K. Then by (2.8.1),

$$x_n := \frac{1}{n + 1}x + (1 - \frac{1}{n + 1})y \in \text{int } K \quad \text{for all } n.$$

Since $x_n \to y$, it follows that $y \in \overline{\text{int } K}$. This proves the first equation in (2.8.2).

Since $K \subset \overline{K}$, we have $\text{int}\,K \subset \text{int}\,\overline{K}$. For the reverse inclusion, let $y \in \text{int}\,\overline{K}$. Then there exists $\epsilon > 0$ such that $B(y,\epsilon) \subset \overline{K}$. Now let $x \in \text{int}\,K \setminus \{y\}$ and set $z := (1+\rho)y - \rho x$, where $\rho := \frac{\epsilon}{2\|x-y\|}$. Since $\|z - y\| = \rho\|y - x\| = \frac{\epsilon}{2}$, we see that $z \in B(y,\epsilon) \subset \overline{K}$. But $y = (1+\rho)^{-1}z + \rho(1+\rho)^{-1}x \in \text{int}\,K$ by (2.8.1). This proves $\text{int}\,\overline{K} \subset \text{int}\,K$ and verifies the second equation in (2.8.2). ∎

The **boundary** of a set S, denoted by $\text{bd}\,S$, is the closure minus the interior of S:

$$\text{bd}\,S = \overline{S} \setminus \text{int}\,S.$$

The following lemma shows that the boundaries of S and $X \setminus S$ are the same, and also that the interior of S is the complement of the closure of the complement of S.

2.9 Lemma. *Let S be any subset of X. Then:*

(1) $\text{int}\,S = X \setminus (\overline{X \setminus S})$,
(2) $\text{int}\,(X \setminus S) = X \setminus \overline{S}$,
(3) $\text{bd}\,S = \text{bd}\,(X \setminus S)$.

Proof. (1) Let $x \in \text{int}\,S$. Then there exists $\epsilon > 0$ such that $B(x,\epsilon) \subset S$. Hence $X \setminus B(x,\epsilon) \supset X \setminus S$, and so

$$X \setminus B(x,\epsilon) = \overline{X \setminus B(x,\epsilon)} \supset \overline{X \setminus S}.$$

This implies that $B(x,\epsilon) \cap (\overline{X \setminus S}) = \emptyset$ and hence $x \in X \setminus (\overline{X \setminus S})$. This proves that $\text{int}\,S \subset X \setminus (\overline{X \setminus S})$.

For the reverse inclusion, let $x \in X \setminus (\overline{X \setminus S})$. Since this set is open, there exists $\epsilon > 0$ such that $B(x,\epsilon) \subset X \setminus (\overline{X \setminus S}) \subset X \setminus (X \setminus S) = S$. That is, $x \in \text{int}\,S$.

(2) This follows from (1) by replacing S with $X \setminus S$.

(3) Let $x \in \text{bd}\,S (= \overline{S} \setminus \text{int}\,S)$. If $x \in \text{int}\,(X \setminus S)$, there exists $\epsilon > 0$ such that $B(x,\epsilon) \subset X \setminus S$, and so $B(x,\epsilon) \cap S = \emptyset$. Hence $x \notin \overline{S}$, which is a contradiction. Thus $x \notin \text{int}\,(X \setminus S)$.

If $x \notin \overline{X \setminus S}$, then there exists $\epsilon > 0$ such that $B(x,\epsilon) \cap (X \setminus S) = \emptyset$. It follows that $B(x,\epsilon) \subset S$ and hence $x \in \text{int}\,S$, a contradiction. This proves that $x \in \overline{X \setminus S}$. That is, $x \in \text{bd}\,(X \setminus S)$. Hence this shows that $\text{bd}\,S \subset \text{bd}\,(X \setminus S)$. Replacing S by $X \setminus S$, we obtain $\text{bd}\,(X \setminus S) \subset \text{bd}\,S$. Combining these results, we obtain that $\text{bd}\,S = \text{bd}\,(X \setminus S)$. ∎

Five Basic Problems Revisited

We conclude this chapter by noting that each of the five problems stated in Chapter 1 can be viewed as a problem of best approximation from a convex subset of an appropriate inner product space.

Problem 1. (Best least-squares polynomial approximation to data) Let $\{(t_j, x(t_j)) \mid j = 1, 2, \ldots, m\}$ be a table of data. For any fixed integer $n < m$, find a polynomial $p(t) = \sum_0^n \alpha_i t^i$, of degree at most n, such that the expression

$$\sum_{k=1}^{m} [x(t_k) - p(t_k)]^2$$

is minimized.

Let $T = \{t_1, t_2, \ldots, t_m\}$, $X = l_2(T)$, and let $\mathcal{P}_n$ denote the subspace of X consisting of all polynomials of degree at most n. Thus the problem may be restated as follows: Minimize $\|x - p\|^2$ over all $p \in \mathcal{P}_n$. That is, *find a best approximation to $x \in X$ from the subspace $\mathcal{P}_n$.*

Problem 2. (Solution to an overdetermined system of equations) Consider the linear system of m equations in the n unknowns $x_1, x_2, \ldots, x_n$:

$$a_{11}x_1 + a_{12}x_2 + \cdots + a_{1n}x_n = b_1$$
$$\ldots$$
$$a_{m1}x_1 + a_{m2}x_2 + \cdots + a_{mn}x_n = b_m,$$

or briefly, $Ax = b$. Find a vector $x \in \mathbb{R}^n$ that minimizes the expression

$$\sum_{i=1}^{m} \left(\sum_{j=1}^{n} a_{ij}x_j - b_i \right)^2.$$

Let $X = l_2(m)$. Then the problem is to find an $x \in \mathbb{R}^n$ that minimizes $\|Ax - b\|^2$. Letting M denote the "range of A," i.e.,

$$M = \{y \in X \mid y = Ax, \ x \in \mathbb{R}^n\},$$

we see that M is a subspace of X. Thus, *if we find a best approximation $y \in M$ to b, then any $x \in \mathbb{R}^n$ with $Ax = y$ solves the problem.*

Problem 3. (Best least-squares polynomial approximation to a function) Let x be a real continuous function on the interval $[a, b]$. Find a polynomial $p(t) = \sum_0^n \alpha_i t^i$ of degree at most n that minimizes the expression

$$\int_a^b [x(t) - p(t)]^2 dt.$$

Let $X = C_2[a, b]$ and $M = \mathcal{P}_n$. Then the problem may be restated as follows: *Find a best approximation to x from the subspace M.*

Problem 4. (A control problem) The position θ of the shaft of a dc motor driven by a variable current source u is governed by the differential equation

(4.1) $\theta''(t) + \theta'(t) = u(t), \qquad \theta(0) = \theta'(0) = 0,$

where $u(t)$ is the field current at time t. Suppose that the boundary condition is given by

(4.2) $\theta(1) = 1, \quad \theta'(1) = 0,$

and the energy is proportional to $\int_0^1 u^2(t)dt$. In the class of all real continuous funtions u on $[0, 1]$ that are related to θ by (4.1), find the one so that (4.2) is satisfied and has minimum energy.

If $w(t)$ denotes the shaft angular velocity at time t, then

(4.3) $$\theta(t) = \int_0^t w(s)ds \text{ and } \theta'(t) = w(t).$$

Thus equations (4.1) and (4.2) may be written in the equivalent form

(4.4) $$w'(t) + w(t) = u(t), \quad w(0) = 0,$$

subject to

(4.5) $$\int_0^1 w(t)dt = 1 \text{ and } w(1) = 0.$$

The solution of (4.4) is easily seen to be given by

(4.6) $$w(t) = \int_0^t u(s)e^{s-t}ds.$$

Thus, using (4.5) and (4.6), our problem reduces to that of finding $u \in C_2[0,1]$ that has minimal norm in the set of all such functions that satisfy

(4.7) $$\int_0^1 \left[\int_0^t u(s)e^{s-t}ds \right] dt = 1$$

and

(4.8) $$\int_0^1 u(s)e^{s-1}ds = 0.$$

Let $\mathcal{X}_{[0,t]}$ denote the characteristic function of the interval $[0,t]$; i.e., $\mathcal{X}_{[0,t]}(s) = 1$ if $s \in [0,t]$, and $\mathcal{X}_{[0,t]}(s) = 0$ otherwise. By interchanging the order of integration, we can rewrite (4.7) as

$$1 = \int_0^1 \left[\int_0^t u(s)e^{s-t}ds \right] dt = \int_0^1 \left[\int_0^1 \mathcal{X}_{[0,t]}(s)u(s)e^{s-t}ds \right] dt$$

$$= \int_0^1 \left[\int_0^1 \mathcal{X}_{[0,t]}(s)u(s)e^{s-t}dt \right] ds = \int_0^1 u(s)e^s \left[\int_s^1 e^{-t}dt \right] ds$$

$$= \int_0^1 u(s)e^s[-e^{-t}|_s^1]ds = \int_0^1 u(s)[1 - e^{s-1}]ds.$$

Finally, letting $X = C_2[0,1]$, $x_1(t) = e^{t-1}$, and $x_2(t) = 1 - e^{t-1}$, we see that the problem may be stated as follows. Find $u \in X$ satisfying

$$\langle u, x_1 \rangle = 0, \quad \langle u, x_2 \rangle = 1$$

and having minimal norm. If we set

$$V = \{u \in X \mid \langle u, x_1 \rangle = 0, \quad \langle u, x_2 \rangle = 1\},$$

we see that V is a convex set and the problem may be restated as follows: *Find the best approximation to 0 from V.*

Problem 5. (Positive constrained interpolation) Let $\{x_1, x_2, \ldots, x_n\}$ be a finite set of functions in $L_2[0, 1]$, and let $b_1, b_2, \ldots, b_n$ be real numbers. In the class of all functions $x \in L_2[0, 1]$ such that

$$(5.1) \qquad\qquad x(t) \geq 0 \quad \text{for all } t \in [0, 1]$$

and

$$\int_0^1 x(t) x_i(t) dt = b_i \quad (i = 1, 2, \ldots, n),$$

find the one such that

$$\int_0^1 x^2(t) dt$$

is minimized.

Letting $X = L_2[0, 1]$ and

$$K = \{x \in X \mid x \geq 0, \ \langle x, x_i \rangle = b_i \ (i = 1, 2, \cdots, n)\},$$

we see that the problem reduces to finding the best approximation to 0 from the convex set K, i.e., determining $P_K(0)$.

Exercises

1. Show that the open ball

$$B(x, r) = \{y \in X \mid \|x - y\| < r\}$$

   is convex.

2. Show that the interval

$$[x, y] = \{\lambda x + (1 - \lambda)y \mid 0 \leq \lambda \leq 1\}$$

   is convex.

3. Show that a set K is convex if and only if $\lambda K + (1 - \lambda)K = K$ for every $0 \leq \lambda \leq 1$.

4. Show that the following statements are equivalent for a set C:
   (1) C is a convex cone;
   (2) $x + y \in C$ and $\rho x \in C$ whenever $x, y \in C$ and $\rho \geq 0$;
   (3) $\alpha C + \beta C = C$ for all $\alpha, \beta \geq 0$ with $\alpha^2 + \beta^2 \neq 0$, and $0 \in C$.

5. Show that the following statements are equivalent for a set M:
   (1) M is a subspace;
   (2) $x + y \in M$ and $\alpha x \in M$ whenever $x, y \in M$ and $\alpha \in \mathbb{R}$;
   (3) $\alpha M + \beta M = M$ for all $\alpha, \beta \in \mathbb{R}$ with $\alpha^2 + \beta^2 \neq 0$.

6. (1) Verify that the sum

$$A + B = \{a + b \mid a \in A, \ b \in B\}$$

   of two convex cones (respectively subspaces) is a convex cone (respectively subspace).
   (2) Verify that the scalar multiple

$$\alpha A := \{\alpha a \mid a \in A\}$$

of a convex set (respectively convex cone, subspace) is a convex set (respectively convex cone, subspace).

7. Show that every subspace is a convex cone, and every convex cone is convex. Give examples in $l_2(2)$ of a convex cone that is not a subspace, and a convex set that is not a convex cone.

8. Give an alternative proof of the uniqueness theorem for best approximation along the following lines. Let K be a convex set. Prove the following statements.

    (a) $P_K(x) = K \cap B[x, d(x, K)]$, and hence $P_K(x)$ is a convex set.

    (b) The only nonempty convex subsets of the sphere

$$S[x, r] := \{y \in X \mid \|x - y\| = r\}$$

are the singletons.

    (c) $P_K(x) \subset S[x, d(x, K)]$.

    (d) $P_K(x)$ is either empty or a singleton for each $x \in X$.

9. Give an alternative proof of the uniqueness theorem using the parallellogram law. [Hint: If $y_1, y_2 \in P_K(x)$, deduce $y_1 = y_2$ from the relation

$$\|y_1 - y_2\|^2 = \|(x - y_2) - (x - y_1)\|^2$$

$$= 2\|x - y_2\|^2 + 2\|x - y_1\|^2 - 4\|x - \frac{1}{2}(y_1 + y_2)\|^2.]$$

10. Let A be a nonempty subset of X. The **convex hull** of A, denoted by $\mathrm{co}(A)$, is the intersection of all convex sets that contain A. Verify the following statements.

    (1) $\mathrm{co}(A)$ is a convex set; hence $\mathrm{co}(A)$ is the smallest convex set that contains A.

    (2) A is convex if and only if $A = \mathrm{co}(A)$.

    (3) $\mathrm{co}(A) = \{\sum_1^n \lambda_i x_i \mid x_i \in A, \lambda_i \geq 0, \sum_1^n \lambda_i = 1, n \in \mathbb{N}\}$.

11. Let A be a nonempty subset of X. The **conical hull** of A, denoted by $\mathrm{con}(A)$, is the intersection of all convex cones that contain A. Verify the following statements.

    (1) $\mathrm{con}(A)$ is a convex cone; hence $\mathrm{con}(A)$ is the smallest convex cone that contains A.

    (2) A is a convex cone if and only if $A = \mathrm{con}(A)$.

    (3) $\mathrm{con}(A) = \{\sum_1^n \rho_i x_i \mid x_i \in A, \rho_i \geq 0, n \in \mathbb{N}\}$.

12. Let A be a nonempty subset of X. Verify the following statements.

    (1) $\mathrm{span}(A)$ is the intersection of all subspaces of X that contain A.

    (2) $\mathrm{span}(A)$ is the smallest subspace of X that contains A.

    (3) A is a subspace if and only if $A = \mathrm{span}(A)$.

13. A set V in X is called **affine** if $\alpha x + (1 - \alpha)y \in V$ whenever $x, y \in V$ and $\alpha \in \mathbb{R}$. The **affine hull** of a set S, denoted by $\mathrm{aff}(S)$, is the intersection of all affine sets that contain S. Prove the following statements.

    (1) V is affine if and only if V is a translate of a subspace. That is, $V = M + v$ for some (unique) subspace M and any $v \in V$.

    (2) The intersection of any collection of affine sets is affine. Thus $\mathrm{aff}(S)$ is the smallest affine set that contains S.

    (3) $\mathrm{aff}(S) = \{\sum_1^n \alpha_i x_i \mid x_i \in S, \alpha_i \in \mathbb{R}, \sum_1^n \alpha_i = 1, n \in \mathbb{N}\}$.

    (4) The set S is affine if and only if $S = \mathrm{aff}(S)$.

14. Prove that if K is a convex cone (respectively a subspace, affine), then its closure $\overline{K}$ is also a convex cone (respectively a subspace, affine).

15. Show that $B[x,r] = \overline{B(x,r)}$.

16. (1) Verify that the sum

$$A + B := \{a + b \mid a \in A,\ b \in B\}$$

of two affine sets is an affine set.

(2) Verify that any scalar multiple

$$\alpha A := \{\alpha a \mid a \in A\}$$

of an affine set A is an affine set.

17. Prove that $\mathcal{P}_n$ is an $(n+1)$-dimensional subspace of $C_2[a,b]$. [Hint: Verify that $\{x_0, x_1, \ldots, x_n\}$ is linearly independent, where $x_i(t) = t^i$.]

18. If K is a convex Chebyshev set in an inner product space X and $x \in X \setminus K$, show that $P_K(x') = P_K(x)$ for every $x' \in \mathrm{co}\{x, P_K(x)\}$. That is, if $\lambda \in [0,1]$ and $x' = \lambda x + (1-\lambda)P_K(x)$, then $P_K(x') = P_K(x)$.

19. Show that the uniqueness theorem (Theorem 2.4) is also valid in a complex inner product space.

Historical Notes

The term "metric projection" goes back at least to Aronszajn and Smith (1954), while the term "Chebyshev set" goes back to Efimov and Stechkin (1958). The term "proximinal" is a combination of the words "proximity" and "minimal" and was coined by Killgrove (see Phelps (1957; p. 790)). That the interior and closure of a convex set is convex (i.e., Theorem 2.8) is classical. (See Steinitz (1913, 1914, 1916), and the more modern classic by Rockafellar (1970) for these and other results on convexity.)

The control problem (Problem 4) can be found in Luenberger (1969; p. 66).

There are several good introductory books on approximation theory that include at least one or two chapters on best approximation in inner product spaces. For example, Achieser (1956), Davis (1963), Cheney (1966), Rivlin (1969), and Laurent (1972). A fairly detailed overview of the general theory of best approximation in normed linear spaces can be found in Singer (1970) and Singer (1974).

CHAPTER 3

EXISTENCE AND UNIQUENESS OF BEST APPROXIMATIONS

Existence of Best Approximations

The main existence theorem of this chapter states that every approximatively compact set is proximinal. This result contains virtually all the existence theorems of interest. In particular, the two most useful existence and uniqueness theorems can be deduced from it. They are: (1) Every finite-dimensional subspace is Chebyshev, and (2) every closed convex subset of a Hilbert space is Chebyshev.

We first observe that every proximinal set must be closed.

3.1 Proximinal Sets are Closed. *Let K be a proximinal subset of X. Then K is closed.*

Proof. If K were not closed, there would exist a sequence $\{x_n\}$ in K such that $x_n \to x$ and $x \notin K$. Then $d(x, K) \leq \|x - x_n\| \to 0$, so that $d(x, K) = 0$. But $\|x - y\| > 0$ for each $y \in K$, since $x \notin K$. This contradicts $P_K(x) \neq \emptyset$. ∎

The converse to Theorem 3.1 is false. Indeed, in every incomplete inner product space there is a closed subspace that is not proximinal. This general fact will follow from results proved in Chapter 6 (see, e.g., Exercise 6(b) at the end of Chapter 6). For the moment, we content ourselves with a particular example.

3.2 A Closed Subspace of $C_2[-1, 1]$ That Is Not Proximinal. *Let*

$$M = \left\{ y \in C_2[-1, 1] \,\middle|\, \int_0^1 y(t)\,dt = 0 \right\}.$$

Then M is a closed subspace of $C_2[-1, 1]$ that is not proximinal.

Proof. To see that M is closed, let $\{y_n\} \subset M$ and $y_n \to y$. Then by Schwarz's inequality

$$\left| \int_0^1 y(t)\,dt \right| = \left| \int_0^1 [y(t) - y_n(t)]\,dt \right| \leq \int_{-1}^1 |y(t) - y_n(t)|\,dt$$

$$\leq \left[\int_{-1}^1 |y(t) - y_n(t)|^2\,dt \right]^{\frac{1}{2}} \left[\int_{-1}^1 1\,dt \right]^{\frac{1}{2}} = \sqrt{2}\|y - y_n\| \to 0.$$

Hence $\int_0^1 y(t)\,dt = 0$, which implies $y \in M$, and M is closed. M is clearly a subspace by linearity of integration.

Next define x on $[-1, 1]$ by $x(t) = 1$ for all t. Then $x \in C_2[-1, 1]$, and for each

$y \in M$,

$$\|x - y\|^2 = \int_{-1}^{1} |x(t) - y(t)|^2 \, dt = \int_{-1}^{0} |1 - y(t)|^2 \, dt + \int_{0}^{1} |1 - y(t)|^2 \, dt$$

$$= \int_{-1}^{0} |1 - y(t)|^2 \, dt + \int_{0}^{1} [1 - 2y(t) + y^2(t)] \, dt$$

$$= \int_{-1}^{0} |1 - y(t)|^2 \, dt + 1 + \int_{0}^{1} y^2(t) \, dt \geq 1,$$

and equality holds if and only if $y(t) = 1$ for $-1 \leq t \leq 0$ and $y(t) = 0$ for $0 \leq t \leq 1$. But such a y is not continuous. This proves that $d(x, M) \geq 1$ and $\|x - y\| > 1$ for all $y \in M$.

Next, given any $0 < \epsilon < 1$, define y_ϵ on $[-1, 1]$ by

$$y_\epsilon(t) = \begin{cases} 1 & \text{if } -1 \leq t \leq -\epsilon, \\ -\epsilon^{-1}t & \text{if } -\epsilon < t < 0, \\ 0 & \text{if } 0 \leq t \leq 1. \end{cases}$$

Then it is easy to check that $y_\epsilon \in M$ and $\|x - y_\epsilon\|^2 = 1 + \epsilon/3$. Hence $d(x, M) \leq 1$. From the preceding paragraph, it follows that

$$d(x, M) = 1 < \|x - y\| \quad \text{for all } y \in M.$$

Thus x has no best approximation in M. ■

3.3 Definition. *A nonempty subset K of X is called:*
(1) **complete** *if each Cauchy sequence in K converges to a point in K;*
(2) **approximatively compact** *if given any $x \in X$, each sequence $\{y_n\}$ in K with $\|x - y_n\| \to d(x, K)$ has a subsequence that converges to a point in K.*

Remarks. One can easily verify (see Exercise 3 at the end of the chapter) the following. (1) Every complete set is closed, and (2) in a Hilbert space, a nonempty set is complete if and only if it is closed. The main reason for introducing complete sets is that often the inner product space X is not complete, but many of the results will nevertheless hold because the set in question *is* complete. An important example of a complete set in any inner product space is a finite-dimensional subspace (see Theorem 3.7 below).

A sequence $\{y_n\}$ in K satisfying $\|x - y_n\| \to d(x, K)$ is called a **minimizing sequence** for x. Thus K is approximatively compact if each minimizing sequence has a subsequence converging to a point in K. It is useful to observe that every minimizing sequence is bounded. This is a consequence of the inequality $\|y_n\| \leq \|y_n - x\| + \|x\| \to d(x, K) + \|x\|$.

3.4 Existence of Best Approximations. (1) *Every approximatively compact set is proximinal.*

(2) *Every complete convex set is a (approximatively compact) Chebyshev set.*

Proof. (1) Let K be approximatively compact, and let $x \in X$. Choose a minimizing sequence $\{y_n\}$ in K for x. Let $\{y_{n_j}\}$ be a subsequence that converges to some $y \in K$. By Theorem 1.12,

$$\|x - y\| = \lim \|x - y_{n_j}\| = d(x, K).$$

Thus y is a best approximation to x from K, and hence K is proximinal.

(2) Let K be a complete convex set and fix any $x \in X$. Suppose that $\{y_n\}$ is minimizing for x: $\|x - y_n\| \to d(x, K)$. Then by the parallelogram law (Theorem 1.5),

$$\|y_n - y_m\|^2 = \|(x - y_m) - (x - y_n)\|^2$$
$$= 2(\|x - y_m\|^2 + \|x - y_n\|^2) - \|2x - (y_m + y_n)\|^2$$
$$= 2(\|x - y_m\|^2 + \|x - y_n\|^2) - 4\|x - \tfrac{1}{2}(y_m + y_n)\|^2.$$

Since K is convex, $\frac{1}{2}(y_m + y_n) \in K$, and hence

$$(3.4.1) \qquad \|y_n - y_m\|^2 \le 2(\|x - y_m\|^2 + \|x - y_n\|^2) - 4d(x, K)^2.$$

Since $\{y_n\}$ is a minimizing sequence for x, it follows that the right side of (3.4.1) tends to zero as n and m tend to infinity. Thus $\{y_n\}$ is a Cauchy sequence. Since K is complete, $\{y_n\}$ converges to some point $y \in K$. This proves that K is approximatively compact.

By the first part, K is proximinal. The conclusion now follows from Theorem 2.4 and the fact that K is convex. ∎

Remarks. (1) The converse of Theorem 3.4(1) is false. That is, there are proximinal sets that are not approximatively compact. For example, let $K = \{y \in l_2 \mid \|y\| = 1\}$ and consider the sequence $\{e_n\}$, where $e_n(i) = \delta_{ni}$ for all n, i. Clearly, $e_n \in K$ and $d(0, K) = 1 = \|e_n\|$ for each n, so $\{e_n\}$ is a minimizing sequence for 0. But $\|e_n - e_m\| = \sqrt{2}$ if $n \ne m$ implies that no subsequence of $\{e_n\}$ converges. Thus K is not approximatively compact. However, it is easy to verify that $P_K(0) = K$ and $P_K(x) = \{x/\|x\|\}$ if $x \ne 0$ (see Exercise 6 at the end of the chapter). Hence K is proximinal. However, if K is convex, this cannot happen.

(2) The converse of Theorem 3.4(2) is false. That is, there are convex Chebyshev sets that are not complete. For suppose

$$X = \{x \in \ell_2 \mid x(n) = 0 \text{ except for at most finitely many } n\}.$$

Then the set $K = \{x \in X \mid x(1) = 1\}$ is clearly convex, and for any $x \in X$, the element y defined by $y(1) = 1$ and $y(i) = x(i)$ for every $i > 1$ is in K and is obviously the best approximation to x from K. However, the sequence of vectors $\{x_n\}$ defined by $x_n(i) = 1/i$ for every $i \le n$ and $x_n(i) = 0$ for every $i > n$ is a Cauchy sequence in K that does not converge to any point in X. (If it converged, its limit would be the vector x defined by $x(i) = 1/i$ for all i, and this vector is not in X.)

Uniqueness of Best Approximations

The uniqueness of best approximations from convex sets was already observed in Theorem 2.4. The next result states that every closed convex subset of Hilbert space is Chebyshev.

3.5 Closed Convex Subsets of Hilbert Space are Chebyshev. *Every closed convex subset of a Hilbert space is a (approximatively compact) Chebyshev set.*

Proof. Let K be a closed convex subset of the Hilbert space X. By Theorem 3.4(2), it suffices to verify that K is complete. Let $\{y_n\}$ be a Cauchy sequence in K. Since X is complete, there exists $y \in X$ such that $y_n \to y$. Since K is closed, $y \in K$. Thus K is complete. ∎

3.6 Corollary. *Every closed convex subset of a Hilbert space has a unique element of minimal norm.*

Proof. Let K be the closed convex set. By Theorem 3.5, 0 has a unique best approximation $y_0 \in K$. That is, $\|y_0\| < \|y\|$ for all $y \in K \backslash \{y_0\}$. ∎

Example 3.2 shows that completeness of X cannot be dropped from the hypothesis of Theorem 3.5 or Corollary 3.6.

The following theorem isolates the key step in the proof of our second main consequence of Theorem 3.4.

3.7 Finite-Dimensional Subspaces are Complete. *Let M be a finite-dimensional subspace of the inner product space X. Then:*

(1) Each bounded sequence in M has a subsequence that converges to a point in M;

(2) M is closed;

(3) M is complete;

(4) Suppose $\{x_1, x_2, \ldots, x_n\}$ is a basis for M, $y_k = \sum_{i=1}^{n} \alpha_{ki} x_i$, and $y = \sum_{1}^{n} \alpha_i x_i$. Then $y_k \to y$ if and only if $\alpha_{ki} \to \alpha_i$ for each $i = 1, 2, \ldots, n$.

Proof. Let $M = \text{span}\{x_1, x_2, \ldots, x_n\}$ be n-dimensional. We proceed to simultaneously prove (1) and (2) by induction on n. For $n = 1$, let $\{y_k\}$ be a bounded sequence in M. Then $y_k = \alpha_k x_1$, so $\{\alpha_k\}$ must be a bounded sequence in $\mathbb{R}$. Thus there exists a subsequence $\{\alpha_{k_j}\}$ and $\alpha_0 \in \mathbb{R}$ such that $\alpha_{k_j} \to \alpha_0$. Setting $y_0 = \alpha_0 x_1 \in M$, we see that

$$\|y_{k_j} - y_0\| = |\alpha_{k_j} - \alpha_0| \, \|x_1\| \to 0.$$

Of course, if $\{y_k\}$ is actually a convergent sequence, then $\{y_k\}$ is bounded (by Lemma 1.15), and the above argument shows that its limit is in M. This proves (1) and (2) when $n = 1$.

Next assume that (1) and (2) are true when the dimension of M is n and suppose $M = \text{span}\{x_1, x_2, \ldots, x_n, x_{n+1}\}$ is $(n+1)$-dimensional. Set $M_n = \text{span}\{x_1, \ldots, x_n\}$, so M_n is n-dimensional. Since M_n is closed by the induction hypothesis and $x_{n+1} \notin M_n$, it follows that $d(x_{n+1}, M_n) > 0$. Let $\{y_k\}$ be a bounded sequence in M. Then $y_k = z_k + \beta_k x_{n+1}$, where $z_k \in M_n$. Using Theorem 2.7 (2) and the fact that $0 \in M_n$, we get

$$\|y_k\| \geq d(y_k, M_n) = d(\beta_k x_{n+1}, M_n) = |\beta_k| d(x_{n+1}, M_n),$$

which implies

$$|\beta_k| \leq \frac{\|y_k\|}{d(x_{n+1}, M_n)}.$$

Thus $\{\beta_k\}$ is a bounded sequence, and

$$\|z_k\| = \|y_k - \beta_k x_{n+1}\| \leq \|y_k\| + |\beta_k| \, \|x_{n+1}\|$$

implies that $\{z_k\}$ is a bounded sequence. By the induction hypothesis, there is a subsequence $\{z_{k_j}\}$ and $z_0 \in M_n$ such that $z_{k_j} \to z_0$. Since $\{\beta_{k_j}\}$ is bounded, by passing to a further subsequence if necessary, we may assume that $\beta_{k_j} \to \beta_0 \in \mathbb{R}$. Hence setting $y_0 = z_0 + \beta_0 x_{n+1}$, it follows that $y_0 \in M$ and

$$\|y_{k_j} - y_0\| \leq \|z_{k_j} - z_0\| + |\beta_{k_j} - \beta_0| \, \|x_{n+1}\| \to 0.$$

Again, if $\{y_k\}$ is actually a convergent sequence, this proves that its limit is in M. Thus (1) and (2) hold for M.

This completes the inductive step and hence the proof statements (1) and (2).

(3) To prove that M is complete, it suffices by (1) to note that every Cauchy sequence is bounded. But this follows by Lemma 1.17(2).

(4) The proof of the "only if" part of (4) has essentially been done in the course of establishing (1). We leave the full details to the reader (see Exercise 13 at the end of the chapter). The proof of the "if" part follows easily from the inequality

$$\|y_k - y\| \le \sum_{i=1}^{n} |\alpha_{ki} - \alpha_i| \|x_i\|. \quad \blacksquare$$

While we needed completeness of the *whole space* to prove Theorem 3.5, it is not necessary to assume completeness of the space for the next result, since it is a consequence of the hypotheses that the *sets* in question are complete, and this is enough.

3.8 Finite-Dimensional Subspaces are Chebyshev. *Let X be an inner product space. Then:*

(1) *Every closed subset of a finite-dimensional subspace of X is proximinal.*

(2) *Every closed convex subset of a finite-dimensional subspace of X is Chebyshev.*

(3) *Every finite-dimensional subspace is Chebyshev.*

Proof. Statement (2) follows from (1) and Theorem 2.4. Statement (3) follows from (2) and Theorem 3.7(2). Thus it suffices to verify (1).

Let K be a closed subset of the finite-dimensional subspace M of X. By Theorem 3.4(1), it suffices to show that K is approximatively compact. Let $x \in X$ and let $\{y_n\}$ be a minimizing sequence in K for x. Then $\{y_n\}$ is bounded. By Theorem 3.7(1), $\{y_n\}$ has a subsequence converging to a point $y_0 \in M$. Since K is closed, $y_0 \in K$. Thus K is approximatively compact. $\blacksquare$

As a specific application of Theorem 3.8, which is typical of the practical applications that are often made, we verify the following proposition.

3.9 An Application. *Let $X = C_2[a,b]$ (or $X = L_2[a,b]$), n a nonnegative integer, and*

$$C := \{p \in \mathcal{P}_n \mid p(t) \ge 0 \text{ for all } t \in [a,b]\}.$$

Then C is a Chebyshev convex cone in X.

Proof. It is clear that C is a convex cone in the finite-dimensional subspace $\mathcal{P}_n$ of X. By Theorem 3.8, it suffices to verify that C is closed. Let $\{x_k\}$ be a sequence in C and $x_k \to x$. Since $\mathcal{P}_n$ is closed by Theorem 3.7, $x \in \mathcal{P}_n$. It remains to verify that $x \ge 0$. If not, then $x(t_0) < 0$ for some $t_0 \in [a,b]$. By continuity, there exists a nontrivial interval $[c,d]$ in $[a,b]$ that contains t_0 and such that $x(t) \le \frac{1}{2}x(t_0) < 0$ for all $t \in [c,d]$. Then

$$\|x_k - x\|^2 = \int_a^b |x_k(t) - x(t)|^2 dt \ge \int_c^d |x_k(t) - x(t)|^2 dt$$

$$\ge \int_c^d |x(t)|^2 dt \ge \int_c^d |\tfrac{1}{2}x(t_0)|^2 dt$$

$$= \frac{1}{4}|x(t_0)|^2(d-c) > 0.$$

Since the right side is a positive constant, this contradicts the fact that $x_k \to x$. Hence $x \geq 0$. ∎

Compactness Concepts

We next introduce two other types of "compactness." These are stronger than approximative compactness, and will also be used in the rest of the book.

3.10 Definition. *A subset K of X is called* **compact** *(respectively* **boundedly compact***) if each sequence (respectively bounded sequence) in K has a subsequence that converges to a point in K.*

The relationship between the various compactness criteria is as follows.

3.11 Lemma. *Consider the following statements about a subset K.*
(1) K is compact.
(2) K is boundedly compact.
(3) K is approximatively compact.
(4) K is proximinal.
Then $(1) \Rightarrow (2) \Rightarrow (3) \Rightarrow (4)$. *In particular, compact convex sets and boundedly compact convex sets are Chebyshev.*

Proof. The implication $(1) \Rightarrow (2)$ is obvious, and $(3) \Rightarrow (4)$ was proved in Theorem 3.4.

To prove $(2) \Rightarrow (3)$, let K be boundedly compact. If $x \in X$ and $\{y_n\}$ is a minimizing sequence in K, then $\{y_n\}$ is bounded. Thus there is a subsequence that converges to a point in K. This proves that K is approximatively compact.

The last statement is a consequence of Theorem 2.4. ∎

It is worth noting that the four properties stated in Lemma 3.11 are distinct. That is, none of the implications is reversible.

An example showing that $(4) \not\Rightarrow (3)$ was given after the proof of Theorem 3.4.

To see that $(3) \not\Rightarrow (2)$, consider the unit ball in l_2: $K = \{y \in l_2 \mid \|y\| \leq 1\}$. Then K is a closed convex subset of l_2, so by Theorem 3.5, K is approximatively compact. Taking the unit vectors $e_n \in l_2$ as defined in the paragraph following Theorem 3.4, we see that $\{e_n\}$ is a bounded sequence in K having no convergent subsequence. Hence K is not boundedly compact.

To see that $(2) \not\Rightarrow (1)$, let $K = \text{span}\{e_1\}$ in l_2. That is, $K = \{\alpha e_1 \mid \alpha \in \mathbb{R}\}$. If $\{x_n\}$ is a bounded sequence in K, then $x_n = \alpha_n e_1$ for a bounded sequence of scalars $\{\alpha_n\}$. Choose a subsequence $\{\alpha_{n_k}\}$ and $\alpha_0 \in \mathbb{R}$ such that $\alpha_{n_k} \to \alpha_0$. Then $x_{n_k} \to \alpha_0 e_1 \in K$, so K is boundedly compact. However, the sequence $y_n = ne_1 \in K$ has the property that $\|y_n - y_m\| = |n - m| \geq 1$ for all $n \neq m$, so it has no convergent subsequence. Thus K is not compact.

Compact subsets of a finite-dimensional subspace can be characterized in a useful alternative way.

3.12 Finite-Dimensional Compacts Sets. *(1) Every compact set is closed and bounded.*
(2) A subset of a finite-dimensional subspace is compact if and only if it is closed and bounded.

Proof. (1) Let K be compact. If $\{x_n\}$ is in K and $x_n \to x$, choose a subsequence converging to a point in K. Since this subsequence must also converge to x, the uniqueness of limits shows that $x \in K$. Hence K is closed. If K were unbounded, then for each n there would exist $y_n \in K$ such that $\|y_n\| > n$. Choose a subsequence $\{y_{n_k}\}$ converging to a point in K. Then $\{y_{n_k}\}$ is bounded by Lemma 1.15, which contradicts $\|y_{n_k}\| > n_k \to \infty$. Thus K is bounded.

(2) Let K be a subset of a finite-dimensional subspace M of X. By (1), it suffices to show that if K is closed and bounded, then it is compact. Let $\{x_n\}$ be a sequence in K. By Theorem 3.7, it has a subsequence converging to a point $x_0 \in M$. Since K is closed, $x_0 \in K$. Thus K is compact. ∎

The converse of (1) is false. That is, there are closed and bounded sets that are not compact. In fact, the set $K = \{x \in l_2 \mid \|x\| \le 1\}$ is not compact since the unit vectors $\{e_n\}$ in K, where $e_n(i) = \delta_{ni}$, satisfy $\|e_n - e_m\| = \sqrt{2}$ whenever $n \ne m$. Thus $\{e_n\}$ has no convergent subsequence.

Exercises

1. For any nonempty set K, show that $|d(x, K) - d(y, K)| \le \|x - y\|$ for all $x, y \in X$. In particular, if $x_n \to x$, then $d(x_n, K) \to d(x, K)$. [Hint: For any $z \in K$, $d(x, K) \le \|x - z\| \le \|x - y\| + \|y - z\|$.]

2. Let K be a convex set, $x \in K$, and let $\{y_n\}$ in K be a minimizing sequence for x. Show that $\{y_n\}$ is a Cauchy sequence, hence bounded. [Hint: Look at the proof of Theorem 3.4.]

3. Verify the following statements.
   (a) Closed subsets of complete sets are complete.
   (b) Every complete set is closed.
   (c) In a Hilbert space, a nonempty subset is complete if and only if it is closed.

4. Prove the following two statements.
   (a) If K is a closed convex subset of an inner product space X that is contained in a *complete* subset of X, then K is Chebyshev.
   (b) Theorems 3.5 and 3.8(2) can be deduced from part (a).

5. Verify the following statements.
   (a) A closed subset of a finite-dimensional subspace of X is boundedly compact.
   (b) Every finite-dimensional subspace is boundedly compact, but never compact.

6. Let K denote the unit sphere in X: $K = \{x \in X \mid \|x\| = 1\}$. Verify:
   (a) K is not convex.
   (b) $P_K(x) = x/\|x\|$ if $x \in X \backslash \{0\}$ and $P_K(0) = K$.
   (c) K is proximinal, but not Chebyshev.

7. Verify the following statements.
   (a) Every finite set is compact.
   (b) The union of a finite number of compact sets is compact.
   (c) A closed subset of a compact set is compact.
   (d) The intersection of any collection of compact sets is compact.

8. If A is compact and B is closed, then $A + B$ is closed. In particular, any translate of a closed set is closed.

9. If X is infinite-dimensional and M is a finite-dimensional subspace, show that there exists $x \in X$ such that $\|x\| = 1$ and $d(x, M) = 1$. [Hint: Take any $x_0 \in X \backslash M$ and set $x = d(x_0, M)^{-1}(x_0 - P_M(x_0))$.]

10. Show that the closed unit ball in X,

$$B[0,1] = \{\, x \in X \mid \|x\| \leq 1 \,\},$$

is compact if and only if X is finite-dimensional. [Hint: If X is infinite-dimensional, use Exercise 9 to inductively construct a sequence $\{x_n\}$ in X such that $\|x_n\| = 1$ and $\|x_n - x_m\| \geq 1$ if $n \neq m$.]

11. Let

$$M = \{ x \in l_2 \mid x(n) = 0 \text{ for all but finitely many } n \}.$$

Show that M is a subspace in l_2 that is not closed. What is $\overline{M}$?

12. Prove that the set

$$M = \{ x \in l_2 \mid x(2n) = 0 \text{ for all } n \}$$

is an infinite-dimensional Chebyshev subspace in l_2. What is $P_M(x)$ for any $x \in l_2$?

13. Let $C = \{\, x \in l_2 \mid x(n) \geq 0 \text{ for all } n \,\}$. Prove that C is a convex Chebyshev subset of l_2, and determine $P_C(x)$ for any $x \in l_2$.

14. Let $\{p_k\} \subset \mathcal{P}_n \subset C_2[a,b]$ and suppose $\|p_k\| \to 0$. Show that $p_k \to 0$ *uniformly* on $[a,b]$. That is, for each $\epsilon > 0$, there exists an integer N such that $\sup_{x \in [a,b]} |p_k(x)| < \epsilon$ whenever $k \geq N$. [Hint: Theorem 3.7(4).]

15. Let
$$X = \{ x \in l_2 \mid x(n) = 0 \text{ for all but finitely many } n \}$$

and $M = \{ x \in X \mid \sum_1^\infty \frac{1}{2^n} x(n) = 0 \}$. Verify the following statements.

(a) M is a closed subspace in X that is not equal to X.

(b) $M^\perp := \{ x \in X \mid x \perp M \} = \{0\}$.
[Hint: If $z \in M^\perp$, then $\langle z, e_n - 2e_{n+1} \rangle = 0$ for every positive integer n. What does this say about z? Here e_j denotes the element in X such that $e_j(n) = 0$ if $n \neq j$ and $e_j(j) = 1$.]

(c) If some $x \in X$ has 0 as a best approximation in M, then $x \in M^\perp$.
[Hint: If not, then $\langle x, y \rangle > 0$ for some $y \in M$. Then for $\lambda > 0$ sufficiently small, the element λy satisfies $\|x - \lambda y\|^2 = \|x\|^2 - \lambda[2\langle x, y \rangle - \lambda\|y\|^2] < \|x\|^2$.]

(d) No element of $X \setminus M$ has a best approximation in M!
[Hint: If some $x \in X \setminus M$ had a best approximation $x_0 \in M$, then $z = x - x_0$ has 0 as a best approximation in M.]

16. In the text, Corollary 3.6 was deduced from Theorem 3.5. Now use Corollary 3.6 to deduce Theorem 3.5. Hence show that Theorem 3.5 and Corollary 3.6 are equivalent. [Hint: A translate of a convex set is convex.]

17. Show that *all* the definitions and results of this chapter are valid in *complex* inner product spaces also.

18. All the definitions and several of the results of this chapter are valid in an arbitrary normed linear space. In particular, verify that the Theorems 3.1, 3.4(1), 3.7, 3.11, and 3.12 hold in any normed linear space X.

Historical Notes

The notion of "compactness" has a long history (see the books by Alexandroff and Hopf (1935) and Kuratowski (1958), (1961)). There is an alternative definition of compactness to the one that we gave that is essential, especially for defining compactness in general topological spaces. A collection S of open subsets of a topological space X is called an *open cover* for the subset K of X if the union of the sets in S contains K. A finite subcollection of an open cover for K is called a *finite subcover* if it is also an open cover for K. The set K is called **compact** (or, in the older literature, **bicompact**) if each open cover of K has a finite subcover. It is well-known that in a metric space this definition of compactness is equivalent to the sequential one we gave in Definition 3.10 (see, e.g., Dunford and Schwartz (1958; Theorem 15, p. 22)).

Bolzano (1817) established the existence of a least upper bound for a bounded sequence of real numbers. In his (unpublished) Berlin lectures in the 1860s, Weierstrass used Bolzano's method, which he duly credited to Bolzano, to prove that *every bounded infinite set of real numbers has a limit point* (see also Grattan–Guiness (1970; p. 74) or Kline (1972; p. 953)). The essence of the method was to divide the bounded interval into two parts and select the part that contained infinitely many points. By repeating this process, he closes down on a limit point of the set. It is an easy step from this result to prove what is now called the "Bolzano–Weierstrass theorem": *A subset of the real line is (sequentially) compact if and only if it is closed and bounded.* Indeed, it is also an easy step to prove that this holds more generally in $\mathbb{R}^n$ (see Theorem 3.12).

Heine (1870) defined uniform continuity for functions of one or several variables and proved that a function that is continuous on a closed bounded interval $[a, b]$ is uniformly continuous. His method used the result that if the interval $[a, b]$ has a cover consisting of a *countable* number of open intervals, then it has a finite subcover. Independently, Borel (1895) recognized the importance of this result and stated it as a separate theorem. Cousin (1895) showed that the open cover in the result of Heine and Borel need not be restricted to be countable, although this fact is often credited to Lebesgue (1904). This result can be easily extended to any closed bounded set in $\mathbb{R}$, and even to $\mathbb{R}^n$. Moreover, the converse holds as well. The consequential result is generally called the "Heine–Borel theorem": *A subset of $\mathbb{R}^n$ is ("open cover") compact if and only if it is closed and bounded.*

The notion of "approximative compactness" goes back at least to Efimov and Stechkin (1961), while that of "bounded compactness" seems to originate with Klee (1953).

The fundamental fact that every closed convex subset of a Hilbert space is Chebyshev dates back to Riesz (1934), who adapted an argument due to Levi (1906). That any finite-dimensional subspace of a normed linear space is proximinal was established by Riesz (1918).

Cauchy (1821; p. 125) gave his criterion for the convergence of a sequence $\{x_n\}$ of real numbers: $\{x_n\}$ converges to a limit x if and only if $|x_{n+r} - x_n|$ can be made smaller than any positive number for all sufficiently large n and all $r > 0$. He proved that this condition is necessary, but only remarked (without proof) that if the condition is satisfied, then the convergence is guaranteed. According to historian Morris Kline (1972), Cauchy may have lacked a complete understanding of the structure of the real numbers at that time to give a proof. That the Cauchy criterion is a necessary and sufficient condition for the convergence of a sequence in $\mathbb{R}^n$ is an easy consequence of the result in $\mathbb{R}$ by arguing in each coordinate separately.

CHAPTER 4

CHARACTERIZATION OF BEST APPROXIMATIONS

Characterizing Best Approximations

We give a characterization theorem for best approximations from convex sets. This result will prove useful over and over again throughout the book. Indeed, it will be the basis for *every* characterization theorem that we give. The notion of a dual cone plays an essential role in this characterization. In the particular case where the convex set is a subspace, we obtain the familiar orthogonality condition, which for finite-dimensional subspaces reduces to a linear system of equations called the "normal equations." When an orthonormal basis of a (finite or infinite-dimensional) subspace is available, the problem of finding best approximations is greatly simplified. The Gram–Schmidt orthogonalization procedure for constructing an orthonormal basis from a given basis is described. An application of the characterization theorem is given to determine best approximations from a translate of a convex cone. Finally, the first three problems stated in Chapter 1 are completely solved.

4.1 Characterization of Best Approximations from Convex Sets. *Let K be a convex subset of the inner product space X, $x \in X$, and $y_0 \in K$. Then $y_0 = P_K(x)$ if and only if*

$$(4.1.1) \qquad \langle x - y_0, y - y_0 \rangle \leq 0 \quad \text{for all } y \in K.$$

Proof. If (4.1.1) holds and $y \in K$, then

$$\|x - y_0\|^2 = \langle x - y_0, x - y_0 \rangle = \langle x - y_0, x - y \rangle + \langle x - y_0, y - y_0 \rangle$$
$$\leq \langle x - y_0, x - y \rangle \leq \|x - y_0\| \, \|x - y\|$$

by Schwarz's inequality. Hence $\|x - y_0\| \leq \|x - y\|$, and so $y_0 = P_K(x)$.

Conversely, suppose (4.1.1) fails. Then $\langle x - y_0, y - y_0 \rangle > 0$ for some $y \in K$. For each $0 < \lambda < 1$, the element $y_\lambda := \lambda y + (1 - \lambda)y_0$ is in K by convexity and

$$\|x - y_\lambda\|^2 = \langle x - y_\lambda, x - y_\lambda \rangle = \langle x - y_0 - \lambda(y - y_0), \; x - y_0 - \lambda(y - y_0) \rangle$$
$$= \|x - y_0\|^2 - 2\lambda \langle x - y_0, y - y_0 \rangle + \lambda^2 \|y - y_0\|^2$$
$$= \|x - y_0\|^2 - \lambda[2 \langle x - y_0, y - y_0 \rangle - \lambda \|y - y_0\|^2].$$

For $\lambda > 0$ sufficiently small, the term in brackets is positive, and thus $\|x - y_\lambda\|^2 < \|x - y_0\|^2$. Hence $y_0 \neq P_K(x)$. ∎

There are two geometric interpretations of Theorem 4.1. The first is that the angle θ between the vectors $x - y_0$ and $y - y_0$ is at least 90 degrees for every $y \in K$

(see Figure 4.1.2). The second interpretation is that the convex set K lies on one side of the hyperplane H that is orthogonal to $x - y_0$ and that passes through y_0 (see Chapter 6 for a discussion of hyperplanes).

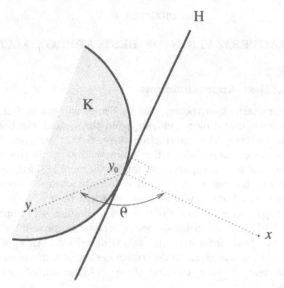

Figure 4.1.2

Dual Cones

There is an alternative way to restate the characterization theorem, Theorem 4.1, that will prove to be quite useful later on. It involves the notion of the "dual cone" of a given set.

4.2 Definition. *Let S be any nonempty subset of the inner product space X. The **dual cone** (or negative polar) of S is the set*

$$S^\circ := \{x \in X \mid \langle x, y \rangle \leq 0 \quad \text{for all} \quad y \in S\}.$$

*The **orthogonal complement** of S is the set*

$$S^\perp := S^\circ \cap (-S^\circ) = \{x \in X \mid \langle x, y \rangle = 0 \quad \text{for all} \quad y \in S\}.$$

Geometrically, the dual cone S° is the set of all vectors in X that make an angle of at least 90 degrees with every vector in S (see Figure 4.2.1).

The following is an obvious restatement of Theorem 4.1 using dual cones.

4.3 Dual Cone Characterization of Best Approximations. *Let K be a convex subset of the inner product space X, $x \in X$, and $y_0 \in K$. Then $y_0 = P_K(x)$ if and only if*

(4.3.1) $$x - y_0 \in (K - y_0)^\circ.$$

Thus the characterization of best approximations requires, in essence, the calculation of dual cones. For certain convex sets (e.g., cones, subspaces, or intersections

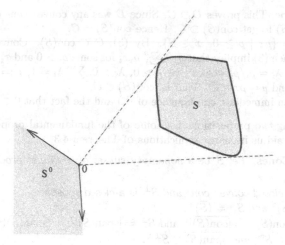

Figure 4.2.1 Dual cone

of such sets), there are substantial improvements in Theorem 4.3 that are possible. Before stating them, it is convenient to list some basic properties of dual cones.

Recall that the **conical hull** of a set S, denoted by $\mathrm{con}(S)$, is the intersection of all convex cones that contain S.

4.4 Conical Hull. *Let S be a nonempty subset of X. Then:*

(1) $\mathrm{con}(S)$ *is a convex cone, the smallest convex cone containing S.*
(2) S *is a convex cone if and only if $S = \mathrm{con}(S)$.*
(3) $\mathrm{con}(S) = \{\sum_1^n \rho_i x_i \mid x_i \in S, \rho_i \geq 0, n \in \mathbb{N}\}$.
(4) *If S is convex, then $\mathrm{con}(S) = \{\rho x \mid \rho \geq 0, x \in S\}$.*
(5) *If S is convex and $0 \in S$, then $\mathrm{con}(S) = \{\rho x \mid \rho > 0, x \in S\}$.*

Proof. The proofs of (1) and (2) are trivial.
(3) Let

$$C := \left\{\sum_1^n \rho_i x_i \mid x_i \in S, \rho_i \geq 0, n \in \mathbb{N}\right\}.$$

It is clear that $C \supset S$ (e.g., if $x \in S$, $x = 1 \cdot x \in C$). Also, let $x, y \in C$ and $\alpha, \beta \geq 0$. Then $x = \sum_1^n \rho_i x_i$ and $y = \sum_1^m \gamma_i y_i$, where $x_i, y_i \in S$ and $\rho_i, \gamma_i \geq 0$. Hence $\alpha x + \beta y = \sum_1^n \alpha \rho_i x_i + \sum_1^m \beta \gamma_i y_i = \sum_1^{n+m} \delta_i z_i$, where $\delta_i = \alpha \rho_i$ and $z_i = x_i$ if $1 \leq i \leq n$, and $\delta_i = \beta \gamma_{i-n}$ and $z_i = y_{i-n}$ if $n+1 \leq i \leq n+m$. Since $z_i \in S$ and $\delta_i \geq 0$, $\alpha x + \beta y \in C$. Thus C is a convex cone containing S. From (1) it follows that $C \supset \mathrm{con}(S)$.

Next let D be any convex cone that contains S. We want to show that $D \supset C$. Let $x \in C$. Then $x = \sum_1^n \rho_i x_i$ for some $x_i \in S$ and $\rho_i \geq 0$. We proceed by induction on n. If $n = 1$, then $x = \rho_1 x_1 \in D$, since $x_1 \in S \subset D$ and D is a cone. Assume $x \in D$ whenever x is a nonnegative linear combination of $n - 1$ elements of S. Then

$$x = \sum_1^n \rho_i x_i = \sum_1^{n-1} \rho_i x_i + \rho_n x_n \in D,$$

since D is a cone. This proves $D \supset C$. Since D was any convex cone containing S, take $D = \text{con}(S)$ to get $\text{con}(S) \supset C$. Hence $\text{con}(S) = C$.

(4) Set $C = \{\rho x \mid \rho \geq 0, \ x \in S\}$. By (3), $C \subset \text{con}(S)$. Conversely, if $y \in \text{con}(S) \setminus \{0\}$, then (3) implies that $y = \sum_1^n \rho_i x_i$ for some $\rho_i > 0$ and $x_i \in S$. Setting $\rho = \sum_1^n \rho_i$ and $\lambda_i = \rho_i / \rho$, we see that $\rho > 0$, $\lambda_i > 0$, $\sum_1^n \lambda_i = 1$, $x := \sum_1^n \lambda_i x_i \in S$ by convexity, and $y = \rho x \in C$. That is, $\text{con}(S) \subset C$.

(5) This is an immediate consequence of (4) and the fact that $0 \in S$. ∎

The following two propositions list some of the fundamental properties of dual cones that will aid us in many applications of Theorem 4.3.

4.5 Dual Cones. *Let S be a nonempty subset of the inner product space X. Then:*

(1) *S° is a closed convex cone and $S^\perp$ is a closed subspace.*

(2) *$S^\circ = (\overline{S})^\circ$ and $S^\perp = (\overline{S})^\perp$.*

(3) *$S^\circ = [\text{con}(S)]^\circ = \overline{[\text{con}(S)]}^\circ$ and $S^\perp = [\text{span}(S)]^\perp = \overline{[\text{span}(S)]}^\perp$.*

(4) *$\overline{\text{con}(S)} \subset S^{\circ\circ}$ and $\overline{\text{span}(S)} \subset S^{\perp\perp}$.*

(5) *If C is a convex cone, then $(C - y)^\circ = C^\circ \cap y^\perp$ for each $y \in C$.*

(6) *If M is a subspace, then $M^\circ = M^\perp$.*

(7) *If C is a Chebyshev convex cone (e.g., a closed convex cone in a Hilbert space), then*
$$C^{\circ\circ} = C.$$

(8) *If M is a Chebyshev subspace (e.g., a closed subspace in a Hilbert space), then*
$$M^{\circ\circ} = M^{\perp\perp} = M.$$

(9) *If X is complete and S is any nonempty subset, then $S^{\circ\circ} = \overline{\text{con}(S)}$, $S^{\perp\perp} = \overline{\text{span}(S)}$, $S^{\circ\circ\circ} = S^\circ$, and $S^{\perp\perp\perp} = S^\perp$.*

Proof. We will prove only the statements about dual cones. The proofs of the analogous statements for orthogonal complements can be easily deduced from these (or verified directly) and are left as an exercise.

(1) Let $x_n \in S^\circ$ and $x_n \to x$. Then for each $y \in S$,
$$\langle x, y \rangle = \lim \langle x_n, y \rangle \leq 0$$

implies $x \in S^\circ$ and S° is closed. Let x, z in S° and $\alpha, \beta \geq 0$. Then, for each $y \in S$,
$$\langle \alpha x + \beta z, y \rangle = \alpha \langle x, y \rangle + \beta \langle z, y \rangle \leq 0,$$

so $\alpha x + \beta z \in S^\circ$ and S° is a convex cone.

(2) Since $S \subset \overline{S}$, then $S^\circ \supset (\overline{S})^\circ$. If $x \in S^\circ$ and $y \in \overline{S}$, choose $y_n \in S$ such that $y_n \to y$. Then
$$\langle x, y \rangle = \lim \langle x, y_n \rangle \leq 0$$

implies $x \in (\overline{S})^\circ$. Thus $S^\circ \subset (\overline{S})^\circ$ and hence $S^\circ = (\overline{S})^\circ$.

(3) Since $\text{con}(S) \supset S$, $[\text{con}(S)]^\circ \subset S^\circ$. Let $x \in S^\circ$ and $y \in \text{con}(S)$. By Theorem 4.4, $y = \sum_1^n \rho_i y_i$ for some $y_i \in S$ and $\rho_i \geq 0$. Then
$$\langle x, y \rangle = \sum_1^n \rho_i \langle x, y_i \rangle \leq 0$$

implies $x \in [\mathrm{con}(S)]^\circ$, so $S^\circ \subset [\mathrm{con}(S)]^\circ$. Thus $S^\circ = [\mathrm{con}(S)]^\circ$.

The second equality of (3) follows from (2).

(4) Let $x \in S$. Then for any $y \in S^\circ$, $\langle x, y \rangle \leq 0$. Hence $x \in S^{\circ\circ}$. That is, $S \subset S^{\circ\circ}$. Since $S^{\circ\circ}$ is a closed convex cone by (1), $\overline{\mathrm{con}(S)} \subset S^{\circ\circ}$.

(5) Now, $x \in (C - y)^\circ$ if and only if $\langle x, c - y \rangle \leq 0$ for all $c \in C$. Taking $c = 0$ and $c = 2y$, it follows that the last statement is equivalent to $\langle x, y \rangle = 0$ and $\langle x, c \rangle \leq 0$ for all $c \in C$. That is, $x \in C^\circ \cap y^\perp$.

(6) If M is a subspace, then $-M = M$ implies

$$M^\circ = M^\circ \cap (-M)^\circ = M^\perp.$$

(7) Let C be a Chebyshev convex cone. By (4) we have $C \subset C^{\circ\circ}$. It remains to verify $C^{\circ\circ} \subset C$. If not, choose $x \in C^{\circ\circ} \backslash C$ and let $y_0 = P_C(x)$. By (5) and Theorem 4.3, we have $x - y_0 \in (C - y_0)^\circ = C^\circ \cap y_0^\perp$. Thus

$$0 < \|x - y_0\|^2 = \langle x - y_0, x - y_0 \rangle = \langle x - y_0, x \rangle \leq 0,$$

which is absurd.

(8) This follows from (6) and (7).

(9) This follows from (3), (7), and (8). ■

The sum of two sets S_1 and S_2 is defined by $S_1 + S_2 = \{ x + y \mid x \in S_1, y \in S_2 \}$. More generally, the sum of a finite collection of sets $\{S_1, S_2, \ldots, S_n\}$, denoted by $S_1 + S_2 + \cdots + S_n$ or $\sum_1^n S_i$, is defined by

$$\sum_1^n S_i := \left\{ \sum_1^n x_i \mid x_i \in S_i \text{ for every } i \right\}.$$

4.6 Dual Cones of Unions, Intersections, and Sums. Let $\{S_1, \ldots, S_m\}$ be a finite collection of nonempty sets in the inner product space X. Then

(1) $\left(\bigcup_1^m S_i \right)^\circ = \bigcap_1^m S_i^\circ$ and $\left(\bigcup_1^m S_i \right)^\perp = \bigcap_1^m S_i^\perp$.

(2) $\overline{\sum_1^m S_i^\circ} \subset \left(\bigcap_1^m S_i \right)^\circ$ and $\overline{\sum_1^m S_i^\perp} \subset \left(\bigcap_1^m S_i \right)^\perp$.

(3) If $0 \in \bigcap_1^m S_i$, then

$$\left(\sum_1^m S_i \right)^\circ = \bigcap_1^m S_i^\circ \quad \text{and} \quad \left(\sum_1^m S_i \right)^\perp = \bigcap_1^m S_i^\perp.$$

(4) If $\{C_1, C_2, \ldots, C_m\}$ is a collection of closed convex cones in a Hilbert space, then

$$\left(\bigcap_1^m C_i \right)^\circ = \overline{\sum_1^m C_i^\circ} \quad \text{and} \quad \left(\bigcap_1^m C_i \right)^\perp = \overline{\sum_1^m C_i^\perp}.$$

(5) If $\{M_1, M_2, \ldots, M_m\}$ is a collection of closed subspaces in a Hilbert space, then

$$\left(\bigcap_1^m M_i \right)^\perp = \overline{\sum_1^m M_i^\perp}.$$

Proof. We will verify the statements concerning dual cones, and leave the proofs of the analogous statements concerning orthogonal complements as an exercise.

(1) $x \in \cap_i S_i^\circ$ if and only if $x \in S_i^\circ$ for each i if and only if $\langle x, y \rangle \leq 0$ for each $y \in S_i$ and all i if and only if $\langle x, y \rangle \leq 0$ for all $y \in \cup_i S_i$ if and only if $x \in (\cup_i S_i)^\circ$.

(2) Let $x \in \sum_1^m S_i^\circ$. Then $x = \sum_1^m x_i$ for some $x_i \in S_i^\circ$. For every $y \in \cap_1^m S_i$, we have $\langle x, y \rangle = \sum_1^m \langle x_i, y \rangle \leq 0$, so $x \in (\cap_1^m S_i)^\circ$.

(3) $x \in (\sum_i S_i)^\circ$ if and only if $\langle x, s \rangle \leq 0$ for each $s \in \sum_i S_i$ if and only if $\langle x, \sum_i s_i \rangle \leq 0$ whenever $s_i \in S_i$ if and only if (since $0 \in \cap_i S_i$) $\langle x, s_i \rangle \leq 0$ for each $s_i \in S_i$ and all i if and only if $x \in S_i^\circ$ for each i if and only if $x \in \cap_i S_i^\circ$.

(4) By Theorem 4.5(7), $C_i^{\circ\circ} = C_i$ for each i. Also, it is easy to check that the sum $\sum_i C_i^\circ$ is a convex cone. Using part (3), we obtain

$$\bigcap_i C_i = \bigcap_i C_i^{\circ\circ} = \bigcap_i (C_i^\circ)^\circ = \left(\sum_i C_i^\circ \right)^\circ.$$

Using Theorem 4.5, it follows that

$$\left(\bigcap_i C_i \right)^\circ = \left(\sum_i C_i^\circ \right)^{\circ\circ} = \left(\overline{\sum_i C_i^\circ} \right)^{\circ\circ} = \overline{\sum_i C_i^\circ}.$$

(5) This is a special case of (4). ∎

Remarks. (1) There is a natural generalization of Theorem 4.6 to the case of *infinitely* many sets. One just needs to make the appropriate definition of the *sum of an infinite number of sets* that does not entail any convergence questions. (See Exercise 23 at the end of the chapter.)

(2) Theorem 4.6 will be particularly useful to us in Chapter 10 when we study the problem of constrained best approximation.

(3) If K_1 and K_2 are two convex sets with $0 \in K_1 \cap K_2$, it is *not* true in general that

(4.6.1) $(K_1 \cap K_2)^\circ = \overline{K_1^\circ + K_2^\circ}.$

In fact, this result is *equivalent* to the condition

(4.6.2) $\overline{con(K_1 \cap K_2)} = \overline{con(K_1)} \cap \overline{con(K_2)}.$

In other words, (4.6.1) holds if and only if the operation of taking the closed conical hull of $K_1 \cap K_2$ *commutes* with the operation of taking the intersection. (See Exercise 24 at the end of the chapter.) Sets that have this property will play an important role in the characterization theorems to be developed in Chapter 10.

In the particular case where the convex set is a convex cone, Theorem 4.3 can be strengthened by using Proposition 4.5(5).

4.7 Characterization of Best Approximations from Convex Cones.
Let C be a convex cone in X, $x \in X$, and $y_0 \in C$. *The following statements are equivalent:*

(1) $y_0 = P_C(x)$;
(2) $x - y_0 \in C^\circ \cap y_0^\perp$;
(3) $\langle x - y_0, y \rangle \leq 0$ for all $y \in C$ and $\langle x - y_0, y_0 \rangle = 0$.

The geometric interpretation of Theorem 4.7 is this: y_0 is the best approximation to x if and only if the error $x - y_0$ is orthogonal to y_0 and makes an angle of at least $90°$ with each vector in C (see Figure 4.7.3).

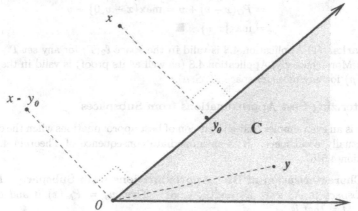

Figure 4.7.3

4.8 An Application: Best Approximation from a Translate of a Convex Cone. Let $X = C_2[a, b]$ (or $L_2[a, b]$), $v \in X$, and

$$K = \{y \in X \mid y(t) \geq v(t) \quad \text{for all} \quad t \in [a, b]\}.$$

Then K is a convex Chebyshev set and

$$(4.8.1) \qquad\qquad P_K(x) = \max\{x, v\}$$

for every $x \in X$. [Here $y = \max\{x, v\}$ denotes the function defined pointwise by $y(t) = \max\{x(t), v(t)\}$.] In particular, if

$$(4.8.2) \qquad\qquad K = \{y \in X \mid y(t) \geq 0 \text{ for all } t \in [a, b]\},$$

we get

$$(4.8.3) \qquad\qquad P_K(x) = x^+ = \max\{x, 0\}.$$

Proof. First note that $K = C + v$, where $C = \{y \in X \mid y \geq 0\}$ is a convex cone. Then

$$C^\circ = \{x \in X \mid \langle x, y \rangle \leq 0 \quad \text{for all} \quad y \geq 0\} = \{x \in X \mid x \leq 0\}.$$

Let $x \in X$ and $y_0 \in C$. Then by Theorem 4.7, $y_0 = P_C(x)$ if and only if $x - y_0 \in C^\circ \cap y_0^\perp$ if and only if $x - y_0 \leq 0$ and $\langle x - y_0, y_0 \rangle = 0$. Since $y_0 \geq 0$, this is obviously equivalent to $x - y_0 \leq 0$ and $(x - y_0)y_0 = 0$, which is equivalent to $x - y_0 \leq 0$ and $y_0(t) = 0$ whenever $x(t) - y_0(t) < 0$. That is, $y_0 = \max\{x, 0\}$. This proves that

$$P_C(x) = \max\{x, 0\}$$

for any $x \in X$. In particular, C is a Chebyshev convex cone in X. By Theorem 2.7 (1) (ii) and (v), we have that $K = C + v$ is a convex Chebyshev set and, for any $x \in X$,

$$\begin{aligned} P_K(x) = P_{C+v}(x) &= P_{C+v}(x - v + v) \\ &= P_C(x - v) + v = \max\{x - v, 0\} + v \\ &= \max\{x, v\}. \quad \blacksquare \end{aligned}$$

Remarks. (1) Application 4.8 is valid in the space $\ell_2(T)$ for any set T.

★ (2) More generally, Application 4.8 (as well as its proof) is valid in the space $L_2(T, \mathcal{S}, \mu)$ for any measure space $(T, \mathcal{S}, \mu)$.

Characterizing Best Approximations from Subspaces

There is an even simpler characterization of best approximations when the convex set is actually a subspace. It is an immediate consequence of Theorem 4.3 and Proposition 4.5(6).

4.9 Characterization of Best Approximations from Subspaces. *Let M be a subspace in X, $x \in X$, and $y_0 \in M$. Then $y_0 = P_M(x)$ if and only if $x - y_0 \in M^\perp$; that is,*

$$(4.9.1) \qquad \langle x - y_0, y \rangle = 0 \quad \text{for all} \quad y \in M.$$

The geometric interpretation of Theorem 4.9 is clear: y_0 is the best approximation to x if and only if the error $x - y_0$ is orthogonal to M (see Figure 4.9.2). This is the reason why $P_M(x)$ is often called the *orthogonal projection* of x onto M.

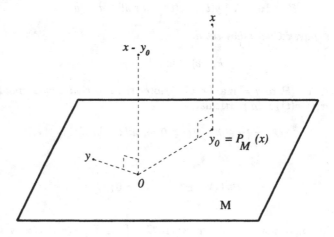

Figure 4.9.2

4.10 Corollary (The Normal Equations). *Let $\{x_1, x_2, \ldots, x_n\}$ be a basis for the n-dimensional subspace M of X. Then M is a Chebyshev subspace, and for each $x \in X$,*

$$(4.10.1) \qquad P_M(x) = \sum_{1}^{n} \alpha_i x_i ,$$

where the scalars α_i are the unique solution to the **normal equations**

$$(4.10.2) \qquad \sum_{i=1}^{n} \alpha_i \langle x_i, x_j \rangle = \langle x, x_j \rangle \qquad (j = 1, 2, \ldots, n) .$$

In particular, if $\{x_1, x_2, \ldots, x_n\}$ is an orthonormal basis for M, then

$$(4.10.3) \qquad P_M(x) = \sum_{1}^{n} \langle x, x_i \rangle x_i \quad \text{for every} \quad x \in X.$$

Proof. Fix $x \in X$ and $y_0 \in M$. Then $y_0 = \sum_1^n \alpha_i x_i$ for some scalars α_i. By Theorem 4.9, $y_0 = P_M(x)$ is equivalent to $\langle x - y_0, y \rangle = 0$ for every $y \in M$. But the latter is clearly equivalent to $\langle x - y_0, x_j \rangle = 0$ for $j = 1, 2, \ldots, n$; i.e., (4.10.2) holds.

If $\{x_1, x_2, \ldots, x_n\}$ is orthonormal, then $\langle x_i, x_j \rangle = \delta_{ij}$, so (4.10.2) implies $\alpha_i = \langle x, x_i \rangle$ for each i and hence (4.10.3) holds. ∎

Gram–Schmidt Orthonormalization

In matrix-vector notation, the normal equations (4.10.2) may be rewritten as

$$(4.10.4) \qquad G(x_1, x_2, \ldots, x_n)\alpha = \beta,$$

where

$$G(x_1, x_2, \ldots, x_n) := \begin{bmatrix} \langle x_1, x_1 \rangle & \langle x_2, x_1 \rangle & \ldots & \langle x_n, x_1 \rangle \\ \langle x_1, x_2 \rangle & \langle x_2, x_2 \rangle & \ldots & \langle x_n, x_2 \rangle \\ \ldots & & & \\ \langle x_1, x_n \rangle & \langle x_2, x_n \rangle & \ldots & \langle x_n, x_n \rangle \end{bmatrix}$$

is called the **Gram matrix** of $\{x_1, x_2, \ldots, x_n\}$, and α and β are the column vectors $\alpha = (\alpha_1, \alpha_2, \ldots, \alpha_n)^T$ and $\beta = (\langle x, x_1 \rangle, \langle x, x_2 \rangle, \ldots, \langle x, x_n \rangle)^T$.

It can be shown (see Exercise 7 at the end of the chapter) that the determinant of the Gram matrix satisfies the inequalities

$$0 \leq \det G(x_1, x_2, \ldots, x_n) \leq \|x_1\|^2 \|x_2\|^2 \cdots \|x_n\|^2 ,$$

and equality holds on the left (respectively right) side if and only if $\{x_1, x_2, \ldots, x_n\}$ is a linearly dependent (respectively orthogonal) set. As a practical matter, it is better if the determinant of the Gram matrix is as large as possible. This is because of the well-known "ill-conditioning" phenomenon: If $\det G(x_1, x_2, \ldots, x_n)$ is close to zero, the inverse of the Gram matrix has some entries that are very large; hence small errors in the calculation of the vector β in (4.10.4) (due to rounding, truncation, etc.) could result in very large errors in the solution vector α.

This suggests that the most desirable situation is to have the determinant of the Gram matrix as large as possible, that is, when the set $\{x_1, x_2, \ldots, x_n\}$ is orthogonal. Moreover, we have just seen in Corollary 4.10 that the normal equations have an especially simple form in this case. For these (and other) reasons, one would like to have an orthogonal basis for any given subspace. The next result describes a *constructive method* for achieving this.

4.11 Gram–Schmidt Orthogonalization Process. Let $\{x_1, x_2, \dots\}$ be a finite or countably infinite set of linearly independent vectors. Let $M_0 = \{0\}$ and $M_n = \operatorname{span}\{x_1, x_2, \dots, x_n\}$ for $n \geq 1$. Define

$$(4.11.1) \qquad y_n = x_n - P_{M_{n-1}}(x_n), \quad e_n = y_n/\|y_n\|$$

for each $n \geq 1$. Then:

(1) $\{y_1, y_2, \dots\}$ is an orthogonal set.
(2) $\{e_1, e_2, \dots\}$ is an orthonormal set.
(3) $\operatorname{span}\{y_1, y_2, \dots, y_n\} = \operatorname{span}\{e_1, e_2, \dots, e_n\} = M_n$ for each $n \geq 1$.
(4) $y_1 = x_1$ and $y_n = x_n - \sum_{i=1}^{n-1} \langle x_n, e_i \rangle e_i$ for $n \geq 2$.

In particular, every finite-dimensional subspace has an orthonormal basis.

Proof. By Theorem 3.8, M_n is Chebyshev for each $n \geq 0$ so y_n is well-defined. Since $\{x_1, x_2, \dots, x_n\}$ is linearly independent for each n, $x_n \notin M_{n-1}$. Hence $y_n \neq 0$ for all n, so e_n is well-defined. By Theorem 4.9, $y_n \in M_{n-1}^\perp$ for every $n \geq 1$. For any $j < n$,

$$y_j = x_j - P_{M_{j-1}}(x_j) \subset M_j \subset M_{n-1},$$

so $y_n \perp y_j$. By symmetry, $y_n \perp y_j$ whenever $n \neq j$. Thus (1) holds.

Since $e_n = y_n/\|y_n\|$, statement (2) and the first equality in (3) follow from (1). Since $\{e_1, e_2, \dots, e_n\} \subset M_n$, M_n is n-dimensional, and orthonormal sets are linearly independent by Theorem 1.7, we obtain that $\operatorname{span}\{e_1, e_2, \dots, e_n\} = M_n$. This completes the proof of (3).

Finally, to verify (4), it suffices to show that $P_{M_{n-1}}(x_n) = \sum_{i=1}^{n-1} \langle x_n, e_i \rangle e_i$ for $n \geq 2$. But this follows from (2), (3), and (4.10.3). ∎

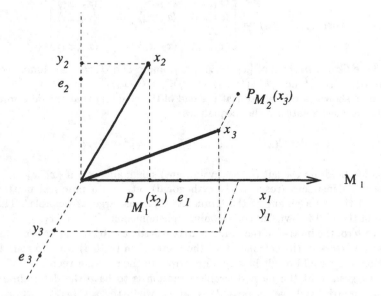

Figure 4.11.2. The Gram–Schmidt theorem

In practice, one constructs the orthonormal set $\{e_1, e_2, \dots\}$ from the linearly independent set $\{x_1, x_2, \dots\}$ inductively as follows (see Figure 4.11.2):

$$y_1 = x_1, \qquad\qquad\qquad e_1 = y_1/\|y_1\|,$$

$$\cdots \qquad\qquad\qquad\qquad \cdots$$

$$y_n = x_n - \sum_{i=1}^{n-1} \langle x_n, e_i \rangle e_i, \qquad e_n = y_n/\|y_n\|,$$

$(n \geq 2)$.

4.12 Examples.

(1) (Legendre polynomials) If one orthonormalizes the set of monomials $\{1, t, t^2, \dots\}$ in $C_2[-1, 1]$, the so-called *Legendre polynomials* are obtained. The first four Legendre polynomials are given by

$$p_0(t) = 1/\sqrt{2}, \qquad\qquad p_1(t) = \frac{\sqrt{6}}{2} t,$$

$$p_2(t) = \frac{\sqrt{10}}{4}(3t^2 - 1), \qquad p_3(t) = \frac{\sqrt{14}}{4}(5t^3 - 3t).$$

In general,

$$p_n(t) = \frac{(-1)^n \sqrt{2n+1}}{2^n \sqrt{2n!}} \frac{d^n}{dt^n}[(1 - t^2)^n]$$

for $n \geq 0$.

(2) (Chebyshev polynomials) Suppose one orthonormalizes the monomials $\{1, t, t^2, \dots\}$ in the weighted space $C_2([-1, 1]; w)$, where $w(t) = (1 - t^2)^{-\frac{1}{2}}$ (recall Example 1.3(4) for the definition of this space). Then the so-called (normalized) *Chebyshev polynomials* of the first kind are obtained. The first four Chebyshev polynomials are given by

$$\tilde{T}_0(t) = \frac{1}{\sqrt{\pi}}, \qquad\qquad \tilde{T}_1(t) = \sqrt{\frac{2}{\pi}} t,$$

$$\tilde{T}_2(t) = \sqrt{\frac{2}{\pi}}(2t^2 - 1), \qquad \tilde{T}_3(t) = \sqrt{\frac{2}{\pi}}(4t^3 - 3t).$$

In general, $\tilde{T}_n(t) = \sqrt{\frac{2}{\pi}} T_n(t)$ for $n \geq 1$, where $T_n(t) := \cos(n \arccos t)$ ($n = 1, 2, \dots$) are the "usual" Chebyshev polynomials.

(3) (Polynomials on discrete sets) Consider a set of $m > 1$ distinct real numbers $T = \{t_1, t_2, \dots, t_m\}$ and the space $l_2(T)$. If one orthonormalizes the monomials $\{1, t, t^2, \dots, t^n\}$, where $n \leq m - 1$, an orthonormal set of polynomials is obtained. For example, if $t_i = i$ for each i, then $l_2(T) = l_2(m)$, and the first three orthonormal polynomials are given by

$$p_0(t) = \frac{1}{\sqrt{m}}, \qquad p_1(t) = \sqrt{\frac{12}{m(m^2 - 1)}}[t - \frac{1}{2}(m + 1)],$$

and $p_2(t) = \lambda[t^2 - (m+1)t + \frac{1}{6}(m+1)(m+2)]$, where the positive scalar λ is chosen so that $\|p_2\| = 1$.

As the above examples show, the Gram–Schmidt orthonormalization procedure can be used to compute orthonormal polynomials of arbitrarily high degree. In practice, however, the following three-term recurrence relation for orthonormal polynomials is more efficient for their actual computation.

4.13 Three-Term Recurrence. *Let* $p_0, p_1, p_2, \ldots$ *be the sequence of orthonormal polynomials obtained by orthonormalizing the monomials* $1, t, t^2, \ldots$ *via the Gram–Schmidt procedure. Then*

$$(4.13.1) \qquad p_{n+1}(t) = (a_n t + b_n) p_n(t) + c_n p_{n-1}(t)$$

for some constants a_n, b_n, *and* c_n $(n = 1, 2, \ldots)$.

Proof. It follows from Theorem 4.11 that $\mathcal{P}_n = \text{span}\{p_0, p_1, \ldots, p_n\}$, $p_n \perp \mathcal{P}_{n-1}$, and $p_n \in \mathcal{P}_n \backslash \mathcal{P}_{n-1}$ for $n = 1, 2, \ldots$. In particular, since $t\, p_n(t) \in \mathcal{P}_{n+1}$, we see that there are scalars α_i with $\alpha_{n+1} \neq 0$ such that $t\, p_n(t) = \sum_{i=0}^{n+1} \alpha_i p_i(t)$. For any $0 \leq j \leq n+1$,

$$\langle t\, p_n(t), p_j(t) \rangle = \left\langle \sum_{i=0}^{n+1} \alpha_i p_i(t), p_j(t) \right\rangle = \sum_{i=0}^{n+1} \alpha_i \langle p_i(t), p_j(t) \rangle = \alpha_j \|p_j\|^2 = \alpha_j.$$

That is,

$$(4.13.2) \qquad \alpha_j = \langle t p_n(t), p_j(t) \rangle \qquad (j = 0, 1, 2, \ldots, n+1).$$

But if $0 \leq j \leq n - 2$, then $t p_j(t) \in \mathcal{P}_{n-1}$ and

$$(4.13.3) \qquad \alpha_j = \langle t p_n(t), p_j(t) \rangle = \langle p_n(t), t p_j(t) \rangle = 0,$$

since $p_n \perp \mathcal{P}_{n-1}$. Thus

$$t p_n(t) = \alpha_{n-1} p_{n-1}(t) + \alpha_n p_n(t) + \alpha_{n+1} p_{n+1}(t).$$

Solving this equation for p_{n+1}, we obtain (4.13.1) with

$$(4.13.4) \qquad a_n = \frac{1}{\alpha_{n+1}}, \quad b_n = -\frac{\alpha_n}{\alpha_{n+1}}, \quad c_n = -\frac{\alpha_{n-1}}{\alpha_{n+1}}. \quad \blacksquare$$

For the actual computation of the orthonormal polynomials using equations (4.13.1)–(4.13.4), the following format is convenient:

$$(4.13.5) \qquad \begin{cases} y_{-1} = 0, \qquad p_{-1} = 0 \\[2mm] y_0 = 1, \qquad p_0 = \dfrac{p_0}{\|p_0\|} \\[2mm] y_{n+1}(t) = [t - \langle t p_n(t), p_n(t) \rangle] p_n(t) - \langle t p_n(t), p_{n-1}(t) \rangle p_{n-1}(t) \\[2mm] p_{n+1} = \dfrac{y_{n+1}}{\|y_{n+1}\|} \qquad (n = 0, 1, 2, \ldots). \end{cases}$$

Fourier Analysis

We will next prove a generalization of formula (4.10.3) for the case where M is not necessarily a finite-dimensional subspace, but possibly infinite-dimensional. It will be convenient to first provide some background material that is of interest in its own right.

4.14 Theorem. *Let* $\{e_1, e_2, \ldots, e_n\}$ *be a finite orthonormal set in the inner product space* X *and let* $M := \operatorname{span}\{e_1, e_2, \ldots, e_n\}$. *Then, for every* $x \in X$,

$$(4.14.1) \qquad\qquad P_M(x) = \sum_1^n \langle x, e_i \rangle e_i$$

and

$$(4.14.2) \qquad\qquad \sum_1^n |\langle x, e_i \rangle|^2 \leq \|x\|^2.$$

Proof. The equality (4.14.1) is just (4.10.3). Moreover, Theorem 4.9 implies that $x - P_M(x) \in M^\perp$. Using the Pythagorean theorem (Theorem 1.9), we get

$$\|x\|^2 = \|P_M(x)\|^2 + \|x - P_M(x)\|^2 \geq \|P_M(x)\|^2$$

and hence

$$(4.14.3) \qquad\qquad \|P_M(x)\| \leq \|x\|.$$

Again by the Pythagorean theorem,

$$(4.14.4) \qquad \|P_M(x)\|^2 = \sum_1^n \|\langle x, e_i \rangle e_i\|^2 = \sum_1^n |\langle x, e_i \rangle|^2.$$

Combining (4.14.3) and (4.14.4), we obtain (4.14.2). ∎

Inequality (4.14.2) states that the sum of the squares of the components of a vector in various orthogonal directions does not exceed the square of the length of the vector. It is this inequality, or more generally the inequality of Theorem 4.16 below, that goes by the name of "Bessel's inequality."

The next lemma is the key to extending Theorem 4.14 to an infinite orthonormal set.

4.15 Lemma. *Let* $\mathcal{O}$ *be an orthonormal set in the inner product space* X, *and let* $x \in X$. *Then the set*

$$(4.15.1) \qquad\qquad \mathcal{O}_x := \{e \in \mathcal{O} \mid \langle x, e \rangle \neq 0\}$$

is countable (i.e., either empty, finite, or countably infinite). Moreover, if we write $\mathcal{O}_x = \{e_1, e_2, \ldots\}$, *then*

$$(4.15.2) \qquad\qquad \sum_i |\langle x, e_i \rangle|^2 \leq \|x\|^2.$$

In particular, the series $\sum_i |\langle x, e_i \rangle|^2$ *converges.*
Proof. For each $n \in \mathbb{N}$, let

$$S_n := \{e \in \mathcal{O} \mid |\langle x, e \rangle|^2 > \|x\|^2/n\}.$$

Since $\mathcal{O}_x = \cup_1^\infty S_n$, it suffices to show that each S_n is finite. But by inequality (4.14.2), S_n contains at most $n-1$ elements. This proves that $\mathcal{O}_x$ is countable. By (4.14.2), we have $\sum_{i=1}^n |\langle x, e_i \rangle|^2 \leq \|x\|^2$ for every n. Since the right side of this inequality is a constant independent of n, it follows by taking the supremum over n that (4.15.2) holds. ∎

To prove Bessel's inequality when there are infinitely many orthonormal elements, we must first explain what the sum $\sum_{e \in \mathcal{O}} |\langle x, e \rangle|^2$ means if $\mathcal{O}$ is infinite and possibly even uncountable. Of course, if $\mathcal{O}$ is finite, say $\mathcal{O} = \{e_1, e_2, \ldots, e_n\}$, then $\sum_{e \in \mathcal{O}} |\langle x, e \rangle|^2 := \sum_1^n |\langle x, e_i \rangle|^2$. Now let $\mathcal{O}$ be an infinite orthonormal set in an inner product space X, and let $x \in X$. By Lemma 4.15, the set $\mathcal{O}_x = \{e \in \mathcal{O} \mid \langle x, e \rangle \neq 0\}$ is countable. Let us index the set $\mathcal{O}_x$ in *some* order, say $\mathcal{O}_x = \{e_1, e_2, \ldots\}$. By Lemma 4.15, the series $\sum_1^\infty |\langle x, e_i \rangle|^2$ is convergent. From the theory of absolutely convergent series, it follows that this series converges to the same sum for every *rearrangement* of the terms of this sum. This shows that the ordering we initially chose in $\mathcal{O}_x$ was unimportant, and we can unambiguously *define*

$$(4.15.3) \qquad \sum_{e \in \mathcal{O}} |\langle x, e \rangle|^2 := \sum_{i=1}^\infty |\langle x, e_i \rangle|^2$$

for *any ordering* $\{e_1, e_2, \ldots\}$ of the elements of $\mathcal{O}_x$.

We have just proved the following generalized version of inequalities (4.14.2) and (4.15.2).

4.16 Bessel's Inequality. *If $\mathcal{O}$ is any orthonormal set in an inner product space X, then*

$$(4.16.1) \qquad \sum_{e \in \mathcal{O}} |\langle x, e \rangle|^2 \leq \|x\|^2 \quad \text{for every } x \in X.$$

This result generalizes the inequality (4.14.2). Before we can generalize (4.14.1) to the case where M is spanned by an infinite orthonormal set, we need to define the expression $\sum_{e \in \mathcal{O}} \langle x, e \rangle e$ in case the orthonormal set $\mathcal{O}$ is infinite.

4.17 Definition (Fourier Series). *Let $\mathcal{O}$ be an orthonormal set in the inner product space X, and for each $x \in X$, let $\mathcal{O}_x$ denote the countable set defined (as in (4.15.1)) by $\mathcal{O}_x = \{e \in \mathcal{O} \mid \langle x, e \rangle \neq 0\}$. We define the **Fourier series** of x relative to $\mathcal{O}$ by*

$$\sum_{e \in \mathcal{O}} \langle x, e \rangle e := \begin{cases} 0 & \text{if } \mathcal{O}_x = \emptyset, \\ \sum_1^n \langle x, e_i \rangle e_i & \text{if } \mathcal{O}_x = \{e_1, e_2, \ldots, e_n\} \text{ is finite.} \end{cases}$$

In the case that $\mathcal{O}_x = \{e_1, e_2, \ldots\}$ is countably infinite, we define

$$\sum_{e \in \mathcal{O}} \langle x, e \rangle e := \lim_n \sum_{i=1}^n \langle x, e_i \rangle e_i,$$

provided that this limit exists. (In the remark following Lemma 4.18, we will see that this limit is *independent* of the order of the terms listed in $\mathcal{O}_x$.) *In each of* these cases, we say that the Fourier series $\sum_{e \in \mathcal{O}} \langle x, e \rangle e$ **converges**. *Otherwise, we say that the Fourier series* **diverges**. *The scalars $\langle x, e \rangle$ are called the* **Fourier coefficients** *of x relative to $\mathcal{O}$.*

We will see below that the Fourier series $\sum_{e \in \mathcal{O}} \langle x, e \rangle e$ *always converges* whenever X is complete or, more generally, whenever the closed subspace $M := \overline{\text{span}\mathcal{O}}$ is complete in the inner product space X. In fact, in either of these cases,

$$\sum_{e \in \mathcal{O}} \langle x, e \rangle e = P_M(x) \quad \text{for each} \quad x \in X$$

(see Theorem 4.22).

The next result shows that the Fourier series of every element in $\overline{\text{span}\mathcal{O}}$ always converges; in fact, it converges to the element itself.

4.18 Lemma. *Let $\mathcal{O}$ be an orthonormal subset of the inner product space X. Then for each $x \in \overline{\text{span}\mathcal{O}}$,*

$$(4.18.1) \qquad x = \sum_{e \in \mathcal{O}} \langle x, e \rangle e.$$

Proof. Fix any $x \in \overline{\text{span}\mathcal{O}}$. If $\mathcal{O}_x = \emptyset$, then $\langle x, e \rangle = 0$ for every $e \in \mathcal{O}$, i.e., $x \in \mathcal{O}^\perp = (\text{span}\mathcal{O})^\perp$. Since $\overline{\text{span}\mathcal{O}} \cap (\overline{\text{span}\mathcal{O}})^\perp = \{0\}$, it follows that $x = 0$ and (4.18.1) holds.

Next let $\mathcal{O}_x = \{e_1, e_2, \ldots, e_n\}$. Given $\epsilon > 0$, choose $y \in \text{span}\mathcal{O}$ such that $\|x - y\| < \epsilon$. Then $y = \sum_{k=1}^{m} \alpha_k s_k$ for some $s_k \in \mathcal{O}$ and $\alpha_k \in \mathbb{R}$. By Theorem 4.14,

$$(4.18.2) \qquad \left\| x - \sum_{1}^{m} \langle x, s_k \rangle s_k \right\| \le \left\| x - \sum_{1}^{m} \alpha_k s_k \right\| = \|x - y\| < \epsilon.$$

Suppose that $\langle x, s_m \rangle = 0$. Then $\sum_{1}^{m} \langle x, s_k \rangle s_k = \sum_{1}^{m-1} \langle x, s_k \rangle s_k$ and

$$\left\| x - \sum_{1}^{m} \alpha_k s_k \right\|^2 = \|x\|^2 - 2 \sum_{1}^{m} \alpha_k \langle x, s_k \rangle + \sum_{1}^{m} \alpha_k^2$$

$$\ge \|x\|^2 - 2 \sum_{1}^{m-1} \alpha_k \langle x, s_k \rangle + \sum_{1}^{m-1} \alpha_k^2$$

$$= \left\| x - \sum_{1}^{m-1} \alpha_k s_k \right\|^2.$$

Thus by replacing y by $\sum_{1}^{m-1} \alpha_k s_k$, we see that (4.18.2) is still valid. By repeating this argument, we see that we may discard those terms in the sums $\sum_{1}^{m} \langle x, s_k \rangle s_k$ and $\sum_{1}^{m} \alpha_k s_k$ such that $\langle x, s_k \rangle = 0$, i.e., such that $s_k \notin \mathcal{O}_x$. That is, we may always choose $y \in \text{span}\mathcal{O}_x = \text{span}\{e_1, \ldots, e_n\}$. For any such y,

$$\left\| x - \sum_{1}^{n} \langle x, e_i \rangle e_i \right\| \le \|x - y\| < \epsilon$$

by Theorem 4.14. Since ϵ was arbitrary, $x = \sum_1^n \langle x, e_i \rangle e_i$.

Finally, suppose $\mathcal{O}_x = \{e_1, e_2, \dots\}$ is countably infinite. Given $\epsilon > 0$, choose $y \in \text{span}\mathcal{O}$ such that $\|x - y\| < \epsilon$. The same argument as in the preceding paragraph shows that y may be chosen from $\text{span}\mathcal{O}_x$. Thus

$$\left\| x - \sum_1^n \langle x, e_i \rangle e_i \right\| \leq \|x - y\| < \epsilon$$

for n sufficiently large. It follows that $x = \lim_n \sum_1^n \langle x, e_i \rangle e_i$. If the elements of $\mathcal{O}_x$ are arranged in any other order, say $\mathcal{O}_x = \{e_1', e_2', \dots\}$, then, for any $\epsilon > 0$, choose N_1 such that $\|x - \sum_1^n \langle x, e_i \rangle e_i\| < \epsilon$ for every $n \geq N_1$. Now choose $N_2 \geq N_1$ such that $\{e_1, e_2, \dots, e_{N_1}\} \subset \{e_1'.e_2', \dots, e_{N_2}'\}$. Then for every $n \geq N_2$,

$$\left\| x - \sum_1^n \langle x, e_i' \rangle e_i' \right\| \leq \left\| x - \sum_1^{N_1} \langle x, e_i \rangle e_i \right\| < \epsilon.$$

Hence $\lim_n \sum_1^n \langle x, e_i' \rangle e_i' = x$. In every case, we have shown that (4.18.1) holds. ∎

Remark. To verify the statement in parentheses that was made in Definition 4.17, suppose that $y = \lim_n \sum_1^n \langle x, e_i \rangle e_i$ exists. Then $y \in \overline{\text{span}\mathcal{O}}$, and by continuity of the inner product we deduce that $\langle y, e \rangle = \langle x, e \rangle$ for every $e \in \mathcal{O}$. Thus $\mathcal{O}_x = \mathcal{O}_y$. By Lemma 4.18, $y = \lim_n \sum_1^n \langle y, e_i' \rangle e_i' = \lim_n \sum_1^n \langle x, e_i' \rangle e_i'$ for any reordering $\{e_1', e_2', \dots\}$ of $\mathcal{O}_x = \mathcal{O}_y$.

While Lemma 4.18 shows that the Fourier series for x converges when $x \in \overline{\text{span}\mathcal{O}}$, it will be seen below that this series often converges for *every* $x \in X$. For example, the next theorem shows that if X is complete (or, more generally, if $\overline{\text{span}\mathcal{O}}$ is complete), then *the Fourier series converges for every* $x \in X$. Before formulating the main theorem concerning Fourier series, we make the following observation. If $x = \sum_{e \in \mathcal{O}} \langle x, e \rangle e$ and $y = \sum_{e \in \mathcal{O}} \langle y, e \rangle e$, then

$$(4.18.3) \qquad \langle x, y \rangle = \sum_{e \in \mathcal{O}} \langle x, e \rangle \langle e, y \rangle,$$

where

$$(4.18.4) \qquad \sum_{e \in \mathcal{O}} \langle x, e \rangle \langle e, y \rangle := \lim_n \sum_{i=1}^n \langle x, e_i \rangle \langle e_i, y \rangle$$

and $\{e_1, e_2, \dots\}$ is any enumeration of the (countable) set $\mathcal{O}_x \cup \mathcal{O}_y$.

To see this, we first note that we can write $x = \lim_n \sum_1^n \langle x, e_i \rangle e_i$ and $y = \lim_n \sum_1^n \langle y, e_i \rangle e_i$. By continuity of the inner product in each argument, we deduce

$$\langle x, y \rangle = \lim_n \sum_{i=1}^n \langle x, e_i \rangle \lim_m \sum_{j=1}^m \langle e_j, y \rangle \langle e_i, e_j \rangle = \lim_n \sum_{i=1}^n \langle x, e_i \rangle \langle e_i, y \rangle.$$

Since the scalar $\langle x, y \rangle$ is independent of the enumeration in $\mathcal{O}_x \cup \mathcal{O}_y$, the result follows.

4.19 Fourier Series. *Let $\mathcal{O}$ be an orthonormal set in an inner product space X. Consider the following statements:*

(1) $\overline{\text{span}\mathcal{O}} = X$.

(2) *Every element can be expanded in a Fourier series. That is,*
$$x = \sum_{e \in \mathcal{O}} \langle x, e \rangle e \quad \text{for each } x \in X.$$

(3) *(Extended Parseval identity)*
$$\langle x, y \rangle = \sum_{e \in \mathcal{O}} \langle x, e \rangle \langle e, y \rangle \quad \text{for every } x, y \in X.$$

(4) *(Parseval identity)*
$$\|x\|^2 = \sum_{e \in \mathcal{O}} |\langle x, e \rangle|^2 \quad \text{for every } x \in X.$$

(5) $\mathcal{O}$ *is a "maximal" orthonormal set. That is, no orthonormal set in X properly contains $\mathcal{O}$.*

(6) $\mathcal{O}^\perp = \{0\}$.

(7) *(Elements are uniquely determined by their Fourier coefficients)*
$$\langle x, e \rangle = \langle y, e \rangle \quad \text{for every} \quad e \in \mathcal{O} \quad \text{implies that} \quad x = y.$$

Then (1) $\Leftrightarrow$ (2) $\Leftrightarrow$ (3) $\Leftrightarrow$ (4) $\Rightarrow$ (5) $\Leftrightarrow$ (6) $\Leftrightarrow$ (7).
Moreover, if $\overline{\text{span}\mathcal{O}}$ is complete (for example, if X is complete), then all seven statements are equivalent.

Proof. (1) $\Rightarrow$ (2). This is a consequence of Lemma 4.18.

(2) $\Rightarrow$ (3). This was established in (4.18.3) and (4.18.4).

(3) $\Rightarrow$ (4). Fix any $x \in X$ and put $y = x$ in (3) to get $\|x\|^2 = \sum_{e \in \mathcal{O}} |\langle x, e \rangle|^2$.

(4) $\Rightarrow$ (1). If (4) holds, let $x \in X$ and $\mathcal{O}_x = \{e_1, e_2, \dots\}$. Then

$$\left\| x - \sum_1^n \langle x, e_i \rangle e_i \right\|^2 = \|x\|^2 - 2\left\langle x, \sum_1^n \langle x, e_i \rangle e_i \right\rangle + \left\| \sum_1^n \langle x, e_i \rangle e_i \right\|^2$$
$$= \|x\|^2 - \sum_1^n |\langle x, e_i \rangle|^2 \to 0 \text{ as } n \to \infty.$$

This proves that $x \in \overline{\text{span}\mathcal{O}}$ and hence that (1) holds.

(4) $\Rightarrow$ (5). If (5) fails, then $\mathcal{O}$ is not maximal, so we can choose an $x \in X \setminus \mathcal{O}$ such that $\mathcal{O} \cup \{x\}$ is an orthonormal set. Then $x \in \mathcal{O}^\perp$ implies that $\langle x, e \rangle = 0$ for every $e \in \mathcal{O}$, so that $\sum_{e \in \mathcal{O}} |\langle x, e \rangle|^2 = 0$, but $\|x\|^2 > 0$. This shows that (4) fails.

(5) $\Rightarrow$ (6). If (6) fails, then there exists $x \in \mathcal{O}^\perp \setminus \{0\}$. Thus the set $\mathcal{O} \cup \{x/\|x\|\}$ is an orthonormal set that is strictly larger that $\mathcal{O}$; hence $\mathcal{O}$ is not maximal, and (5) fails.

(6) $\Rightarrow$ (7). If $\langle x, e \rangle = \langle y, e \rangle$ for every $e \in \mathcal{O}$, then $x - y \in \mathcal{O}^\perp = \{0\}$.

(7) $\Rightarrow$ (5). If (5) fails, $\mathcal{O}$ is not maximal, and we can choose an $x \in X$ such that $\mathcal{O} \cup \{x\}$ is orthonormal. Then $\langle x, e \rangle = 0 = \langle 0, e \rangle$ for every $e \in \mathcal{O}$; but $\|x\| = 1$, so (7) fails.

Finally, assume that $\overline{\text{span}\mathcal{O}}$ is complete. If (1) fails, choose $x \in X \setminus \overline{\text{span}\mathcal{O}}$. Since $M := \overline{\text{span}\mathcal{O}}$ is complete, it follows by Theorems 3.4 and 4.9 that M is a Chebyshev subspace and $0 \neq x - P_M(x) \in M^\perp \subset \mathcal{O}^\perp$. Thus (6) fails. This proves that (6) implies (1), and hence verifies the last statement of the theorem. ∎

Since each closed subspace M of an inner product space is an inner product space in its own right, we can apply Theorem 4.19 to M instead of X to obtain the following corollary.

4.20 Corollary. *Let M be a closed subspace of the inner product space X and let $\mathcal{O}$ be an orthonormal subset of M. Consider the following statements.*

(1) $\overline{\text{span}\mathcal{O}} = M$.
(2) $x = \sum_{e \in \mathcal{O}} \langle x, e \rangle e$ *for every* $x \in M$.
(3) $\langle x, y \rangle = \sum_{e \in \mathcal{O}} \langle x, e \rangle \langle e, y \rangle$ *for every* $x, y \in M$.
(4) $\|x\|^2 = \sum_{e \in \mathcal{O}} |\langle x, e \rangle|^2$ *for every* $x \in M$.
(5) $\mathcal{O}$ *is a maximal orthonormal subset of* M.
(6) $M \cap \mathcal{O}^\perp = \{0\}$.
(7) $x, y \in M$ *and* $\langle x, e \rangle = \langle y, e \rangle$ *for every* $e \in \mathcal{O}$ *implies* $x = y$.

Then $(1) \Leftrightarrow (2) \Leftrightarrow (3) \Leftrightarrow (4) \Rightarrow (5) \Leftrightarrow (6) \Leftrightarrow (7)$.
Moreover, if M is complete, then all seven statements are equivalent.

Remark. An orthonormal set $\mathcal{O}$ with the property that $M = \overline{\text{span}\mathcal{O}}$ is often called an **orthonormal basis** for M. By the preceding corollary, $\mathcal{O}$ is an orthonormal basis for M if and only if $x = \sum_{e \in \mathcal{O}} \langle x, e \rangle e$ for every $x \in M$. Clearly, when $\mathcal{O}$ is finite, this is just the usual definition of (orthonormal) basis for M. It is natural to ask what closed subspaces actually have orthonormal bases. By the Gram–Schmidt theorem, every finite-dimensional subspace has an orthonormal basis. We can show, more generally, that every complete subspace of an inner product space has an orthonormal basis. By the above corollary, it suffices to show that every inner product space contains a maximal orthonormal set.

4.21 Existence of Maximal Orthonormal Subsets. *Every inner product space $X \neq \{0\}$ contains a maximal orthonormal set.*

Proof. The proof uses Zorn's lemma (see Appendix 1). Let $\mathcal{C}$ denote the collection of all orthonormal subsets of X. Now, $\mathcal{C} \neq \emptyset$, since $\{x/\|x\|\} \in \mathcal{C}$ for any nonzero $x \in X$. Order $\mathcal{C}$ by containment: $\mathcal{O}_1 \succ \mathcal{O}_2$ if and only if $\mathcal{O}_1 \supset \mathcal{O}_2$. Let $\mathcal{T}$ be any totally ordered subset of $\mathcal{C}$. It is clear that $\cup \{\mathcal{O} \mid \mathcal{O} \in \mathcal{T}\}$ is an upper bound for $\mathcal{T}$. By Zorn's lemma, $\mathcal{C}$ contains a maximal element $\mathcal{O}$, and this must be a maximal orthonormal set in X. ∎

4.22 Approximating from Infinite-Dimensional Subspaces. *Let M be a complete subspace of the inner product space X. Then:*

(1) *M is a Chebyshev subspace.*
(2) *M has an orthonormal basis.*
(3) *If $\mathcal{O}$ is any orthonormal basis for M, then*

$$(4.22.1) \qquad P_M(x) = \sum_{e \in \mathcal{O}} \langle x, e \rangle e \quad \text{for every } x \in X.$$

Proof. (1). By Theorem 3.4(2), M is a Chebyshev subspace.

(2). By Corollary 4.20 and Theorem 4.21, M has an orthonormal basis.

(3). For any $x \in X$, let $x_0 = P_M(x)$. By Corollary 4.20,

$$(4.22.2) \qquad x_0 = \sum_{e \in \mathcal{O}} \langle x_0, e \rangle e.$$

Since $x - x_0 \in M^\perp$ by Theorem 4.9 and since $M = \overline{\text{span}\mathcal{O}}$, it follows that $\langle x - x_0, e \rangle = 0$ for every $e \in \mathcal{O}$. That is, $\langle x, e \rangle = \langle x_0, e \rangle$ for every $e \in \mathcal{O}$. Substitute this into (4.22.2) to obtain (4.22.1). ∎

The preceding theorem shows that if an orthonormal basis for a complete subspace M is available, then it is easy (in principle) to compute best approximations from M by using formula (4.22.1). Now we give an application of this theorem.

4.23 Application. *Let Γ be a nonempty set and $X = \ell_2(\Gamma)$. For each $j \in \Gamma$, define $e_j \in X$ by*

$$e_j(i) = \delta_{ij} \quad (i, j \in \Gamma).$$

That is, e_j is the function in X that is 1 at j and 0 elsewhere. Fix any nonempty subset $\Gamma_0 \subset \Gamma$. Then the subspace

$$M_{\Gamma_0} := \{x \in X \mid x(j) = 0 \quad \text{for every } j \in \Gamma \setminus \Gamma_0\}$$

is a Chebyshev subspace of X,

$$\mathcal{O}_{\Gamma_0} := \{e_j \mid j \in \Gamma_0\}$$

is an orthonormal basis for M_{Γ_0}, and

$$(4.23.1) \qquad P_{M_{\Gamma_0}}(x) = \sum_{e \in \mathcal{O}_{\Gamma_0}} \langle x, e \rangle e = \sum_{j \in \Gamma_0} x(j) e_j \quad \text{for every} \quad x \in X.$$

In particular, when $\Gamma_0 = \Gamma$, we have that $\mathcal{O}_\Gamma = \{e_j \mid j \in \Gamma\}$ and $M_\Gamma = X$. That is, $\mathcal{O}_\Gamma$ is an orthonormal basis for X.

To see this, first observe that $\mathcal{O}_{\Gamma_0}$ is clearly an orthonormal subset of M_{Γ_0}, and M_{Γ_0} is a closed, hence complete, subspace of the Hilbert space X. Next note that if $x \in M_{\Gamma_0}$, then $\langle x, e_j \rangle = x(j) = 0$ for every $j \in \Gamma \setminus \Gamma_0$, while if $x \in \mathcal{O}_{\Gamma_0}^\perp$, then $x(j) = \langle x, e_j \rangle = 0$ for every $j \in \Gamma_0$. Thus if $x \in M_{\Gamma_0} \cap \mathcal{O}_{\Gamma_0}^\perp$, it follows that $x(j) = 0$ for every $j \in \Gamma$, or $x = 0$. That is, $M_{\Gamma_0} \cap \mathcal{O}_{\Gamma_0}^\perp = \{0\}$. By Corollary 4.20, $\mathcal{O}_{\Gamma_0}$ is an orthonormal basis for M_{Γ_0}. Formula (4.23.1) now follows from Theorem 4.22.

Solutions to the First Three Basic Problems

We conclude this chapter by giving *complete solutions* to the first three of the five problems initially posed in Chapter 1. The fourth (respectively fifth) problem will be solved in Chapter 7 (respectively Chapter 10).

Problem 1. (Best least-squares polynomial approximation to data) Let $\{(t_j, x(t_j)) \mid j = 1, 2, \ldots, m\}$ be a table of data. For any fixed integer $n < m$, find a polynomial $p(t) = \sum_0^n \alpha_i t^i$, of degree at most n, such that the expression

$$\sum_{k=1}^m [x(t_k) - p(t_k)]^2$$

is minimized.

We saw in Chapter 2 that letting $T = \{t_1, t_2, \ldots, t_m\}$, $X = l_2(T)$, and $M = \mathcal{P}_n$, the problem may be restated as follows: Find the best approximation to $x \in X$ from the $(n + 1)$-dimensional subspace M.

By Corollary 4.10, $p(t) = \sum_0^n \alpha_i t^i$ is the best approximation to x if and only if the α_i satisfy the normal equations

$$\sum_{i=0}^n \alpha_i \langle t^i, t^j \rangle = \langle x, t^j \rangle \qquad (j = 0, 1, \ldots, n) ,$$

where

$$\langle t^i, t^j \rangle = \sum_{k=1}^m t_k^i t_k^j = \sum_{k=1}^m t_k^{i+j}$$

and

$$\langle x, t^j \rangle = \sum_{k=1}^m x(t_k) t_k^j .$$

Hence the normal equations can be written as

$$(1.1) \qquad \sum_{i=0}^n \alpha_i \left(\sum_{k=1}^m t_k^{i+j} \right) = \sum_{k=1}^m t_k^j x(t_k) \qquad (j = 0, 1, \ldots, n) .$$

In particular, when we are seeking the best *constant* approximation to x (i.e., $n = 0$), the normal equations (1.1) reduce to

$$\alpha_0 \sum_{k=1}^m 1 = \sum_{k=1}^m x(t_k),$$

or

$$\alpha_0 = \frac{1}{m} \sum_{k=1}^m x(t_k) .$$

That is, *the best constant approximation to x is its mean.*

Similarly, *the best linear approximation to x* (i.e., $n = 1$) is given by $p_1(t) = \alpha_0 + \alpha_1 t$, where the α_i are given by

$$\alpha_0 = \frac{\left(\sum_1^m t_k^2 \right) \left(\sum_1^m x(t_k) \right) - \left(\sum_1^m t_k \right) \left(\sum_1^m t_k x(t_k) \right)}{m \left(\sum_1^m t_k^2 \right) - \left(\sum_1^m t_k \right)^2}$$

and

$$\alpha_1 = \frac{m \left(\sum_1^m t_k x(t_k) \right) - \left(\sum_1^m t_k \right) \left(\sum_1^m x(t_k) \right)}{m \left(\sum_1^m t_k^2 \right) - \left(\sum_1^m t_k \right)^2}$$

Problem 2. (Solution to an overdetermined system of equations) Consider the linear system of m equations in the n unknowns $x_1, x_2, \ldots, x_n$:

$$a_{11}x_1 + a_{12}x_2 + \cdots + a_{1n}x_n = b_1$$

$$\cdots$$

$$a_{m1}x_1 + a_{m2}x_2 + \cdots + a_{mn}x_n = b_m$$

or briefly, $Ax = b$. Find a vector $x = (x_1, x_2, \ldots, x_n) \in \mathbb{R}^n$ that minimizes the expression

$$\sum_{i=1}^{m} \left(\sum_{j=1}^{n} a_{ij}x_j - b_i \right)^2.$$

We saw in Chapter 2 that letting $X = l_2(m)$ and $M = \{ y \in X \mid y = Ax,$ $x \in \mathbb{R}^n \}$, the problem may be restated as follows: Find the best approximation y_0 to b from the subspace M. Then choose any $x_0 \in \mathbb{R}^n$ such that $Ax_0 = y_0$.

By Theorem 4.9, $y_0 = P_M(b)$ if and only if $b - y_0 \in M^\perp$ if and only if $\langle b - y_0, Ax \rangle = 0$ for every $x \in \mathbb{R}^n$. At this point we make the following observation. If A^* denotes the *transpose* matrix of A (i.e., the (i, j) entry of A^* is the (j, i) entry of $A : a_{ij}^* = a_{ji}$), then, for each $x \in \mathbb{R}^n$ and $y \in \mathbb{R}^m$, we have

$$\langle y, Ax \rangle = \sum_{i=1}^{m} y_i \left(\sum_{j=1}^{n} a_{ij}x_j \right) = \sum_{j=1}^{n} \left(\sum_{i=1}^{m} a_{ji}^* y_i \right) x_j = \langle A^* y, x \rangle$$

(where the inner product on the left corresponds to the space $l_2(m)$, while that on the right corresponds to the space $l_2(n)$).

Continuing from the top of the preceding paragraph, we see that $y_0 = P_M(b)$ is equivalent to $\langle A^*(b - y_0), x \rangle = 0$ for every $x \in \mathbb{R}^n$, which is equivalent to $A^*(b - y_0) = 0$; i.e., $A^* b = A^* y_0$. Now, $y_0 \in M$, and so $y_0 = Ax_0$ for some $x_0 \in \mathbb{R}^n$. Thus we can finally conclude that $x_0 \in \mathbb{R}^n$ is a solution to the problem if and only if

$$(2.1) \qquad A^* A x_0 = A^* b.$$

In particular, if the matrix $A^* A$ is nonsingular, there is a unique solution of (2.1) given by

$$(2.2) \qquad x_0 = (A^* A)^{-1} A^* b.$$

However, if $A^* A$ is singular, there will always be more than one solution x_0 of (2.1). We should emphasize that this does *not* contradict the uniqueness guaranteed by Theorem 2.4. For although the best approximation $y_0 \in M$ to b *is* unique (by Theorem 2.4), there may be more than one $x_0 \in \mathbb{R}^n$ with $Ax_0 = y_0$. That is, A may not be one-to-one.

Problem 3. (Best least-squares polynomial approximation to a function) Let x be a real continuous function on the interval $[a, b]$. Find a polynomial $p(t) = \sum_0^n \alpha_i t^i$, of degree at most n, that minimizes the expression

$$\int_a^b [x(t) - p(t)]^2 dt.$$

Letting $X = C_2[a, b]$ and $M = \mathcal{P}_n$, the problem is to find the best approximation to x from the subspace M. By Corollary 4.10, $p(t) = \sum_0^n \alpha_i t^i$ is the best approximation to x *if and only if* the α_i satisfy the normal equations

$$\sum_{i=0}^n \alpha_i \int_a^b t^{i+j} dt = \int_a^b t^j x(t) dt \qquad (j = 0, 1, \ldots, n),$$

or equivalently,

(3.1) $$\sum_{i=0}^n \frac{\alpha_i}{i+j+1} (b^{i+j+1} - a^{i+j+1}) = \int_a^b t^j x(t) dt \qquad (j = 0, 1, \ldots, n).$$

In particular, the best *constant* approximation to x (i.e., $n = 0$) is given by its integral mean, or average:

$$\alpha_0 = \frac{1}{b-a} \int_a^b x(t) dt.$$

Exercises

1. (**Approximation by translates of cones**). Let C be a convex cone, $z \in X$, and $K = C + z$. Let $x \in X$ and $y_0 \in K$. Verify that $y_0 = P_K(x)$ if and only if $\langle x - y_0, y \rangle \leq 0$ for all $y \in C$ and $\langle x - y_0, y_0 - z \rangle = 0$.
2. (**Approximation by affine sets, i.e., translates of subspaces**). Let M be a subspace, $v \in X$, and $V = M + v$.
    (a) Let $x \in X$ and $y_0 \in V$. Show that $y_0 = P_V(x)$ if and only if $x - y_0 \in M^\perp$.
    (b) Prove that $P_V(x + z) = P_V(x)$ for every $x \in X$ and $z \in M^\perp$.
3. (**Strong uniqueness**) Let K be a convex Chebyshev set in X. Prove that for any $x \in X$,

$$\|x - y\|^2 \geq \|x - P_K(x)\|^2 + \|y - P_K(x)\|^2$$

for every $y \in K$.
    This is a type of *strong uniqueness* result for best approximations: It gives a quantitative estimate of how much larger $\|x - y\|$ is than $\|x - P_K(x)\|$ in terms of $\|y - P_K(x)\|$. [Hint: Theorem 4.1.]
4. Let $C = \{p \in C_2[a, b] \mid p \in \mathcal{P}_n, p \geq 0\}$.
    (a) Show that C is a Chebyshev convex cone.
    (b) When $n = 0$, show that for every $x \in C_2[a, b]$,

$$P_C(x) = \max \left\{ 0, \frac{1}{b-a} \int_a^b x(t) dt \right\}.$$

    (c) If $n \geq 1$, can you exhibit a formula for $P_C(x)$?

5. (a) Let $C = \{x \in l_2(I) \mid x(i) \geq 0 \text{ for all } i \in I\}$. Show that C is a Chebyshev convex cone in $l_2(I)$ and

$$P_C(x) = x^+ := \max\{x, 0\}$$

for every $x \in l_2(I)$.

★ (b) Let $C = \{x \in L_2(\mu) \mid x \geq 0\}$. Show that C is a Chebyshev convex cone in $L_2(\mu)$ and $P_C(x) = x^+$ for each $x \in L_2(\mu)$. (This generalizes part (a).)

6. (**Distance to finite-dimensional subspace**). Let $\{x_1, x_2, \ldots, x_n\}$ be linearly independent in X and $M = \operatorname{span}\{x_1, x_2, \ldots, x_n\}$. Show that

$$d(x, M)^2 = \frac{g(x, x_1, x_2, \ldots, x_n)}{g(x_1, x_2, \ldots, x_n)}$$

for every $x \in X$, where $g(y_1, y_2, \ldots, y_m)$ is the determinant of the $m \times m$ **Gram matrix**

$$G(y_1, y_2, \ldots, y_m) = \begin{bmatrix} \langle y_1, y_1 \rangle & \langle y_1, y_2 \rangle & \cdots & \langle y_1, y_m \rangle \\ \langle y_2, y_1 \rangle & \langle y_2, y_2 \rangle & \cdots & \langle y_2, y_m \rangle \\ \cdots & & & \\ \langle y_m, y_1 \rangle & \langle y_m, y_2 \rangle & \cdots & \langle y_m, y_m \rangle \end{bmatrix}.$$

[Hint: Adjoin the equation $d(x, M)^2 = \|x - P_M(x)\|^2 = \langle x - P_M(x), x \rangle = \langle x, x \rangle - \sum_1^n \alpha_i \langle x_i, x \rangle$ to the normal equations (4.10.2) and solve for $d(x, M)^2$ by Cramer's rule.]

7. (**Gram determinants**). Let $\{x_1, x_2, \ldots, x_n\}$ be a set of n vectors in X and $g(x_1, x_2, \ldots, x_n)$ the Gram determinant as defined in Exercise 6. Verify the following statements.

(a) $g(x_1, x_2, \ldots, x_n)$ is a symmetric function of the n arguments x_i.
[Hint: What happens when *two* of the x_i's are interchanged?]
(b) $0 \leq g(x_1, x_2, \ldots, x_n) \leq \|x_1\|^2 \|x_2\|^2 \cdots \|x_n\|^2$.
[Hint: Exercise 6.]
(c) The equality $g(x_1, x_2, \ldots, x_n) = 0$ holds if and only if the set $\{x_1, \ldots, x_n\}$ is linearly dependent.
(d) The equality $g(x_1, x_2, \ldots, x_n) = \|x_1\|^2 \|x_2\|^2 \cdots \|x_n\|^2$ holds if and only if $\{x_1, x_2, \ldots, x_n\}$ is an orthogonal set.
(e) The inequality $g(x_1, x_2) \geq 0$ is just Schwarz's inequality. (Hence, the inequality on the left in part (b) generalizes Schwarz's inequality.)

8. (**Hadamard's determinant inequality**). Let A be an $n \times n$ real matrix:

$$A = \begin{bmatrix} a_{11} & a_{12} & \ldots & a_{1n} \\ a_{21} & a_{22} & \ldots & a_{2n} \\ \cdots & & & \\ a_{n1} & a_{n2} & \ldots & a_{nn} \end{bmatrix}, \text{ or } A = \begin{bmatrix} a_1 \\ a_2 \\ \cdots \\ a_n \end{bmatrix},$$

where $a_i = (a_{i1}, a_{i2}, \ldots, a_{in})$ denotes the ith row vector of A. We regard each a_i as an element in $l_2(n)$.

(a) Show that

$$|\det(A)| \leq \|a_1\| \, \|a_2\| \cdots \|a_n\|.$$

[Hint: The matrix AA^* is a Gram matrix.]

(b) If $|a_{ij}| \leq c$ for each i, j, then

$$|\det(A)| \leq (c\sqrt{n})^n.$$

9. Let $\{x_1, x_2, \ldots, x_n\}$ be a basis for the subspace M. Show that for each $x \in X$,

$$P_M(x) = x - \frac{1}{g(x_1, x_2, \ldots, x_n)} \begin{vmatrix} x & x_1 & \cdots & x_n \\ \langle x_1, x \rangle & \langle x_1, x_1 \rangle & \cdots & \langle x_1, x_n \rangle \\ \langle x_2, x \rangle & \langle x_2, x_1 \rangle & \cdots & \langle x_2, x_n \rangle \\ & \cdots & & \\ \langle x_n, x \rangle & \langle x_n, x_1 \rangle & \cdots & \langle x_n, x_n \rangle \end{vmatrix}.$$

(The determinant on the right is, of course, understood to be the linear combination of $x, x_1, \ldots, x_n$ that one obtains by formally expanding this determinant by cofactors of the first row.)

10. (**Cauchy determinant formula**). Let a_i, b_i be real numbers such that $a_i + b_j \neq 0$ for all $i, j = 1, 2, \ldots, n$. Show that if

$$D = \begin{vmatrix} \frac{1}{a_1+b_1} & \frac{1}{a_1+b_2} & \cdots & \frac{1}{a_1+b_n} \\ \frac{1}{a_2+b_1} & \frac{1}{a_2+b_2} & \cdots & \frac{1}{a_2+b_n} \\ & \cdots & & \\ \frac{1}{a_n+b_1} & \frac{1}{a_n+b_2} & \cdots & \frac{1}{a_n+b_n} \end{vmatrix},$$

then

$$D = \frac{\prod\limits_{1 \leq i < j \leq n} (a_i - a_j)(b_i - b_j)}{\prod\limits_{i,j=1}^{n} (a_i + b_j)}.$$

11. In the space $C_2[0, 1]$, consider the monomials $x_i(t) = t^{i-1}$ $(i = 1, 2, \ldots, n)$.

(a) Show that $\langle x_i, x_j \rangle = (i + j - 1)^{-1}$.

(b) In this case the Gram matrix $G(x_1, x_2, \ldots, x_n)$ is also called the **Hilbert matrix**. Show that the determinant of the Hilbert matrix is given by

$$g(x_1, x_2, \ldots, x_n) = \frac{[1!2!\cdots(n-1)!]^4}{1!2!\cdots(2n-1)!}.$$

[Hint: Exercise 10 with $a_i = b_i = i - \frac{1}{2}$.]

(c) Verify that

$$\lim_{n \to \infty} g(x_1, x_2, \ldots, x_n) = 0.$$

What practical implications does this have for solving the normal equations when n is large?

12. Find the best approximation to $x \in C_2[-1, 1]$ from the subspace $\mathcal{P}_2$ in the cases where

(a) $x(t) = t^3$,

(b) $x(t) = e^t$,

(c) $x(t) = \sin 2\pi t$.

13. (a) Compute the first four Legendre polynomials (see Example 4.12 (1)).

(b) Compute the first four Chebyshev polynomials of the first kind (see Example 4.12 (2)).

(c) Compute the first three orthonormal polynomials in $l_2(m)$, $m \geq 4$ (see Example 4.12 (3)).

14. In the weighted space $C_2([a, b]; w)$, let $\{p_0, p_1, p_2, \dots\}$ denote the set of polynomials obtained by orthonormalizing the set of monomials $\{1, t, t^2, \dots\}$. Show that the best approximation to t^n from $\mathcal{P}_{n-1}$ is given by $q_{n-1}(t) := t^n - \alpha_n p_n(t)$, where α_n is a constant chosen such that

$$\alpha_n p_n(t) = t^n + \text{lower-order terms.}$$

15. Show that the zeros of the orthonormal polynomials obtained in Exercise 14 are real, simple, and lie in $[a, b]$.

16. Prove that a finite-dimensional inner product space is complete by using the fact that it has an orthonormal basis.

17. If $\{x_1, x_2, \dots, x_n\}$ is an orthonormal basis for the subspace M, show that

$$x = \sum_1^n \langle x, x_i \rangle x_i$$

for every $x \in M$.

18. (a) Verify that for a table of "constant data," i.e., $\{(t_j, c) \mid j = 1, \dots, m\}$, both the best constant approximation and the best linear approximation to this data are given by $p(t) \equiv c$. What about the best approximation from $\mathcal{P}_n$ when $n > 1$?

(b) Verify that for a table of "linear data," i.e., $\{(t_j, at_j + b) \mid j = 1, \dots, m\}$, the best constant approximation is given by

$$p_0(t) = a \left(\frac{1}{m} \sum_1^m t_k \right) + b,$$

and the best linear approximation is given by

$$p_1(t) = at + b.$$

What about the best approximation from $\mathcal{P}_n$ when $n > 1$?

19. Let $\{x_1, x_2, \dots, x_n\} \subset X$ and let $K = \{\sum_1^n \lambda_i x_i \mid \lambda_i \geq 0, \sum_1^n \lambda_i = 1\}$. (That is, K is the "convex hull" of $\{x_1, x_2, \dots, x_n\}$.) Let $y_0 \in K$. Show that $y_0 = P_K(0)$ if and only if

$$\langle y_0, x_i \rangle \geq \|y_0\|^2 \qquad (i = 1, 2, \dots, n).$$

20. An alternative approach to obtaining characterization theorems for best approximations is via differentiation. For example, finding the best approximation to x from the subspace $M = \text{span}\{x_1, x_2, \dots, x_n\}$ is equivalent to minimizing the function of n real variables

$$f(\alpha_1, \alpha_2, \dots, \alpha_n) := \left\| x - \sum_1^n \alpha_i x_i \right\|^2.$$

By expanding this in terms of inner products, show that f is a differentiable function of the α_i. Deduce the normal equations (4.10.2) again from the conditions

(20.1) $$\frac{\partial f}{\partial \alpha_i} = 0 \qquad (i = 1, 2, \ldots, n) .$$

Can you justify why the necessary condition (20.1) is also sufficient? [Hint: When $\sum_1^n \alpha_i^2$ is large, so is $f(\alpha_1, \alpha_2, \ldots, \alpha_n)$.] Finally, the solution to (20.1) is unique. This can be established using the nonsingularity of the Gram matrix $G(x_1, x_2, \ldots, x_n)$ (see Exercise 7), or it can be verified by showing that f is a "strictly convex" function, and such functions have a unique minimum.

21. For any collection of nonempty sets $\{S_1, S_2, \ldots, S_m\}$ in an inner product space, show that $(\cap_1^m S_i)^\circ \supset \overline{\sum_1^m S_i^\circ}$.

22. In Theorems 4.5 and 4.6, prove the statements concerning orthogonal complements.

23. **(Infinite sum of sets)** Suppose I is any index set, and for each $i \in I$, let S_i be a nonempty subset of the inner product space X such that $0 \in S_i$ for all except possibly finitely many $i \in I$. We define the **sum** of the sets S_i, denoted by $\sum_{i \in I} S_i$, by

$$\left\{ \sum_{i \in I} s_i \mid s_i \in S_i \text{ for all } i \in I, \ s_i = 0 \text{ for all except finitely many } i \right\}.$$

(Note that if I is finite, this reduces to the usual definition of the sum of a finite collection of sets.) Prove a generalization of Theorem 4.6 valid for *any* indexed collection of nonempty sets, not necessarily a finite collection.

24. Let X be a Hilbert space and C and D closed convex subsets with $0 \in C \cap D$. Show that the following statements are equivalent.
    (1) $(C \cap D)^\circ = \overline{C^\circ + D^\circ}$;
    (2) $(C \cap D)^\circ \subset \overline{C^\circ + D^\circ}$;
    (3) $\overline{\text{con}(C)} \cap \overline{\text{con}(D)} \subset \overline{\text{con}(C \cap D)}$;
    (4) $\overline{\text{con}(C)} \cap \overline{\text{con}(D)} = \overline{\text{con}(C \cap D)}$.
    While Theorem 4.6(3) shows that (1) (and hence all four statements above) hold when C and D are Chebyshev convex *cones*, the next exercise shows that these conditions do *not* always hold for general Chebyshev convex sets.

25. Consider the two sets in the Hilbert space $\ell_2(2)$ defined by

$$C := \{x \in \ell_2(2) \mid x^2(1) + [x(2) - 1]^2 \le 1\} \quad \text{and} \quad D := -C.$$

(a) Verify that both C and D are convex Chebyshev sets.
(b) Show that $C \cap D = \{0\}$, $\overline{\text{con}(C)} = \{x \in \ell_2(2) \mid x(2) \ge 0\}$, $\overline{\text{con}(D)} = \{x \in \ell_2(2) \mid x(2) \le 0\}$, $\overline{\text{con}(C \cap D)} = \{0\}$, and $\overline{\text{con}(C)} \cap \overline{\text{con}(D)} = \{x \in \ell_2(2) \mid x(2) = 0\}$.
(c) Show that $(C \cap D)^\circ \ne \overline{C^\circ + D^\circ}$. [Hint: Either a direct proof or appeal to (b) and Exercise 24.] This proves that Theorem 4.6(3) is *not* valid if the

Chebyshev convex cones are replaced by more general convex Chebyshev sets.

26. Let M and N denote the following subsets of $\ell_2(4)$: $M = \text{span}\{e_1, e_2 + e_3\}$ and $N = \text{span}\{e_1 + e_2, e_3 + e_4\}$, where $\{e_1, e_2, e_3, e_4\}$ is the canonical orthonormal basis for $\ell_2(4)$, i.e., $e_i(j) = \delta_{ij}$. Verify the following statements.
   (a) M and N are 2-dimensional Chebyshev subspaces of $\ell_2(4)$.
   (b) $M \cap N = \{0\}$.
   (c) For each element $x = (x(1), x(2), x(3), x(4)) \in \ell_2(4)$, we have

$$P_M(x) = x(1)e_1 + \frac{1}{2}[x(2) + x(3)](e_2 + e_3)$$

and

$$P_N(x) = \frac{1}{2}[x(1) + x(2)](e_1 + e_2) + \frac{1}{2}[x(3) + x(4)](e_3 + e_4).$$

27. Let $X = \ell_2$ and let $\{e_n \mid n = 1, 2, \dots\}$ denote the canonical orthonormal basis of X, i.e., $e_n(i) = \delta_{ni}$ for all positive integers n and i. Let $M_o = \text{span}\{e_{2n-1} \mid n = 1, 2, \dots\}$ and $M_e = \text{span}\{e_{2n} \mid n = 1, 2, \dots\}$. Show that M_o and M_e are Chebyshev subspaces, $M_e = M_o^{\perp}$, $M_o + M_e = \ell_2$, $P_{M_o}(x) = \sum_1^{\infty} x(2n-1)e_{2n-1}$, and $P_{M_e}(x) = \sum_1^{\infty} x(2n)e_{2n}$ for all $x \in \ell_2$.

28. An inner product space X is called **separable** if it contains a countable dense subset D, that is, if $\overline{D} = X$.
   (a) If D is countable and $\overline{\text{span}}D = X$, show that X is separable. [Hint: The subset of $\text{span}D$ consisting of those linear combinations whose coefficients are all *rational* numbers is countable and dense in $\text{span}D$.]
   (b) Every finite-dimensional inner product space X is separable.
   (c) Show that every orthonormal subset of a separable inner product space is countable. [Hint: If e and e' are distinct members of an orthonormal set, then $\|e - e'\| = \sqrt{2}$.]

29. Let X be a *complex* inner product space. Verify the following statements.
   (a) The characterization theorem (Theorem 4.1) is valid if the inequality (4.1.1) is replaced by

(4.1.1′) $\qquad \text{Re}\langle x - y_0, y - y_0 \rangle \leq 0 \quad$ for every $y \in K$.

   (b) The characterization theorem (Theorem 4.3) is valid if the dual cone of a set S is defined by $S^{\circ} := \{x \in X \mid \text{Re}\langle x, y \rangle \leq 0 \text{ for every } y \in S\}$.
   (c) Theorems 4.5, 4.6, 4.9, 4.10, 4.11, 4.16, 4.18, 4.19, 4.20, 4.21, and 4.22 are valid as stated.

Historical Notes

The characterization of best approximations (Theorem 4.1) can be traced back at least to Aronszajn (1950a) (see also Cheney and Goldstein (1959)), although it may be older. The characterization of best approximations from convex cones (Theorem 4.7), which is a strengthening of Theorem 4.1 in the special case of cones, is essentially contained in Moreau (1962).

The conical hull and dual cone of a set and many of their basic properties can be found in Steinitz (1914) (see also Fenchel (1953)). Theorem 4.4(5) can be found in

Rockafellar (1970; Corollary 2.6.3, p. 14). Dual cones can be defined even in more general normed linear spaces, but in this case they must lie in the dual space. See Kato (1984) for some facts about dual cones in normed linear spaces.

In 1883, Gram (1883) generalized a problem of Chebyshev for finite sums by considering the problem of minimizing the integral

$$(4.25.1) \qquad \int_a^b \rho(t)[f(t) - \sum_{j=1}^n a_j \psi_j(t)]^2 dt$$

over all possible choices of the scalars a_j. By cleverly applying the "orthogonalization process" to the ψ_j's (now recognized as the forerunner of the Gram–Schmidt orthogonalization procedure, Theorem 4.11), he was able to reduce the problem to the case where the ψ_j's form an orthonormal set, and the a_j's that minimize (4.25.1) are the Fourier coefficients $a_j = \int_a^b \rho(t)f(t)\psi_j(t)\, dt$. Gram also considered an infinite orthonormal set $\{\psi_1, \psi_2, \dots\}$ and was able to see that the condition that the expression in (4.25.1) converged to zero as n tends to infinity was connected to the maximality of the orthonormal set $\{\psi_1, \psi_2, \dots\}$.

The fact that all orthonormal bases of a Hilbert space have the same cardinality was shown by Löwig (1934; p. 31) and Rellich (1935; p. 355). Löwig (1934; p. 27) proved that two Hilbert spaces are "equivalent" if and only if they have the same dimension. (The dimension of a Hilbert space is the cardinality of a maximal orthonormal basis. See Exercise 4 of Chapter 6 for the definition of "equivalent.")

The equivalence of the first three statements of Theorem 4.19, when X is a separable Hilbert space, was proved by von Neumann (1929a; Satz 7).

A constructive approach to computing the best approximation $P_M(x)$ as given in (4.22.1) was given by Davis, Mallat, and Zhang (1994). (See also Cheney and Light (2000).)

CHAPTER 5

THE METRIC PROJECTION

Metric Projections onto Convex Sets

In this chapter we shall study the various properties of the metric projection onto a convex Chebyshev set K. It is always true that P_K is nonexpansive and, if K is a subspace, even linear. There are a substantial number of useful properties that P_K possesses when K is a subspace or a convex cone. For example, every inner product space is the direct sum of any Chebyshev subspace and its orthogonal complement. More generally, a useful duality relation holds between the metric projections onto a Chebyshev convex cone and onto its dual cone. The practical advantage of such a relationship is that determining best approximations from a convex cone is equivalent to determining them from the dual cone. The latter problem is often more tractable than the former. Finally, we record a reduction principle that allows us to replace one approximation problem by another one that is often simpler.

We begin by first observing the equivalence of the "$\epsilon - \delta$" definition of continuity with the sequential definition.

5.1 Lemma. *Let X and Y be nonempty subsets of inner product spaces, F a mapping from X into Y, and $x_0 \in X$. Then the following statements are equivalent:*
(1) *$F(x_n) \to F(x_0)$ whenever $x_n \in X$ and $x_n \to x_0$;*
(2) *For each $\epsilon > 0$, there exists $\delta = \delta(\epsilon, x_0) > 0$ such that $\|F(x) - F(x_0)\| < \epsilon$ whenever $x \in X$ and $\|x - x_0\| < \delta$.*

Proof. (1) $\Rightarrow$ (2). Suppose (2) fails. Then there is an $\epsilon > 0$ such that for each integer $n \geq 1$, there exists $x_n \in X$ with $\|x_n - x_0\| < 1/n$ and $\|F(x_n) - F(x_0)\| \geq \epsilon$. Then $x_n \to x_0$ but $F(x_n) \not\to F(x_0)$. Thus (1) fails.

(2) $\Rightarrow$ (1). Suppose (2) holds, $x_n \in X$, and $x_n \to x_0$. For each $\epsilon > 0$, choose $\delta > 0$ such that $\|F(x) - F(x_0)\| < \epsilon$ whenever $x \in X$ and $\|x - x_0\| < \delta$. Choose an integer N such that $\|x_n - x_0\| < \delta$ for $n \geq N$. Then $\|F(x_n) - F(x_0)\| < \epsilon$ for $n \geq N$. That is, $F(x_n) \to F(x_0)$, and (1) holds. ∎

5.2 Definition. *Let X and Y be (nonempty subsets of) inner product spaces, $F: X \to Y$, and $x_0 \in X$. F is said to be **continuous** at x_0 if both the equivalent conditions (1) and (2) of Lemma 5.1 are satisfied. F is called (i) **continuous** (on X) if it is continuous at each point of X; (ii) **uniformly continuous** (on X) if for each $\epsilon > 0$, there exists $\delta = \delta(\epsilon) > 0$ such that $\|F(x) - F(y)\| < \epsilon$ whenever $x, y \in X$ and $\|x - y\| < \delta$; (iii) **Lipschitz continuous** if there exists a constant $c \geq 0$ such that $\|F(x) - F(y)\| \leq c\|x - y\|$ for all x, y in X; (iv) **nonexpansive** if $\|F(x) - F(y)\| \leq \|x - y\|$ for all $x, y \in X$.*

Clearly, every nonexpansive mapping is Lipschitz continuous, every Lipschitz continuous mapping is uniformly continuous, and every uniformly continuous mapping is continuous. Let us next observe that the distance functional $F(x) = d(x, K)$ is nonexpansive.

5.3 Distance is Nonexpansive. *Let K be a nonempty subset of X. Then for every pair x, y in X,*

$$(5.3.1) \qquad |d(x, K) - d(y, K)| \le \|x - y\|.$$

In particular, the function $x \mapsto d(x, K)$ is nonexpansive, hence uniformly continuous.

Proof. For any $z \in K$,

$$d(x, K) \le \|x - z\| \le \|x - y\| + \|y - z\|.$$

Taking the infimum over all $z \in K$ yields

$$d(x, K) \le \|x - y\| + d(y, K).$$

Combining this with the inequality obtained by interchanging the roles of x and y, we obtain the result. ■

While a continuous function is not generally uniformly continuous (e.g., $f(t) = 1/t$ on $(0, 1)$), this will be the case when the domain is compact.

5.4 Theorem. *Let X be a compact subset of an inner product space, Y an inner product space, and suppose $F : X \to Y$ is continuous. Then F is uniformly continuous.*

Proof. If not, then there exists $\epsilon > 0$ such that for each $n \in \mathbb{N}$ there are points $x_n, x'_n \in X$ with $\|x_n - x'_n\| < 1/n$ and

$$(5.4.1) \qquad \|F(x_n) - F(x'_n)\| \ge \epsilon.$$

Since X is compact, there is a subsequence $\{x_{n_k}\}$ and $x_0 \in X$ such that $x_{n_k} \to x_0$. It follows that $x'_{n_k} \to x_0$ also. Since F is continuous at x_0, there exists $\delta > 0$ such that $\|F(x) - F(x_0)\| < \epsilon/2$ whenever $\|x - x_0\| < \delta$. Then for k large, $\|x_{n_k} - x_0\| < \delta$ and $\|x'_{n_k} - x_0\| < \delta$, so that

$$\|F(x_{n_k}) - F(x'_{n_k})\| \le \|F(x_{n_k}) - F(x_0)\| + \|F(x_0) - F(x'_{n_k})\| < \epsilon,$$

which contradicts (5.4.1). ■

Some useful properties of the metric projection onto a convex Chebyshev set are recorded next.

5.5 Metric Projections are Monotone and Nonexpansive. *Let K be a convex Chebyshev set. Then:*
(1) P_K is "idempotent":

$$(5.5.1) \qquad P_K(P_K(x)) = P_K(x) \text{ for every } x.$$

Briefly, $P_K^2 = P_K$.

(2) P_K is "firmly nonexpansive":

(5.5.2) $\langle x - y, P_K(x) - P_K(y) \rangle \geq \|P_K(x) - P_K(y)\|^2$ for all x, y.

(3) P_K is "monotone":

(5.5.3) $\langle x - y, P_K(x) - P_K(y) \rangle \geq 0$ for all x, y.

(4) P_K is "strictly nonexpansive":

(5.5.4) $\|x - y\|^2 \geq \|P_K(x) - P_K(y)\|^2 + \|x - P_K(x) - [y - P_K(y)]\|^2$ for all x, y.

(5) P_K is nonexpansive:

(5.5.5) $\|P_K(x) - P_K(y)\| \leq \|x - y\|$ for all x, y.

(6) P_K is uniformly continuous.

Proof. (1) This follows because each $y \in K$ is its own best approximation in K: $P_K(y) = y$.

(2) We have

$$\langle x - y, P_K(x) - P_K(y) \rangle = \langle x - P_K(x), P_K(x) - P_K(y) \rangle$$
$$+ \langle P_K(x) - P_K(y), P_K(x) - P_K(y) \rangle$$
$$+ \langle P_K(y) - y, P_K(x) - P_K(y) \rangle.$$

The first and third terms on the right are nonnegative by Theorem 4.1, and the second term is $\|P_K(x) - P_K(y)\|^2$. This verifies (2).

(3) This is an immediate consequence of (2).

(4) Using (2), we obtain for each x, y in X that

$$\|x - y\|^2 = \|[x - P_K(x)] + [P_K(x) - P_K(y)] + [P_K(y) - y]\|^2$$
$$= \|P_K(x) - P_K(y)\|^2 + \|[x - P_K(x)] - [y - P_K(y)]\|^2$$
$$+ 2\langle P_K(x) - P_K(y), x - P_K(x) - [y - P_K(y)] \rangle$$
$$= \|P_K(x) - P_K(y)\|^2 + \|x - P_K(x) - [y - P_K(y)]\|^2$$
$$+ 2\langle P_K(x) - P_K(y), x - y \rangle - 2\|P_K(x) - P_K(y)\|^2$$
$$\geq \|P_K(x) - P_K(y)\|^2 + \|x - P_K(x) - [y - P_K(y)]\|^2.$$

This proves (4).

Finally, note that (5) follows immediately from (4), and (6) from (5). ∎

In practice, the element x being approximated may not always be specified exactly (because of round-off, truncation, or other types of error). Thus, instead of approximating the "ideal" element x, we will actually be approximating the element $x + \delta x$, where δx is the error vector involved. The inequality (5.5.5) (see also Figure 5.5.6) implies that $\|P_K(x + \delta x) - P_K(x)\| \leq \|\delta x\|$. In other words, the error incurred in the best approximation is *no larger* than the error in the original vector. This is obviously a desirable property.

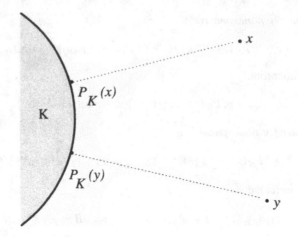

Figure 5.5.6

The metric projection onto a Chebyshev convex cone or Chebyshev subspace has additional, and stronger, properties.

5.6 Metric Projections onto Convex Cones. *Let C be a Chebyshev convex cone in the inner product space X (e.g., a closed convex cone in a Hilbert space). Then:*

(1) C° is a Chebyshev convex cone.

(2) For each $x \in X$,

$$x = P_C(x) + P_{C^\circ}(x) \quad \text{and} \quad P_C(x) \perp P_{C^\circ}(x).$$

Moreover, this representation is unique in the sense that if $x = y + z$ for some $y \in C$ and $z \in C^\circ$ with $y \perp z$, then $y = P_C(x)$ and $z = P_{C^\circ}(x)$.

(3) $\|x\|^2 = \|P_C(x)\|^2 + \|P_{C^\circ}(x)\|^2$ for all $x \in X$. Hence, $\|x\|^2 = d(x,C)^2 + d(x,C^\circ)^2$.

(4) $C^\circ = \{x \in X \mid P_C(x) = 0\}$ and $C = \{x \in X \mid P_{C^\circ}(x) = 0\} = \{x \in X \mid P_C(x) = x\}$.

(5) $\|P_C(x)\| \leq \|x\|$ for every $x \in X$; moreover, $\|P_C(x)\| = \|x\|$ if and only if $x \in C$.

(6) $C^{\circ\circ} = C$.

(7) P_C is "positively homogeneous":

$$P_C(\lambda x) = \lambda P_C(x) \quad \text{for all } x \in X, \ \lambda \geq 0.$$

Proof. To prove (1) and (2), let $x \in X$ and $c_0 = x - P_C(x)$. By Theorem 4.7, $c_0 \in C^\circ$ and $c_0 \perp (x - c_0)$. For every $y \in C^\circ$,

$$\langle x - c_0, y \rangle = \langle P_C(x), y \rangle \leq 0.$$

Hence $x - c_0 \in (C^\circ)^\circ$. By Theorem 4.7 (applied to C° rather than C), we get that $c_0 = P_{C^\circ}(x)$. This proves that C° is a Chebyshev convex cone, $x = P_C(x) + P_{C^\circ}(x)$, and $P_{C^\circ}(x) \perp P_C(x)$. To complete the proof of (2), we must verify the uniqueness of this representation.

Let $x = y + z$, where $y \in C$, $z \in C^\circ$, and $y \perp z$. For each $c \in C$,

$$\langle x - y, c \rangle = \langle z, c \rangle \leq 0$$

and

$$\langle x - y, y \rangle = \langle z, y \rangle = 0.$$

By Theorem 4.7, $y = P_C(x)$. Similarly, $z = P_{C^\circ}(x)$.

(3) The first statement follows from (2) and the Pythagorean theorem. Moreover, using this and (2), we obtain

$$\|x\|^2 = \|x - P_{C^\circ}(x)\|^2 + \|x - P_C(x)\|^2 = d(x, C^\circ)^2 + d(x, C)^2,$$

which verifies the last statement.

(4) This follows using (2). For example, $x \in C^\circ$ if and only if $x = P_{C^\circ}(x)$ if and only if $P_C(x) = 0$.

(5) From (3), $\|P_C(x)\| \leq \|x\|$ is clear. Also from (3), $\|P_C(x)\| = \|x\|$ if and only if $P_{C^\circ}(x) = 0$, which from (4) is equivalent to $x \in C$.

(6) This was already verified in Theorem 4.5. It also follows by applying (4) to the Chebyshev convex cone C° rather than C. That is,

$$C^{\circ\circ} = (C^\circ)^\circ = \{x \in X \mid P_{C^\circ}(x) = 0\} = C.$$

(7) Let $x \in X$ and $\lambda \geq 0$. Then by (2), $x = P_C(x) + P_{C^\circ}(x)$ and $\lambda x = \lambda P_C(x) + \lambda P_{C^\circ}(x)$. Since both C and C° are convex cones, $\lambda P_C(x) \in C$ and $\lambda P_{C^\circ}(x) \in C^\circ$. By (2) and the uniqueness of the representation for λx, we see that $P_C(\lambda x) = \lambda P_C(x)$. ∎

5.7 Remark. The representation 5.6(2) can be also expressed by the operator equation

(5.7.1) $$I = P_C + P_{C^\circ},$$

where I is the identity mapping: $I(x) = x$ for all x. In particular, this relationship implies that to find the best approximation to x from *either* set C or C°, it is enough to find the best approximation to x from *any one* of the sets. What this means in practice is that we can compute whichever of the best approximations $P_C(x)$ or $P_{C^\circ}(x)$ is more tractable; the other one is then automatically obtained from the relation $x = P_C(x) + P_{C^\circ}(x)$ (see Figure 5.7.2). This "duality" idea will be more fully exploited in later chapters.

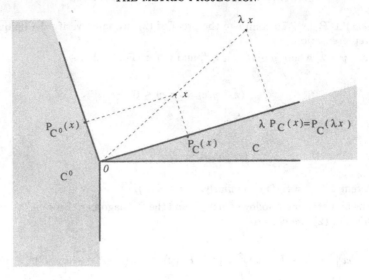

Figure 5.7.2

Since $M^\circ = M^\perp$ when M is a subspace, we obtain the following corollary of Theorem 5.6.

5.8 Metric Projections onto Subspaces. *Let M be a Chebyshev subspace of the inner product space X (e.g., a closed subspace of a Hilbert space). Then:*

(1) $M^\perp$ *is a Chebyshev subspace.*

(2) $x = P_M(x) + P_{M^\perp}(x)$ *for each $x \in X$. Briefly, $I = P_M + P_{M^\perp}$.*

Moreover, this representation is unique in the sense that if $x = y + z$, where $y \in M$ and $z \in M^\perp$, then $y = P_M(x)$ and $z = P_{M^\perp}(x)$.

(3) $\|x\|^2 = \|P_M(x)\|^2 + \|P_{M^\perp}(x)\|^2$ *for all x. Hence, $\|x\|^2 = d(x, M)^2 + d(x, M^\perp)^2$.*

(4) $M^\perp = \{x \in X \mid P_M(x) = 0\}$ *and $M = \{x \in X \mid P_{M^\perp}(x) = 0\} = \{x \in X \mid P_M(x) = x\}$.*

(5) $\|P_M(x)\| \le \|x\|$ *for all $x \in X$; $\|P_M(x)\| = \|x\|$ if and only if $x \in M$.*

(6) $M^{\perp\perp} = M$.

Just as in Remark 5.7, property 5.8(2) allows us to choose whichever one of $P_M(x)$ or $P_{M^\perp}(x)$ is the "easiest" to compute, and the other is automatically obtained. For example, suppose M were infinite-dimensional, but $M^\perp$ was finite-dimensional. In principle, it is easy to compute $P_{M^\perp}(x)$. This amounts to solving the normal equations of Theorem 4.10 (for $M^\perp$ instead of M) to obtain $P_{M^\perp}(x)$, and hence also $P_M(x) = x - P_{M^\perp}(x)$. This "duality" idea will also be pursued in more detail in later chapters.

If A and B are nonempty subsets of the inner product space X, we say that X is the **orthogonal sum** of A and B, and write $X = A \boxplus B$, if each $x \in X$ has a unique representation of the form $x = a + b$, where $a \in A$, $b \in B$, and $a \perp b$. We say that X is the **direct sum** of A and B, and write $X = A \oplus B$, if each $x \in X$ has a unique representation in the form $x = a + b$, where $a \in A$ and $b \in B$. Note that if A and B are actually *subspaces*, then $X = A \oplus B$ if and only if $X = A + B$ and $A \cap B = \{0\}$.

There are simple examples in $l_2(2)$ of a closed convex cone C such that $l_2(2) = C \boxplus C^\circ$, but $l_2(2) \neq C \oplus C^\circ$. (For example, take $C = \{x \in l_2(2) \mid x(1) \geq 0\}$.) Also, it is easy to see that if M and N are any pair of distinct 1-dimensional subspaces of $l_2(2)$ that are not orthogonal, then $l_2(2) = M \oplus N$, but $l_2(2) \neq M \boxplus N$. In contrast to these examples, if $B \subset A^\perp$, then $X = A \boxplus B$ if and only if $X = A \oplus B$.

Using this terminology, we can now state a generalization for part (2) of both Theorems 5.6 and 5.8. This result actually characterizes the convex cones or subspaces that are Chebyshev.

5.9 The Projection Theorem. *Let M (respectively C) be a subspace (respectively a convex cone) in the inner product space X. Then M (respectively C) is Chebyshev if and only if*

$$(5.9.1) \qquad X = M \oplus M^\perp \ (respectively \ X = C \boxplus C^\circ).$$

In particular, if M (respectively C) is a closed subspace (respectively closed convex cone) in a Hilbert space X, then (5.9.1) holds.

Proof. The last statement of the theorem follows from the first and Theorem 3.5.

If C is Chebyshev, then (5.9.1) follows from Theorem 5.6 (2).

Conversely, suppose $X = C \boxplus C^\circ$. Then for each $x \in X$, there exists a unique $c \in C$ and $c_0 \in C^\circ$ such that $x = c + c_0$ and $c \perp c_0$. Hence $x - c \in C^\circ$ and $x - c \perp c$. By Theorem 4.7, $c = P_C(x)$. Hence C is Chebyshev.

Similarly, $X = M \oplus M^\perp$ if and only if M is Chebyshev. ∎

Note that the proof also showed that if $X = C \boxplus C^\circ$, then $x = c + c_0$, where $c = P_C(x)$ and $c_0 = P_{C^\circ}(x)$.

Linear Metric Projections

Actually, further useful properties of the metric projection onto a Chebyshev *subspace* are known besides those listed in Theorem 5.8. The most important of these is that the metric projection is "linear." We now give the definition of this fundamental concept.

5.10 Definition. *If X and Y are inner product spaces, a mapping $F: X \to Y$ is called **linear** if and only if*

$$(5.10.1) \qquad F(\alpha x + \beta y) = \alpha F(x) + \beta F(y)$$

for all x, y in X and α, β in $\mathbb{R}$.

Equivalently, F is linear if and only if it is additive,

$$(5.10.2) \qquad F(x + y) = F(x) + F(y),$$

and homogeneous,

$$(5.10.3) \qquad F(\alpha x) = \alpha F(x),$$

for all x, y in X and $\alpha \in \mathbb{R}$.

Linear mappings are also commonly called **linear operators**. A linear operator F is said to be **bounded** if there exists a constant c such that

$$(5.10.4) \qquad \|F(x)\| \leq c\|x\| \quad \text{for all } x \in X.$$

(To be more precise, we should have written $\|F(x)\|_Y \leq c\|x\|_X$ to differentiate between the norms involved. However, no confusion should ever result by omitting the space indices.)

In the special case where $Y = \mathbb{R}$ is the scalar field, a linear operator $F : X \to \mathbb{R}$ is also called a **linear functional**.

The property of linearity is a rather strong one. Indeed, for such mappings, continuity and boundedness are equivalent, as is continuity at a *single* point!

5.11 Boundedness is Equivalent to Continuity. *Let $F : X \to Y$ be a linear mapping. Then the following statements are equivalent:*
(1) *F is bounded;*
(2) *F is Lipschitz continuous;*
(3) *F is uniformly continuous;*
(4) *F is continuous;*
(5) *F is continuous at some point.*

Proof. (1) $\Rightarrow$ (2). If F is bounded, there exists $c > 0$ such that $\|F(x)\| \leq c\|x\|$ for all x. By linearity,

$$\|F(x) - F(y)\| = \|F(x - y)\| \leq c\|x - y\|$$

for all $x, y \in X$. That is, F is Lipschitz continuous.

The implications (2) $\Rightarrow$ (3) $\Rightarrow$ (4) $\Rightarrow$ (5) are obvious.

(5) $\Rightarrow$ (1). Suppose F is continuous at the point $x_0 \in X$. If F were not bounded, then for each integer $n \geq 1$, there exists $x_n \in X$ such that $\|F(x_n)\| > n\|x_n\|$. Set $z_n = x_n/(n\|x_n\|)$. Then $\|z_n\| = 1/n$ and

$$(5.11.1) \qquad \|F(z_n)\| = \frac{1}{n\|x_n\|}\|F(x_n)\| > 1$$

for every n. But $z_n \to 0$ implies $z_n + x_0 \to x_0$. By continuity of F at x_0, we obtain

$$\|F(z_n)\| = \|F(z_n + x_0) - F(x_0)\| \to 0,$$

which contradicts (5.11.1). Thus F must be bounded. ∎

For a bounded linear operator F, the smallest constant c that works in (5.10.4) is called the **norm** of F and is denoted by $\|F\|$. Thus

$$(5.11.2) \qquad \|F\| := \inf\{c \mid \|F(x)\| \leq c\|x\| \text{ for all } x \in X\}.$$

5.12 Norm of Operator. *Let X and Y be inner product spaces with $X \neq \{0\}$ and let $F : X \to Y$ be a bounded linear operator. Then*

$$(5.12.1) \qquad \|F(x)\| \leq \|F\| \, \|x\| \quad \text{for all } x \in X,$$

and

(5.12.2) $$\|F\| = \sup_{x \neq 0} \frac{\|F(x)\|}{\|x\|} = \sup_{\|x\| \leq 1} \|F(x)\| = \sup_{\|x\| = 1} \|F(x)\|.$$

Proof. The relation (5.12.1) follows from the first equality in (5.12.2). To verify (5.12.2), first note that

$$\|F\| = \inf \left\{ c \mid \frac{\|F(x)\|}{\|x\|} \leq c \text{ for all } x \neq 0 \right\} = \sup_{x \neq 0} \frac{\|F(x)\|}{\|x\|}.$$

Also,

$$\sup_{\|x\| = 1} \|F(x)\| \leq \sup_{\|x\| \leq 1} \|F(x)\| \leq \sup_{0 < \|x\| \leq 1} \frac{\|F(x)\|}{\|x\|} \leq \sup_{x \neq 0} \frac{\|F(x)\|}{\|x\|}$$

$$= \sup_{x \neq 0} \left\| F \left(\frac{x}{\|x\|} \right) \right\| = \sup_{\|x\| = 1} \|F(x)\|.$$

Hence equality must hold throughout this string of inequalities. ∎

Next we consider a few examples of bounded linear operators. One of the most important, from the point of view of approximation theory, is the metric projection onto a Chebyshev subspace.

5.13 Linearity of Metric Projection. *Let M be a Chebyshev subspace in the inner product space X (e.g., a closed subspace of a Hilbert space). Then:*

(1) P_M *is a bounded linear operator and* $\|P_M\| = 1$ *(unless $M = \{0\}$, in which case $\|P_M\| = 0$).*

(2) P_M *is idempotent:* $P_M^2 = P_M$.

(3) P_M *is "self-adjoint":*

(5.13.1) $$\langle P_M(x), y \rangle = \langle x, P_M(y) \rangle \text{ for all } x, y \text{ in } X.$$

(4) *For every $x \in X$,*

(5.13.2) $$\langle P_M(x), x \rangle = \|P_M(x)\|^2.$$

(5) P_M *is "nonnegative":*

(5.13.3) $$\langle P_M(x), x \rangle \geq 0 \text{ for every } x.$$

Proof. (1) Let x, y in X and α, β in $\mathbb{R}$. By Theorem 4.9, $x - P_M(x)$ and $y - P_M(y)$ are in $M^\perp$. Since $M^\perp$ is a subspace,

$$\alpha x + \beta y - [\alpha P_M(x) + \beta P_M(y)] = \alpha(x - P_M(x)) + \beta(y - P_M(y)) \in M^\perp.$$

Since $\alpha P_M(x) + \beta P_M(y) \in M$, Theorem 4.9 implies that $\alpha P_M(x) + \beta P_M(y) = P_M(\alpha x + \beta y)$. Thus P_M is linear. Using part (5) of Theorem 5.8, we get $\|P_M(x)\| \leq \|x\|$ for all x. Thus P_M is bounded and $\|P_M\| \leq 1$. Since $P_M(y) = y$ for every $y \in M$, $\|y\| = \|P_M(y)\| \leq \|P_M\| \|y\|$ implies that $\|P_M\| \geq 1$ and hence $\|P_M\| = 1$.

(2) This follows from (1) of Theorem 5.5.

(3) For any x, y in X, Theorem 4.9 implies that $\langle P_M(x), y - P_M(y) \rangle = 0$, and hence

$$(5.13.4) \qquad \langle P_M(x), y \rangle = \langle P_M(x), P_M(y) \rangle.$$

By interchanging the roles of x and y in (5.13.4), we obtain

$$\langle x, P_M(y) \rangle = \langle P_M(y), x \rangle = \langle P_M(y), P_M(x) \rangle = \langle P_M(x), P_M(y) \rangle = \langle P_M(x), y \rangle.$$

(4) Take $y = x$ in (5.13.4).

(5) This clearly follows from (4). ∎

The metric projection on a Chebyshev subspace M is commonly called the *orthogonal projection* onto M for obvious reasons.

The Reduction Principle

There is a useful "reduction principle" that can be stated as a commuting property for a pair of metric projections. It allows us, for example, to "reduce" a problem of best approximation in a given convex subset K of X from the whole space X to any Chebyshev subspace M that contains K. That is, we need to consider best approximation only to points in the "reduced" space M rather than from all points in X.

In the next theorem, and throughout the remainder of the book, we will denote the composition of two operators by juxtaposition. For example, the operator $P_A P_B$ is defined by

$$P_A P_B(x) := P_A \circ P_B(x) = P_A(P_B(x)) \quad \text{for all } x.$$

5.14 Reduction Principle. *Let K be a convex subset of the inner product space X and let M be any Chebyshev subspace of X that contains K. Then:*
(1) $P_K P_M = P_K = P_M P_K$,
(2) $d(x, K)^2 = d(x, M)^2 + d(P_M(x), K)^2$ *for every* $x \in X$.

Proof. The statement $P_K = P_M P_K$ is obvious, since $K \subset M$. For any $y \in K$, we have that $y \in M$ and $x - P_M(x) \in M^\perp$ by Theorem 4.9, so that

$$(5.14.1) \qquad \|x - y\|^2 = \|x - P_M(x)\|^2 + \|P_M(x) - y\|^2.$$

Since the first term on the right of (5.14.1) is independent of y, it follows that $y \in K$ minimizes $\|x - y\|^2$ if and only if it minimizes $\|P_M(x) - y\|^2$. This proves that $P_K(x)$ exists if and only if $P_K(P_M(x))$ exists and $P_K(x) = P_K(P_M(x))$. Hence $P_K(x) = \emptyset$ if and only if $P_K(P_M(x)) = \emptyset$. In either case, (1) is verified. From this and (5.14.1) we obtain (2). ∎

The geometric interpretation of the reduction principle is the following: To find the best approximation to x from K, we first project x onto M and then project $P_M(x)$ onto K. The result is $P_K(x)$ (see Figure 5.14.2).

Remark. The convexity of K is not essential in the reduction principle. That is, there is a generalization of Theorem 5.14 that is valid for nonconvex sets in which P_K is a "set-valued" mapping (see Exercise 20 at the end of the chapter).

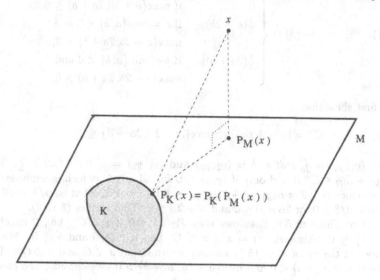

Figure 5.14.2. The reduction principle

One example of the practical value of the reduction principle is this. Suppose we are seeking best approximations to some $x \in X$ from a closed convex subset K of a *finite*-dimensional subspace M of X. It is easy—at least in principle— to compute $P_M(x)$ (and $d(x, M)$). This amounts to solving the normal equations (4.10.2). By the reduction principle, the problem of finding $P_K(x)$ (and $d(x, K)$) is thus reduced to that of finding $P_K(P_M(x))$ (and $d(P_M(x), K)$). Thus, by replacing x with $P_M(x)$, *we may assume that* $x \in M$, *i.e., that* $X = M$ *is finite-dimensional.* We give such an application next.

5.15 Application of Reduction Principle. Let $x \in C_2[0, 1]$ and $C = \{y \in \mathcal{P}_1 \mid y \geq 0 \text{ on } [0, 1]\}$. What is $P_C(x)$? That is, we seek the best approximation to x from the convex cone of nonnegative linear functions. By the reduction principle, this reduces to finding the best approximation to $P_{\mathcal{P}_1}(x)$ from C. Since $P_{\mathcal{P}_1}(x)$ can be obtained from the normal equations (4.10.2) by solving two equations in two unknowns, we assume that this has already been done. That is, we assume that $x \in \mathcal{P}_1$ and we seek $P_C(x)$. Thus we can restate the problem as follows:

Let $X = C_2[0, 1] \cap \mathcal{P}_1$ and $C = \{y \in X \mid y \geq 0\}$. Find $P_C(x)$ for any $x \in X$.

First observe that if $y_1(t) = t$ and $y_2(t) = 1 - t$, then $X = \text{span}\{y_1, y_2\}$ and C is the conical hull of the y_i; i.e.,

$$C = \{\alpha y_1 + \beta y_2 \mid \alpha \geq 0, \ \beta \geq 0\}.$$

To verify the latter statement, we need only observe that if $y = \alpha y_1 + \beta y_2$, then $y \in C$ if and only if $\beta = y(0) \geq 0$ and $\alpha = y(1) \geq 0$.

Let $x \in X$; hence $x = ay_1 + by_2$. We will verify that

(5.15.1)
$$P_C(x) = \begin{cases} x & \text{if } \min\{a, b\} \geq 0, \\ 0 & \text{if } \max\{a + 2b, 2a + b\} \leq 0, \\ \frac{1}{2}(a + 2b)y_2 & \text{if } a = \min\{a, b\} < 0 \text{ and} \\ & \quad \max\{a + 2b, 2a + b\} > 0, \\ \frac{1}{2}(2a + b)y_1 & \text{if } b = \min\{a, b\} < 0 \text{ and} \\ & \quad \max\{a + 2b, 2a + b\} > 0. \end{cases}$$

We first show that

(5.15.2) $C^\circ = \{\alpha y_1 + \beta y_2 \mid \max\{\alpha + 2\beta, 2\alpha + \beta\} \leq 0\}.$

Now, $\langle y_1, y_1 \rangle = \int_0^1 t^2 dt = \frac{1}{3} = \langle y_2, y_2 \rangle$ and $\langle y_1, y_2 \rangle = \int_0^1 t(1 - t)dt = \frac{1}{6}$. Thus $y = \alpha y_1 + \beta y_2 \in C^\circ$ if and only if $\langle y, z \rangle \leq 0$ for all $z \in C$, which is equivalent to $\langle y, y_i \rangle \leq 0$ for $i = 1, 2$ or $\alpha \langle y_1, y_i \rangle + \beta \langle y_2, y_i \rangle \leq 0$ for $i = 1, 2$. That is, $\alpha/3 + \beta/6 \leq 0$ and $\alpha/6 + \beta/3 \leq 0$, or $2\alpha + \beta \leq 0$ and $\alpha + 2\beta \leq 0$. This verifies (5.15.2).

By (4) of Theorem 5.6, it follows that $P_C(x) = 0$ if $x \in C^\circ$, i.e., if $\max\{a + 2b, 2a + b\} \leq 0$. Also, $P_C(x) = x$ if $x \in C$, i.e., if $a \geq 0$ and $b \geq 0$. For the remainder of the proof of (5.15.1) we may assume that $x \notin C$ and $x \notin C^\circ$. Thus we have that $\min\{a, b\} < 0$ and $\max\{a + 2b, 2a + b\} > 0$. We consider two cases.

Case 1. $a = \min\{a, b\} < 0$.

Since $\max\{a + 2b, 2a + b\} > 0$, it follows that $b > 0$ and $a + 2b = \max\{a + 2b, 2a + b\} > 0$. Let $y_0 = \frac{1}{2}(a + 2b)y_2$. Then $y_0 \in C$, $x - y_0 = \frac{a}{2}(2y_1 - y_2)$ and $\langle x - y_0, y_0 \rangle = 0$. By (5.15.2) we see that $x - y_0 \in C^\circ$. By Theorem 4.7, $y_0 = P_C(x)$.

Case 2. $b = \min\{a, b\} < 0$.

A proof similar to that of Case 1 shows that $\frac{1}{2}(2a + b)y_1 = P_C(x)$. This completes the verification of (5.15.1). ∎

We conclude this chapter with a few more examples of bounded linear operators and an example of an *unbounded* linear functional.

A linear operator whose domain is finite-dimensional is always bounded.

5.16 Theorem. *Let $F : X \to Y$ be linear and suppose X is finite-dimensional. Then F is bounded.*

Proof. By Theorem 4.11, we can choose an orthonormal basis $\{e_1, e_2, \ldots, e_n\}$ for X. Then each $x \in X$ has the unique representation $x = \sum_1^n \langle x, e_i \rangle e_i$ by (4.10.3). Let $c = \max\{\|F(e_i)\| \mid i = 1, 2, \ldots, n\}$. Then by Schwarz's inequality (in $l_2(n)$),

$$\|F(x)\| = \left\| \sum_1^n \langle x, e_i \rangle F(e_i) \right\| \leq \sum_1^n |\langle x, e_i \rangle| \, \|F(e_i)\|$$

$$\leq c \sum_1^n |\langle x, e_i \rangle| \leq c\sqrt{n} \left[\sum_1^n |\langle x, e_i \rangle|^2 \right]^{\frac{1}{2}} = c\sqrt{n}\|x\|.$$

Thus F is bounded and $\|F\| \leq c\sqrt{n}$. ∎

The next example is of a bounded linear operator from $C_2[a, b]$ into itself that includes several examples of interest as special cases.

5.17 Bounded Linear Operator on $C_2[a, b]$. Let $k(s, t)$ be a continuous real function of two variables s, t in $[a, b]$. For any $x \in C_2[a, b]$, define $F(x)$ at $s \in [a, b]$ by

$$(5.17.1) \qquad\qquad F(x)(s) = \int_a^b k(s, t)x(t)\, dt.$$

Then F is a bounded linear operator from $C_2[a, b]$ into itself such that

$$(5.17.2) \qquad\qquad \|F\| \le \left[\int_a^b \int_a^b |k(s, t)|^2\, dt\, ds \right]^{\frac{1}{2}}.$$

Proof. It is clear from (5.17.1) that F is linear. It remains to show that $F(x) \in C_2[a, b]$ and to verify (5.17.2). Fix any $s_0 \in [a, b]$. By Theorem 5.4, k is uniformly continuous on $[a, b] \times [a, b]$. Hence, for each $\epsilon > 0$, there exists $\delta = \delta(\epsilon) > 0$ such that

$$|k(s, t) - k(s_0, t)| < \epsilon \quad \text{whenever } |s - s_0| < \delta \text{ and } t \in [a, b].$$

Thus for $|s - s_0| < \delta$,

$$|F(x)(s) - F(x)(s_0)| = \left| \int_a^b [k(s, t) - k(s_0, t)]x(t)dt \right|$$

$$\le \int_a^b |k(s, t) - k(s_0, t)|\, |x(t)|dt$$

$$\le \epsilon \int_a^b |x(t)|dt \le \epsilon\sqrt{b - a} \left[\int_a^b |x(t)|^2 dt \right]^{\frac{1}{2}}$$

$$= \epsilon\sqrt{b - a}\, \|x\|$$

using Schwarz's inequality. Since $\epsilon > 0$ was arbitrary, $F(x)$ is continuous at s_0. Since s_0 was arbitrary, $F(x) \in C_2[a, b]$.

Applying Schwarz's inequality to (5.17.1), we obtain, for each $s \in [a, b]$,

$$|F(x)(s)|^2 = \left| \int_a^b k(s, t)x(t)\, dt \right|^2 \le \left[\int_a^b |k(s, t)|^2 dt \right] \int_a^b |x(t)|^2 dt$$

$$= \left[\int_a^b |k(s, t)|^2 dt \right] \|x\|^2.$$

Integrating this over all $s \in [a, b]$ yields

$$\|F(x)\|^2 = \int_a^b |F(x)(s)|^2 ds \le \int_a^b \int_a^b |k(s, t)|^2\, dt\, ds\, \|x\|^2.$$

This shows that F is bounded and verifies (5.17.2). ∎

In the next chapter we will make a detailed study of the bounded linear *functionals* on an inner product space. For the present, we only show how easy it is to generate a large collection of bounded linear functionals on any inner product space.

5.18 Bounded Linear Functionals. *Given any $z \in X$, define the function $f = f_z$ on X by*

$$(5.18.1) \qquad\qquad f(x) := \langle x, z \rangle \quad \text{for all } x \in X.$$

Then f is a bounded linear functional on X and

$$(5.18.2) \qquad\qquad \|f\| = \|z\|.$$

Proof. f is clearly linear, since the inner product is linear in its first argument. By Schwarz's inequality, $|f(x)| \leq \|x\| \, \|z\|$ for all $x \in X$. Hence f is bounded and $\|f\| \leq \|z\|$. Since $\|z\|^2 = \langle z, z \rangle = f(z) \leq \|f\| \, \|z\|$, it follows that $\|f\| \geq \|z\|$, and hence (5.18.2) holds. ∎

In the next chapter we will see that if X is a Hilbert space, the converse of Theorem 5.18 holds. That is, *every* bounded linear functional f on X has the representation (5.18.1) for *some* $z \in X$. For spaces that are not complete, a representation such as (5.18.1) is not valid for every bounded linear functional. But we will see that every bounded linear functional is the *limit* of such functionals.

We conclude this chapter by giving an example of an *unbounded* (hence discontinuous) linear functional on $C_2[a, b]$.

5.19 An Unbounded Linear Functional on $C_2[a,b]$. Fix any point $t_0 \in [a, b]$ and define F on $C_2[a, b]$ by

$$(5.19.1) \qquad\qquad F(x) = x(t_0), \qquad x \in C_2[a, b].$$

(F is called "point evaluation at t_0.") Then F is easily seen to be a linear functional on $C_2[a, b]$. Consider any sequence $\{x_n\}$ in $C_2[a, b]$ having the property that $0 \leq x_n(t) \leq \sqrt{n}$, $x_n(t_0) = \sqrt{n}$, and $x_n(t) = 0$ whenever $|t - t_0| \geq 1/(2n)$. Then $\|x_n\| \leq 1$ for all n, but $F(x_n) = \sqrt{n}$.

Thus $\sup_{\|x\| \leq 1} |F(x)| = \infty$, and so F is not bounded.

Although we shall not do so here, it is possible to prove the existence of an unbounded linear functional on every *infinite*-dimensional space.

Exercises

1. If K is any set with $0 \in K$, show that $K \cap K^\circ = \{0\}$.
2. If $K = \{x \in X \mid \|x\| \leq 1\}$, show that $K^\circ = K^\perp = \{0\}$. More generally, show that if K is any set with $0 \in \mathrm{int}(K)$, then $K^\circ = K^\perp = \{0\}$.
3. For any set K and scalar $\alpha \neq 0$, $(\alpha K)^\circ = (\operatorname{sgn}\alpha)K^\circ$, where $\operatorname{sgn}\alpha := \alpha/|\alpha|$
4. Fix any $z \in X := C_2[a, b]$ and define F on X by

$$F(x) := z \cdot x \qquad \text{for every } x \in X.$$

   (The mapping F is called "multiplication by z.") Show that
   (a) $F : X \mapsto X$ is linear.
   (b) F is bounded and $\|F\| = \|z\|_\infty$, where $\|z\|_\infty := \max_{a \leq t \leq b} |z(t)|$.
5. In the space $C_2[0, 1]$, consider the function $x(t) = \sin t$. Suppose the truncated Taylor series $\tilde{x}(t) = t - \frac{1}{6}t^3$ is used instead of $x(t)$. If K is any convex

Chebyshev set in $C_2[0,1]$, find the largest error incurred if $P_K(\tilde{x})$ is used instead of $P_K(x)$. That is, obtain an upper bound for $\|P_K(x) - P_K(\tilde{x})\|$.

6. Fix an integer $n \geq 2$, and let $M = \{x \in l_2(n) \mid \sum_i^n x(i) = 0\}$.

   (a) Show that M is a Chebyshev subspace of $l_2(n)$ having dimension $n - 1$.

   (b) Find the best approximation in M to each of the unit basis vectors $e_1, e_2, \ldots, e_n$ in $l_2(n)$. (Here $e_i(j) = \delta_{ij}$.)

   (c) Find the best approximation in M to any $x \in l_2(n)$. [Hint: It suffices to use (b) and linearity of P_M.]

7. Let $\{x_1, x_2, \ldots, x_n\}$ be a linearly independent set in X and define F on X by

$$F(x) = \sum_1^n \langle x, x_i \rangle x_i.$$

   (a) Show that F is a bounded linear operator from X into itself.

   (b) Obtain an upper bound for $\|F\|$.

8. Show that the mapping F on l_2 defined by

$$F(x)(n) = x(n+1) \ (n = 1, 2, \ldots)$$

   is a bounded linear operator from l_2 into itself with $\|F\| = 1$. Is F one-to-one or onto? [F is called the (backward) shift operator.]

9. Let K be an approximatively compact Chebyshev set (not necessarily convex). Show that P_K is continuous. [Hint: If $x_n \to x_0$, then $\{P_K(x_n)\}$ is a minimizing sequence.]

10. **(The cone of positive functions)** Let $C = \{y \in C_2[a,b] \mid y \geq 0\}$. Verify the following:

    (a) C is a Chebyshev convex cone and
    $$P_C(x) = x^+ := \max\{x, 0\} \quad \text{for each} \ x \in C_2[a,b].$$

    (b) $C^\circ = -C$.

    (c) C° is a Chebyshev convex cone and
    $$P_{C^\circ}(x) = -x^- := -\max\{-x, 0\} \quad \text{for each} \ x \in C_2[a,b].$$

    (d) $x = x^+ - x^-$ for every $x \in C_2[a,b]$.

11. Let $X = C_2[a,b]$ and

$$C = \{y \in X \mid y \in \mathcal{P}_0, \ y \geq 0\}.$$

    (a) What is C°?

    (b) What is $P_C(x)$ for any $x \in X$?

12. **(Isotone regression)** Let $X = l_2(n)$ and

$$C = \{x \in X \mid x(1) \leq x(2) \leq \cdots \leq x(n)\}.$$

    (a) Show that C is a Chebyshev convex cone.

    (b) Show that for every x, y in X,

$$\langle x, y \rangle := \sum_1^n x(i)y(i)$$

$$= x(1) \sum_1^n y(i) + [x(2) - x(1)] \sum_2^n y(i)$$

$$+ [x(3) - x(2)] \sum_3^n y(i) + \cdots + [x(n) - x(n-1)]y(n).$$

(c) Verify that $C^\circ = \{y \in X \mid \sum_1^n y(i) = 0, \sum_k^n y(i) \le 0 \text{ for } k \ge 2\}$.

(d) When $n = 2$, show that $y_0 = P_C(x)$ if and only if $y_0 = x$ when $x \in C$ and $y_0(i) = a$ $(i = 1, 2)$, where $a = \frac{1}{2}[x(1) + x(2)]$, when $x \notin C$. [Hint: Part (c).]

(e) Characterize $P_C(x)$ for any $x \in l_2(n)$. [Hint: Part (c).]

13. Let $M_1 \supset M_2 \supset \cdots \supset M_n$ be Chebyshev subspaces. Show that

$$P_{M_1} P_{M_2} \cdots P_{M_n} = P_{M_n} = P_{M_n} P_{M_{n-1}} \cdots P_{M_1}.$$

14. Let M and N be orthogonal subspaces that are Chebyshev. Show that $M + N$ is a Chebyshev subspace and

$$P_{M+N} = P_M + P_N.$$

15. Let M and N be Chebyshev subspaces. Show that $N \subset M$ if and only if $P_N P_M = P_N = P_M P_N$.

16. A **projection** onto a closed subspace M is a bounded linear operator P from X onto M such that $P^2 = P$. Show that a bounded linear operator P from a Hilbert space X into itself is a projection onto M if and only if $P = F + (I - F)P_M$ for some bounded linear operator $F : X \to M$.

17. Let $X = \mathcal{P}_n$ with the inner product $\langle x, y \rangle = \int_0^1 x(t)y(t)dt$. Let

$$C_n = \{p \in X \mid p \ge 0\}.$$

What is C_0° ? (C_1° was described in Application 5.15.) Can you describe C_n° when $n > 1$?

18. Let K be a convex Chebyshev set in X. Show that for each pair x, y in X, either $\|P_K(x) - P_K(y)\| < \|x - y\|$ or $P_K(x) - P_K(y) = x - y$. [Hint: Theorem 5.5(2) and the Schwarz inequality.]

19. Let $\mathcal{L}(X, Y)$ denote the set of all bounded linear operators from X into Y. With the pointwise operations of addition and scalar multiplication of operators and the norm (5.11.2), show that $\mathcal{L}(X, Y)$ is a normed linear space.

20. Prove the following generalization of the reduction principle (Theorem 5.14). *Let S be a nonempty subset of the inner product space X and let M be a Chebyshev subspace of X that contains S. Then for every $x \in X$,*
(1) $P_S(P_M(x)) = P_S(x) = P_M(P_S(x))$,
(2) $d(x, S)^2 = d(x, M)^2 + d(P_M(x), S)^2$.
[Hint: Look at the proof of Theorem 5.14.]

21. If X is a *complex* inner product space, the definition of dual cone and linear operator are modified as follows:

$$K^\circ := \{x \in X \mid Re\langle x, y \rangle \le 0 \text{ for all } y \in K\},$$

$$F(\alpha x + \beta y) = \alpha F(x) + \beta F(y) \quad \text{for all } x, \ y \text{ in } X, \ \alpha, \ \beta \in \mathbb{C}.$$

Verify the following statements for a *complex* inner product space X.
(a) Lemma 5.1 is valid.

   (b) Theorem 5.5 is valid provided that $\langle x - y, P_K(x) - P_K(y) \rangle$ is replaced
by its real part where it appears in (2) and (3) of Theorem 5.5.
   (c) Theorems 5.6, 5.8, and 5.9 are valid.
   (d) Theorems 5.11 through 5.14 and Theorems 5.16 and 5.18 are all valid.
22. Definitions 5.2 and 5.10 make sense in any normed linear space. Verify that
    Lemmas 5.1, 5.3, 5.11, 5.12, and 5.16 are all valid in arbitrary normed linear
    spaces.

Historical Notes

   The term *metric projection* was apparently first used in a paper of Aronszajn
and Smith (1954).
   For results concerning the metric projection onto a closed convex subset of a
Hilbert space, perhaps the best single reference is the first part of the long two-
part paper of Zarantonello (1971). (In the second part of this same paper, he even
developed a "spectral theory" that was based on projections onto closed convex
cones that extended the known theory that was based on projections onto closed
subspaces.)
   The monotonicity of the metric projection (i.e., Theorem 5.5(3)) was observed
by Nashed (1968). Parts (2) and (6) of Theorem 5.6 (in a Hilbert space setting)
are due to Moreau (1962) (see also Zarantonello (1971; p. 256) and Aubin (1979;
p. 18)). The projection theorem (Theorem 5.9), for closed subspaces of a separable
Hilbert space (respectively for closed convex cones in a Hilbert space), goes back
at least to von Neumann (1929a) (respectively Moreau (1962)). The "if" part of
the projection theorem, for both the subspace and cone cases, seems new.
   That boundedness is equivalent to continuity for linear operators (i.e., Theorem
5.11), and that the orthogonal projection onto a closed subspace of a Hilbert space
is linear, idempotent, and self-adjoint, are both classical results (see von Neumann
(1929a)).
   The reduction principle (Theorem 5.14) can be traced to Deutsch (1965) (see
also Laurent (1972; Prop. (2.2.6), p. 46)).

CHAPTER 6

BOUNDED LINEAR FUNCTIONALS
AND BEST APPROXIMATION
FROM HYPERPLANES AND HALF-SPACES

Bounded Linear Functionals

In this chapter we study the dual space of all bounded linear functionals on an inner product space. Along with the metric projections onto Chebyshev subspaces, these functionals are the most important linear mappings that arise in our work. We saw in the last chapter that every element of the inner product space X naturally generates a bounded linear functional on X (see Theorem 5.18). Here we give a general representation theorem for *any* bounded linear functional on X (Theorem 6.8). This implies the Fréchet–Riesz representation theorem (Theorem 6.10), which states that in a Hilbert space every bounded linear functional has a "representer." This fact can be used to show that every Hilbert space is equivalent to its dual space. In general (incomplete) inner product spaces we can characterize the bounded linear functionals that are generated by elements, or representers, in X. They are precisely those that "attain their norm" (Theorem 6.12). We should mention that many of the results of this chapter—particularly those up to Theorem 6.12—can be substantially simplified or omitted entirely if the space X is assumed *complete*, i.e., if X is a Hilbert space. Because many of the important spaces that arise in practice are *not* complete, however (e.g., $C_2[a, b]$), we have deliberately avoided the simplifying assumption of completeness. This has necessitated developing a somewhat more involved machinery to handle the problems arising in incomplete spaces.

Hyperplanes are the level sets of bounded linear functionals. They play an important role in the geometric interpretation of various phenomena (e.g., the "strong separation principle", Theorem 6.23). Many problems of best approximation can be given "dual" reformulations in terms of problems concerning best approximation from certain hyperplanes. The usefulness of such a reformulation stems from the fact that the distance from a point to a hyperplane and the best approximation from a hyperplane to a point have *simple explicit formulas*!

Finally, we study the problem of best approximation from "half-spaces" and, more generally, from a "polyhedron" (i.e., an intersection of finitely many half-spaces). A half-space is a set determined by all the points on one side of a hyperplane. As with hyperplanes, there are simple formulas for computing best approximations from half-spaces as well as the distance to half-spaces. Moreover, half-spaces are the fundamental building blocks for closed convex sets, since *every* closed convex set in an inner product space is an intersection of half-spaces (Theorem 6.29).

In this paragraph we state explicitly some basic properties of linear functionals that follow as particular consequences of results proved in Chapter 5 for arbitrary linear operators.

Recall that a **linear functional** on the inner product space X is a real function f on X such that

$$(6.1.1) \qquad f(\alpha x + \beta y) = \alpha f(x) + \beta f(y)$$

for all x, y in X and α, β in $\mathbb{R}$. A linear functional f is **bounded** if there exists a constant c such that

$$(6.1.2) \qquad |f(x)| \le c\|x\| \text{ for all } x \in X.$$

The smallest of these constants c is called the **norm** of f and has the alternative expressions

$$(6.1.3) \qquad \|f\| = \sup_{x \ne 0} \frac{|f(x)|}{\|x\|} = \sup_{\|x\| \le 1} |f(x)| = \sup_{\|x\| = 1} |f(x)|.$$

Finally, we recall the following facts.

6.1 Theorem. *A linear functional f is bounded if and only if it is continuous (Theorem 5.11). Moreover, f is always bounded if X is finite-dimensional (Theorem 5.16).*

6.2 Definition. *The **dual space** of an inner product space X, denoted by X^*, is the set of all bounded (= continuous) linear functionals on X. Further, addition and scalar multiplication in X^* are defined pointwise:*

$$(f + g)(x) := f(x) + g(x),$$
$$(\alpha f)(x) := \alpha f(x) \qquad \text{for all } x \in X, \quad \alpha \in \mathbb{R}.$$

It is easy to verify that if f and g are linear functionals on X, then so are $f + g$ and αf. Moreover, we have the following theorem.

6.3 The Dual Space is a Normed Linear Space. *With the norm of each $f \in X^*$ defined by (6.1.3), X^* is a normed linear space.*
   Proof. Let f, g be in X^*. Then, for every $x \in X$,

$$|(f + g)(x)| = |f(x) + g(x)| \le |f(x)| + |g(x)| \le (\|f\| + \|g\|)\|x\|.$$

Hence $f + g \in X^*$ and $\|f + g\| \le \|f\| + \|g\|$. Also,

$$\sup_{\|x\|=1} |(\alpha f)(x)| = \sup_{\|x\|=1} |\alpha f(x)| = |\alpha| \sup_{\|x\|=1} |f(x)| = |\alpha| \, \|f\|.$$

Thus $\alpha f \in X^*$ and $\|\alpha f\| = |\alpha| \, \|f\|$. Thus X^* is a linear space. Clearly, $\|f\| \ge 0$ and $\|f\| = 0$ if and only if $f(x) = 0$ for all $x \in X$, i.e., $f = 0$. Thus X^* is a normed linear space. ∎

Actually, it is possible to define an inner product on X^* (that generates the same norm) so that X^* is even a Hilbert space! This will follow easily from a result to be established shortly (see also Exercises 7 and 8).

From now on, we will usually use the more suggestive notation x^*, y^*, etc. (rather than f, g, etc.) to denote elements of the dual space X^*.

In Chapter 5 we showed how the elements of X generate bounded linear functionals on X. We can say substantially more.

6.4 Definition. *Let* ˆ *denote the "hat" mapping on X defined by* $x \mapsto \hat{x}$, *where*

(6.4.1) $\hat{x}(y) := \langle y, x \rangle$ *for every* $y \in X$.

Using Theorem 5.18 we see that $\hat{x} \in X^*$ and $\|\hat{x}\| = \|x\|$.

6.5 X Generates Elements in X*. *The hat mapping* ˆ : $X \to X^*$ *defined as in Definition 6.4 is a one-to-one bounded linear operator, having norm one, such that* $\|\hat{x}\| = \|x\|$ *for every* $x \in X$.

Proof. Let $F(x) = \hat{x}$; i.e., $F(x)(y) = \langle y, x \rangle$ for all y. Then, for any $z \in X$,

$$F(\alpha x + \beta y)(z) = \langle z, \alpha x + \beta y \rangle = \alpha \langle z, x \rangle + \beta \langle z, y \rangle$$
$$= \alpha F(x)(z) + \beta F(y)(z) = [\alpha F(x) + \beta F(y)](z).$$

Thus $F(\alpha x + \beta y) = \alpha F(x) + \beta F(y)$ and F is linear.

As already noted, $\|F(x)\| = \|\hat{x}\| = \|x\|$ for all x, so F has norm one.

Finally, since F is linear and norm-preserving, we have $\|F(x_1) - F(x_2)\| = \|F(x_1 - x_2)\| = \|x_1 - x_2\|$. From this it follows that F is one-to-one. ∎

We call x the "representer" of the functional $\hat{x} \in X^*$. More generally, an element $x \in X$ is called a **representer** for a functional $x^* \in X^*$ if $x^* = \hat{x}$, i.e., if

$$x^*(y) = \langle y, x \rangle \quad \text{for every } y \in X.$$

By Theorem 6.5, a functional $x^* \in X^*$ can have at most one representer, and the set

$$\hat{X} := \{\hat{x} \mid x \in X\}$$

of all functionals in X^* that have representers is a subspace of X^*.

It is natural to ask whether *every* bounded linear functional on X has a representer in X. That is, must $\hat{X} = X^*$? We shall see shortly that this is the case precisely when X is complete. However, "most" bounded linear functionals do have representers.

The following technical lemma contains the key step in proving the general representation theorem for bounded linear functionals, and it is the basis for some other interesting results as well. Loosely speaking, it states that every bounded linear functional is close to one that has a representer.

6.6 Key Lemma. *Let* $x^* \in X^*$ *with* $\|x^*\| = 1$ *and* $x \in X$ *with* $\|x\| \leq 1$. *Then*

(6.6.1) $\|x^* - \hat{x}\| \leq 8[1 - x^*(x)]^{1/2}$.

In particular, if $\epsilon > 0$ *and* $x^*(x) > 1 - (\epsilon/8)^2$, *then*

(6.6.2) $\|x^* - \hat{x}\| < \epsilon$.

Proof. To verify (6.6.1), it suffices to show that for each $y \in X$ with $\|y\| = 1$,

(6.6.3) $|x^*(y) - \langle y, x \rangle| \leq 8[1 - x^*(x)]^{1/2}$.

Fix any $y \in X$ with $\|y\| = 1$, and set $z = x^*(y)x - x^*(x)y$ and $\sigma = \text{sgn}\langle z, x \rangle$. Then $x^*(z) = 0$, and for any $\delta > 0$,

$$\|\sigma z\| \leq \|x^*(y)x\| + \|x^*(x)y\| \leq 2,$$
$$0 \leq 1 - [x^*(x)]^2 = [1 + x^*(x)][1 - x^*(x)] \leq 2[1 - x^*(x)],$$

and

$$\|x - \sigma\delta z\| \geq |x^*(x - \sigma\delta z)| = |x^*(x)|.$$

It follows that

$$|x^*(y) - \langle y, x \rangle| \leq |x^*(y) - x^*(y)\|x\|^2| + |\langle x^*(y)x, x \rangle - \langle x^*(x)y, x \rangle|$$
$$+ |\langle x^*(x)y, x \rangle - \langle y, x \rangle|$$

$$\leq |x^*(y)|[1 - \|x\|^2] + \frac{1}{\delta}\langle \sigma\delta z, x \rangle + [1 - x^*(x)]\|y\|\,\|x\|$$

$$\leq 1 - \|x\|^2 + \frac{1}{2\delta}[\|\sigma\delta z\|^2 + \|x\|^2 - \|\sigma\delta z - x\|^2] + 1 - x^*(x)$$

$$\leq 1 - [x^*(x)]^2 + \frac{1}{2\delta}[4\delta^2 + 1 - |x^*(x)|^2] + 1 - x^*(x)$$

$$\leq 3[1 - x^*(x)] + 2\delta + \frac{1}{\delta}[1 - x^*(x)]$$

$$= \left(3 + \frac{1}{\delta}\right)[1 - x^*(x)] + 2\delta.$$

If $x^*(x) = 1$, then $|x^*(y) - \langle y, x \rangle| \leq 2\delta$, and since δ was arbitrary, $|x^*(y) - \langle y, x \rangle| = 0$. If $x^*(x) < 1$, take $\delta = [1 - x^*(x)]^{1/2}$ to obtain $0 < \delta \leq \sqrt{2}$ and

$$|x^*(y) - \langle y, x \rangle| \leq 3\delta^2 + 3\delta \leq 3\sqrt{2}\delta + 3\delta < 8\delta = 8[1 - x^*(x)]^{1/2}.$$

Thus, in either case (6.6.3) holds. If $x^*(x) > 1 - (\epsilon/8)^2$, then $[1 - x^*(x)]^{1/2} < \epsilon/8$, and (6.6.2) follows from (6.6.1). ∎

A subset D of a normed linear space X is called **dense** in X if for each $x \in X$ and $\epsilon > 0$, there exists $y \in D$ such that $\|x - y\| < \epsilon$. For example, the rational numbers are dense in the set of all real numbers.

As a consequence of Lemma 6.6, we now show that the set of functionals that have representers in X is a dense subspace of X^*.

6.7 Denseness of $\hat{X}$ in X^*. *Let $x^* \in X^* \backslash \{0\}$. If $\{x_n\}$ is any sequence in X with $\|x_n\| \leq 1$ and $\lim x^*(x_n) = \|x^*\|$, then $\{x_n\}$ is a Cauchy sequence,*

(6.7.1)
$$\lim \|x^* - (\|x^*\|\hat{x}_n)\| = 0,$$

and

(6.7.2)
$$x^*(x) = \lim\langle x, \|x^*\|x_n \rangle, \quad x \in X.$$

In particular, $\hat{X}$ is a dense subspace of X^*, and the convergence in (6.7.2) is uniform over bounded subsets of X.

Proof. By replacing x^* with $\|x^*\|^{-1}x^*$, we may assume $\|x^*\| = 1$. For each $\epsilon > 0$, $x^*(x_n) > 1 - (\epsilon/8)^2$ for n large. By Lemma 6.6, $\|x^* - \hat{x}_n\| \le \epsilon$ for n large, so (6.7.1) is verified. Using Theorem 6.5, we obtain that

$$\|x_n - x_m\| = \|\hat{x}_n - \hat{x}_m\| \le \|\hat{x}_n - x^*\| + \|x^* - \hat{x}_m\| \to 0$$

as $n, m \to \infty$. That is, $\{x_n\}$ is a Cauchy sequence. Suppose B is a bounded subset of X, say $\|x\| \le c$ for all $x \in B$. For each $\epsilon > 0$, there exists an integer n_0 such that $c\|x^* - \hat{x}_n\| < \epsilon$ for each $n \ge n_0$. Thus for all $x \in B$ and for all $n \ge n_0$, we have

$$|x^*(x) - \langle x, x_n\rangle| = |x^*(x) - \hat{x}_n(x)| \le \|x^* - \hat{x}_n\| \, \|x\| \le c\|x^* - \hat{x}_n\| < \epsilon.$$

This verifies that the convergence in (6.7.2) is uniform over bounded sets.

That $\hat{X}$ is dense in X^* follows from (6.7.1). Finally, $\hat{X}$ is a subspace of X^* by Theorem 6.5. ∎

Representation of Bounded Linear Functionals

Next we give a general representation theorem for bounded linear functionals on X. It will prove quite useful throughout the remainder of the book. In particular, it will be used in the Fréchet–Riesz representation theorem (Theorem 6.10) and the strong separation theorem (Theorem 6.23).

6.8 Representation and Generation of Bounded Linear Functionals. *If X is an inner product space and $x^* \in X^*$, then there exists a Cauchy sequence $\{x_n\}$ in X such that*

(6.8.1) $$x^*(x) = \lim \langle x, x_n\rangle \quad \text{for each} \quad x \in X,$$

and

(6.8.2) $$\|x^*\| = \lim \|x_n\|.$$

Conversely, if $\{x_n\}$ is a Cauchy sequence in X, then (6.8.1) defines a bounded linear functional x^* on X whose norm is given by (6.8.2).

Proof. Let $x^* \in X^*$. If $x^* = 0$, set $x_n = 0$ for all n. Then (6.8.1) and (6.8.2) hold. If $x^* \neq 0$, choose any sequence $\{y_n\}$ in X with $\|y_n\| = 1$ and $\lim x^*(y_n) = \|x^*\|$. Set $x_n = \|x^*\|y_n$. Then $\|x_n\| = \|x^*\|$ for all n, and by Theorem 6.7, $\{x_n\}$ is a Cauchy sequence such that (6.8.1) and (6.8.2) hold.

Conversely, let $\{x_n\}$ be a Cauchy sequence in X. For any $x \in X$,

$$|\langle x, x_n\rangle - \langle x, x_m\rangle| = |\langle x, x_n - x_m\rangle| \le \|x\| \, \|x_n - x_m\|,$$

so $\{\langle x, x_n\rangle\}$ is a Cauchy sequence in $\mathbb{R}$. Thus the limit

$$x^*(x) := \lim \langle x, x_n\rangle$$

exists for each $x \in X$ and defines a real function x^* on X. Moreover, for each $x, y \in X$ and $\alpha, \beta \in \mathbb{R}$,

$$x^*(\alpha x + \beta y) = \lim \langle \alpha x + \beta y, x_n\rangle = \lim [\alpha\langle x, x_n\rangle + \beta\langle y, x_n\rangle]$$
$$= \alpha \lim\langle x, x_n\rangle + \beta \lim\langle y, x_n\rangle = \alpha x^*(x) + \beta x^*(y)$$

implies that x^* is linear.

Further, $\lim \|x_n\|$ exists by (1) of Theorem 1.12, since $\{x_n\}$ is a Cauchy sequence, and

$$|x^*(x)| = \lim |\langle x, x_n \rangle| \leq (\lim \|x_n\|)\|x\|$$

for each $x \in X$. Thus x^* is bounded and $\|x^*\| \leq \lim \|x_n\|$. To obtain the reverse inequality, let $c = \sup\limits_{n}\|x_n\|$. Given any $\epsilon > 0$, choose an integer N such that $c\|x_n - x_m\| < \epsilon$ for all $n, m \geq N$. Then if $m \geq N$,

$$\|x^*\|\,\|x_m\| \geq x^*(x_m) = \lim_{n}\langle x_m, x_n \rangle = \lim_{n}\left[\langle x_m, x_n - x_m \rangle + \|x_m\|^2\right]$$
$$\geq \liminf_{n}\left[-\|x_m\|\,\|x_n - x_m\| + \|x_m\|^2\right] \geq -\epsilon + \|x_m\|^2.$$

Letting $m \to \infty$ in this inequality, we obtain

$$\|x^*\| \lim_{m}\|x_m\| \geq \lim_{m}\|x_m\|^2 - \epsilon.$$

Since ϵ was arbitrary, it follows that $\|x^*\| \geq \lim \|x_m\|$. Thus (6.8.2) holds. ∎

As an easy consequence of Theorems 6.7 and 6.8 we can give a proof of the celebrated Hahn–Banach extension theorem for linear functionals. It is valid in a more general normed linear space setting, but the particular proof given here seems to be unique to the inner product space case and, unlike the standard proofs of the normed linear space case, does not make any appeal to the axiom of choice.

6.9 Hahn–Banach Extension Theorem. *Let M be a subspace of the inner product space X and $y^* \in M^*$. That is, y^* is a bounded linear functional on M. Then there exists $x^* \in X^*$ such that $x^*|_M = y^*$ and $\|x^*\| = \|y^*\|$.*

Proof. By Theorem 6.7, we can choose $\{y_n\}$ in M with $\|y_n\| = 1$ such that $\|y^*\|\hat{y}_n \to y^*$. In particular, $\{y_n\}$ is a Cauchy sequence and

$$y^*(y) = \lim\langle y, \|y^*\|y_n \rangle, \quad y \in M.$$

Defining x^* on X by

$$x^*(x) = \lim\langle x, \|y^*\|y_n \rangle, \quad x \in X,$$

it follows from Theorem 6.8 that $x^* \in X^*$, $\|x^*\| = \lim \|(\|y^*\|y_n)\| = \|y^*\|$, and, obviously, $x^*|_M = y^*$. ∎

When X is complete, *every* bounded linear functional on X has a representer in X.

6.10 Fréchet–Riesz Representation Theorem. *Let X be a Hilbert space. Then $X^* = \hat{X}$. That is, for each $x^* \in X^*$, there is a unique element $x \in X$ such that*

(6.10.1) $$x^*(y) = \langle y, x \rangle \quad \text{for each} \quad y \in X,$$

and $\|x^*\| = \|x\|$.

Proof. Let $x^* \in X^*$. By Theorem 6.8, there is a Cauchy sequence $\{x_n\}$ in X such that

$$x^*(y) = \lim \langle y, x_n \rangle, \quad y \in X,$$

and $\|x^*\| = \lim \|x_n\|$. Since X is complete, $x := \lim x_n$ exists. By (2) (a) and (b) of Theorem 1.12, $\|x\| = \lim \|x_n\|$ and (6.10.1) holds. The uniqueness of x follows, since if $\langle y, x' \rangle = \langle y, x \rangle$ for all $y \in X$, then $\langle y, x' - x \rangle = 0$ for all y, so $x' - x = 0$, or $x' = x$. ∎

This result actually *characterizes* Hilbert space among all inner product spaces. That is, if X is not complete, there exists an $x^* \in X^*$ that fails to have a representer (see Exercise 6 at the end of the chapter).

Left unanswered thus far has been the question of precisely *which* functionals in X^* have representers in X. We answer this next.

6.11 Definition. *A functional $x^* \in X^*$ is said to* **attain its norm** *if there exists $x \in X$ with $\|x\| = 1$ and $x^*(x) = \|x^*\|$.*

From (6.1.3), we see that

$$\|x^*\| = \sup_{\|x\|=1} |x^*(x)| = \sup_{\|x\|=1} x^*(x).$$

Thus x^* attains its norm if and only if the "supremum" can be replaced by "maximum" in the definition of the norm of x^*.

6.12 Norm-Attaining is Equivalent to Having a Representer. *Let $x^* \in X^*$. Then x^* has a representer in X if and only if x^* attains its norm.*

Proof. If $x^* = 0$, then $x = 0$ is the representer for x^*, and x^* attains its norm at any $y \in X$ with $\|y\| = 1$. Thus we may assume $x^* \neq 0$. By replacing x^* with $x^*/\|x^*\|$, we may further assume that $\|x^*\| = 1$.

Let x^* have the representer x. Then $\|x\| = 1$ and

$$x^*(x) = \langle x, x \rangle = 1 = \|x^*\|.$$

Thus x^* attains its norm at x.

Conversely, suppose x^* attains its norm at x. That is, $\|x\| = 1$ and $x^*(x) = 1$. By Lemma 6.6, $x^* = \hat{x}$. That is, x is the representer for x^*. ∎

We now consider some particular examples of functionals that exhibit the various representations thus far discussed.

6.13 Examples.

(1) If $x^* \in l_2(I)^*$, then (by Theorem 6.10) there exists a unique $x \in l_2(I)$ such that

$$x^*(y) = \sum_{i \in I} y(i)x(i) \quad \text{for every } y \in l_2(I),$$

and $\|x^*\| = \|x\|$.

(2) As special cases of (1), we have that if $x^* \in l_2^*$ (respectively $x^* \in l_2(n)^*$); then there exists a unique $x \in l_2$ (respectively $x \in l_2(n)$) such that

$$x^*(y) = \sum_1^\infty y(i)x(i) \quad \text{for every } y \in l_2$$

(respectively $x^*(y) = \sum_1^n y(i)x(i)$ for every $y \in l_2(n)$) and

$$\|x^*\| = \|x\|.$$

(3) If $x \in C_2[a, b]$, then (by Theorem 5.18 or Theorem 6.5)

$$x^*(y) := \int_a^b y(t)x(t)\, dt \quad \text{for all } y \in C_2[a, b]$$

defines a bounded linear functional on $C_2[a, b]$ with $\|x^*\| = \|x\|$.

(4) If $x^* \in C_2[a, b]^*$ has no representer in $C_2[a, b]$ (or, equivalently, x^* fails to attain its norm), then (by Theorem 6.7) for any sequence $\{x_n\}$ in $C_2[a, b]$ with $\|x_n\| \leq 1$ and $x^*(x_n) \to \|x^*\|$, we have the representation

$$x^*(y) = \lim \|x^*\| \int_a^b y(t)x_n(t)dt \quad \text{for every } y \in C_2[a, b].$$

(5) Consider the space $X = C_2[-1, 1]$ and the functional x^* on X defined by

$$x^*(y) = \int_0^1 y(t)dt \quad \text{for all } y \in X.$$

It is clear that x^* is linear and, for any $y \in X$,

$$|x^*(y)| = \left| \int_0^1 y(t)dt \right| \leq \left[\int_0^1 |y(t)|^2 dt \right]^{\frac{1}{2}} \leq \left[\int_{-1}^1 |y(t)|^2 dt \right]^{\frac{1}{2}} = \|y\|.$$

Hence x^* is bounded and $\|x^*\| \leq 1$. Consider the sequence $\{x_n\}$ in $C_2[-1, 1]$ defined by

$$x_n(t) = \begin{cases} 0 & \text{if } -1 \leq t \leq 0, \\ nt & \text{if } 0 \leq t \leq 1/n, \\ 1 & \text{if } 1/n \leq t \leq 1. \end{cases}$$

Clearly, $\|x_n\| \leq 1$ and

$$x^*(x_n) = \int_0^1 x_n(t)\, dt = \int_0^{1/n} nt\, dt + \int_{1/n}^1 1\, dt = 1/(2n) + (1 - 1/n) \to 1.$$

Thus $\|x^*\| = 1$ and $x^*(x_n) \to \|x^*\|$. By Theorem 6.7, x^* has the representation

$$x^*(y) = \lim \int_{-1}^1 y(t)x_n(t)dt \quad \text{for every } y \in X.$$

To see that x^* does not have a representer, it suffices by Theorem 6.12 to show that x^* does not attain its norm. If x^* attained its norm, then there would exist $x \in C_2[-1, 1]$ such that $\|x\| = 1$ and $x^*(x) = 1$. Then (by Schwarz's inequality in $C_2[0, 1]$),

$$1 = \int_0^1 x(t)\, dt \leq \left[\int_0^1 x^2(t)\, dt \right]^{\frac{1}{2}} \leq \left[\int_{-1}^1 x^2(t)\, dt \right]^{\frac{1}{2}} = \|x\| = 1.$$

Hence equality must hold throughout this string of inequalities. It follows that $x(t) = 0$ for $t \in [-1, 0)$ and $x(t) = 1$ on $(0, 1]$. But such an x is not continuous. This contradiction shows that x^* does not attain its norm.

Best Approximation from Hyperplanes

The level sets of bounded linear functionals are called "hyperplanes" and are useful for providing geometric interpretations of various phenomena.

6.14 Definition. *A* **hyperplane** *in X is any set of the form*

$$H = \{y \in X \mid x^*(y) = c\},$$

where $x^* \in X^* \setminus \{0\}$ and $c \in \mathbb{R}$. The **kernel**, *or null space, of a functional $x^* \in X^*$ is the set*

$$\ker x^* := \{y \in X \mid x^*(y) = 0\}.$$

Note that a hyperplane is never empty. For since $x^* \neq 0$, choose any $x_0 \in X$ with $x^*(x_0) \neq 0$. Then the element $y = [c/x^*(x_0)]x_0$ satisfies $x^*(y) = c$.

We next show that if $x^* \in X^* \setminus \{0\}$, then $\ker x^*$ is a **maximal subspace** in X; that is, X is the only subspace that properly contains $\ker x^*$. Moreover, each hyperplane $\{y \in X \mid x^*(y) = c\}$ is a translate of $\ker x^*$.

If M and N are two subspaces of X, recall that X is the **direct sum** of M and N, denoted by $X = M \oplus N$, provided that $X = M + N$ and $M \cap N = \{0\}$. Equivalently, $X = M \oplus N$ if and only if each $x \in X$ has a unique representation of the form $x = y + z$, where $y \in M$ and $z \in N$ (see Exercise 3 at the end of the chapter). Recall that Theorem 5.9 established that if M is a Chebyshev subspace, then $X = M \oplus M^\perp$.

6.15 Hyperplanes Are Translates of Maximal Subspaces. *Let X be an inner product space, $x^* \in X^* \setminus \{0\}$, $c \in \mathbb{R}$, $M = \ker x^*$, and $H = \{y \in X \mid x^*(y) = c\}$. Then:*

(1) *For any $x_1 \in X \setminus M$, $X = M \oplus \operatorname{span}\{x_1\}$.*
(2) *For any $x_0 \in H$, $H = M + x_0$.*
(3) *M is a closed maximal subspace in X.*
(4) *H is a closed convex subset of X.*

In particular, if X is complete, then H is Chebyshev.

Proof. (1) Fix $x_1 \in X \setminus M$. For each $x \in X$, set $y = x - x^*(x)[x^*(x_1)]^{-1}x_1$. Then $y \in M$ and $x = y + x^*(x)[x^*(x_1)]^{-1}x_1$. Thus $X = M + \operatorname{span}\{x_1\}$. If $z \in M \cap \operatorname{span}\{x_1\}$, then $z = \alpha x_1$. If $\alpha \neq 0$, then $x_1 = (1/\alpha)z \in M$, which is a contradiction. Thus $\alpha = 0$, and so $z = 0$. That is, $M \cap \operatorname{span}\{x_1\} = \{0\}$, and this verifies (1).

(2) Fix any $x_0 \in H$. If $x \in M + x_0$, then $x = y + x_0$ for some $y \in M$ implies $x^*(x) = x^*(x_0) = c$. Thus $x \in H$. Conversely, if $x \in H$, then $y = x - x_0$ is in M and $x = y + x_0 \in M + x_0$.

(3) It is clear that M is a subspace, since x^* is linear; and M is closed, since x^* is continuous. It remains to show that M is maximal. If Y is a subspace in X with $Y \supset M$ and $Y \neq M$, choose any $x_1 \in Y \setminus M$. By (1),

$$X = M + \operatorname{span}\{x_1\} \subset Y \subset X,$$

so $Y = X$.

(4) H is the translate of a closed subspace, so it is closed by Exercise 8 at the end of Chapter 3, and convex by (5) of Proposition 2.3. The last statement follows from Theorem 3.5. ∎

In particular, in an n-dimensional space X, a hyperplane is the translate of an $(n-1)$-dimensional subspace of X. For example, in Euclidean 3-space (respectively 2-space) a hyperplane is a "plane" (respectively a "line").

A hyperplane has the pleasant property that there are simple formulas for its metric projection, as well as for its distance from any point.

6.16 Distance to Hyperplanes. *Let X be an inner product space, $x^* \in X^* \setminus \{0\}$, $c \in \mathbb{R}$, and $H = \{y \in X \mid x^*(y) = c\}$. Then*

$$(6.16.1) \qquad d(x, H) = \frac{1}{\|x^*\|} |x^*(x) - c|$$

for each $x \in X$.

Proof. Assume first $c = 0$. Then $H = \{y \in X \mid x^*(y) = 0\}$, and we must show that

$$(6.16.2) \qquad d(x, H) = \frac{1}{\|x^*\|} |x^*(x)|.$$

Fix any $x \in X$. For any $y \in H$,

$$\frac{1}{\|x^*\|} |x^*(x)| = \frac{1}{\|x^*\|} |x^*(x - y)| \leq \|x - y\|$$

implies that

$$(6.16.3) \qquad \frac{1}{\|x^*\|} |x^*(x)| \leq d(x, H).$$

Conversely, given any ϵ with $0 < \epsilon < \|x^*\|$, choose $z \in X$ with $\|z\| = 1$ and $x^*(z) > \|x^*\| - \epsilon$. Then $y := x - x^*(x)[x^*(z)]^{-1} z$ is in H and

$$\|x - y\| = \frac{|x^*(x)|}{|x^*(z)|} \leq \frac{|x^*(x)|}{\|x^*\| - \epsilon}$$

implies $d(x, H) \leq |x^*(x)|(\|x^*\| - \epsilon)^{-1}$. Since ϵ was arbitrary, it follows that $d(x, H) \leq |x^*(x)| \|x^*\|^{-1}$. Combining this with (6.16.3) we obtain (6.16.2).

Now suppose $c \neq 0$. Choose any $x_0 \in H$ and define

$$H_0 := H - x_0 = \{y \in X \mid x^*(y) = 0\}.$$

By the first part of the proof and the invariance by translation (Theorem 2.7(1)(i)), we obtain

$$d(x, H) = d(x - x_0, H - x_0) = d(x - x_0, H_0)$$

$$= \frac{1}{\|x^*\|} |x^*(x - x_0)| = \frac{1}{\|x^*\|} |x^*(x) - c|,$$

which completes the proof. ∎

We can now characterize the hyperplanes that are Chebyshev. The functionals having representers once again play the essential role.

6.17 Chebyshev Hyperplanes. *Let X be an inner product space, $x^* \in X^* \setminus \{0\}$, $c \in \mathbb{R}$, and $H = \{y \in X \mid x^*(y) = c\}$. Then the following statements are equivalent:*

(1) *H is Chebyshev;*
(2) *H is proximinal;*
(3) *Some $x \in X \setminus H$ has a best approximation in H;*
(4) *x^* attains its norm;*
(5) *x^* has a representer in X.*

Moreover, if H is Chebyshev, then

$$(6.17.1) \qquad P_H(x) = x - \|z\|^{-2}(\langle x, z \rangle - c)z$$

for every $x \in X$, where z is the representer for x^*.

Proof. The implications $(1) \implies (2) \implies (3)$ are obvious.

$(3) \implies (4)$. If some $x \in X \setminus H$ has a best approximation $y_0 \in H$, then by Theorem 6.16,

$$\|x - y_0\| = \frac{1}{\|x^*\|} |x^*(x) - c| = \frac{1}{\|x^*\|} |x^*(x - y_0)|.$$

Setting $z = \sigma(x - y_0)$, where $\sigma = \|x - y_0\|^{-1} \operatorname{sgn} x^*(x - y_0)$, we see that $\|z\| = 1$ and $x^*(z) = \|x^*\|$. Thus x^* attains its norm (at z).

The equivalence of (4) and (5) is just Theorem 6.12.

$(5) \implies (1)$. Let $z \in X$ be the representer of x^*. Then $\hat{z} = x^*$ and $\|z\| = \|x^*\|$. For any $x \in X$, set $y = x - \|z\|^{-2}(\langle x, z \rangle - c)z = x - \|x^*\|^{-2}[x^*(x) - c]z$. Then $x^*(y) = c$, so that $y \in H$, and

$$\|x - y\| = \|x^*\|^{-2}|x^*(x) - c|\|z\| = \|x^*\|^{-1}|x^*(x) - c| = d(x, H)$$

(using Theorem 6.16). Thus $y = P_H(x)$ by Theorem 2.4. This proves that H is Chebyshev. In the process, we have also verified (6.17.1). ∎

As a consequence of (6.17.1) and Theorem 4.9, observe that the representer of a functional x^* is always *orthogonal* to $\ker x^* = \{x \in X \mid x^*(x) = 0\}$.

As a simple application of Theorems 6.16 and 6.17 in Euclidean 2-space $l_2(2)$, we obtain the well-known distance formula from a point to a line and the (not as well-known) formula for the best approximation in the line to the point. Specifically, fix any scalars a, b, and c, where $a^2 + b^2 \neq 0$, and consider the line

$$H = \{y \in l_2(2) \mid ay(1) + by(2) = c\} = \{y \in l_2(2) \mid \langle y, z \rangle = c\},$$

where $z = (a, b)$. Then for any $x \in l_2(2)$,

$$d(x, H) = (a^2 + b^2)^{-\frac{1}{2}} |ax(1) + bx(2) - c|$$

and

$$P_H(x) = x - (a^2 + b^2)^{-1}[ax(1) + bx(2) - c]z$$

(see Figure 6.17.2).

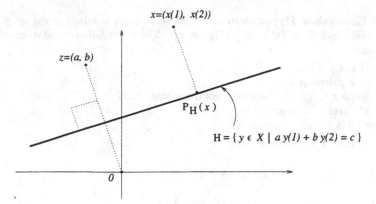

Figure 6.17.2

Next we give an immediate consequence of Theorem 6.17 in the space $C_2[a, b]$.

6.18 Best Approximation from Hyperplanes in $C_2[a, b]$. *Let*
$z \in C_2[a, b] \setminus \{0\}, c \in \mathbb{R}$, *and*

$$H = \left\{ y \in C_2[a, b] \,\Big|\, \int_a^b y(t)z(t)dt = c \right\}.$$

Then H is a Chebyshev hyperplane in $C_2[a, b]$ and

(6.18.1) $$P_H(x) = x - \|z\|^{-2} \left[\int_a^b x(t)z(t)dt - c \right] z$$

for all $x \in C_2[a, b]$.

As a specific example of Corollary 6.18, consider $C_2[0, 1]$ and

$$H = \left\{ y \in C_2[0, 1] \,\Big|\, \int_0^1 y(t)t^2 dt = 1 \right\}.$$

Then $z(t) = t^2$, $\|z\|^2 = \int_0^1 t^4 dt = \frac{1}{5}$, and

$$P_H(x) = x - 5 \left[\int_0^1 x(t)t^2 dt - 1 \right] z$$

for every $x \in C_2[0, 1]$. In particular, if $x(t) = e^{t^3}$, we get

$$P_H(e^{t^3}) = e^{t^3} - (\tfrac{5}{3})(e - 4)t^2.$$

From Theorem 6.17 we saw that a hyperplane H is Chebyshev if a *single* point that is not in H has a best approximation in H. In particular, if H is not Chebyshev, *no* point that is not in H has a best approximation in H. Sets with this latter property are called **anti-proximinal**. Using the characterization of Chebyshev hyperplanes (Theorem 6.17), it is easy to give examples of anti-proximinal hyperplanes. Of course, owing to part (4) of Theorem 6.15, these examples must be in *incomplete* inner product spaces.

6.19 Two Examples of Anti-Proximinal Hyperplanes.
(1) Recall that in Example 3.2 we showed that the closed subspace

$$M = \left\{ y \in C_2[-1,1] \,\middle|\, \int_0^1 y(t)\,dt = 0 \right\}$$

is not proximinal in $C_2[-1,1]$. In Example (5) of 6.13, we also showed that the functional x^* defined on $C_2[-1,1]$ by

$$x^*(y) := \int_0^1 y(t)\,dt$$

is in $C_2[-1,1]^*$, but fails to have a representer; hence M is a hyperplane. By Theorem 6.17, *no* point in $C_2[-1,1] \setminus M$ has a best approximation in M.
(2) *Let*

$$X = \{x \in l_2 \mid x(n) = 0 \text{ except for at most finitely many } n\}$$

and

$$H = \left\{ y \in X \,\middle|\, \sum_1^\infty 2^{-n/2} y(n) = 0 \right\}.$$

Then H is a hyperplane in X, and no point in $X \setminus H$ has a best approximation in H.

To verify this, define x^* on X by

$$x^*(y) = \sum_1^\infty 2^{-n/2} y(n) \text{ for every } y \in X.$$

(This sum is actually a finite sum, since $y(n) = 0$ for n sufficiently large.) Clearly, x^* is a linear functional on X. For any $y \in X$ with $\|y\| = 1$, Schwarz's inequality (in l_2) implies

$$(6.19.1) \qquad |x^*(y)| \le \left[\sum_1^\infty 2^{-n} \right]^{1/2} \|y\| = 1.$$

Thus $x^* \in X^*$, $\|x^*\| \le 1$, and $H = \ker x^*$ is a hyperplane. Next we show that $\|x^*\| = 1$. For each integer $N \ge 1$, define $y_N(n) = 2^{-n/2}$ for all $n \le N$ and $y_N(n) = 0$ if $n > N$. Then $y_N \in X$, $\|y_N\| \le 1$ for all N, and

$$x^*(y_N) = \sum_1^N 2^{-n} = 1 - 2^{-N} \to 1.$$

Thus $\|x^*\| = 1$.

If x^* attained its norm, then $x^*(y) = 1$ for some $y \in X$ with $\|y\| = 1$. Thus equality holds in (6.19.1) for this y. By the condition of equality in Schwarz's inequality, we must have $y = \lambda z$ for some $\lambda > 0$, where $z(n) = 2^{-n/2}$ for all n. But $z \notin X$ (since $z(n) \ne 0$ for every n) implies $y = \lambda z \notin X$ as well. This proves that x^* does not attain its norm. By Theorem 6.17, no point in $X \setminus H$ has a best approximation in H.

Strong Separation Theorem

Hyperplanes are useful in describing certain geometric phenomena.

6.20 Definition. *A hyperplane $H = \{y \in X \mid x^*(y) = c\}$ is said to* **separate** *the sets K and L if*

$$(6.20.1) \qquad \sup x^*(K) \leq c \leq \inf x^*(L),$$

where $\sup x^*(K) := \sup\{x^*(y) \mid y \in K\}$ and $\inf x^*(L) := \inf\{x^*(y) \mid y \in L\}$.

Geometrically, H separates K and L if K lies in one of the two closed "half-spaces" determined by H, and L lies in the other. (The two closed **half-spaces** determined by H are the sets $H^+ := \{y \in X \mid x^*(y) \geq c\}$ and $H^- := \{y \in X \mid x^*(y) \leq c\}$.) In this chapter we will be interested only in the case where one of the sets is a single point. In Chapter 10 we will consider the more general situation.

The next lemma implies the obvious geometric fact that if H separates K and the point x, then x is closer to H than K (see Figure 6.20.2).

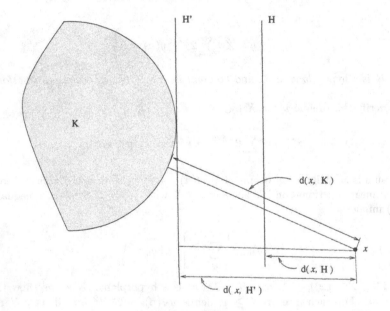

Figure 6.20.2

6.21 Lemma. *Suppose the hyperplane $H = \{y \in X \mid x^*(y) = c\}$ separates the set K and the point x; i.e., $\sup x^*(K) \leq c \leq x^*(x)$. Then the hyperplane $H' = \{y \in X \mid x^*(y) = \sup x^*(K)\}$ also separates K and x and*

$$(6.21.1) \qquad d(x,H) \leq d(x,H') \leq d(x,K).$$

Proof. Using Theorem 6.16, we obtain

$$d(x,H) = \|x^*\|^{-1}|x^*(x) - c| = \|x^*\|^{-1}[x^*(x) - c]$$
$$\leq \|x^*\|^{-1}[x^*(x) - \sup x^*(K)] = d(x,H').$$

Also, for any $y \in K$,

$$\|x^*\|^{-1}[x^*(x) - \sup x^*(K)] \le \|x^*\|^{-1}[x^*(x) - x^*(y)] \le \|x - y\|.$$

This implies $d(x, H') \le d(x, K)$.   ∎

It is useful to know that the dual X^* of an inner product space X is itself a complete inner product space, even if X is not complete. This can be deduced from the following lemma that shows that the parallelogram law holds in X^* (see, e.g., Exercise 7 at the end of the chapter).

6.22 Parallelogram Law in X*. *Let X be an inner product space. For every x^*, y^* in X^*,*

(6.22.1) $$\|x^* + y^*\|^2 + \|x^* - y^*\|^2 = 2\|x^*\|^2 + 2\|y^*\|^2.$$

In particular, if $\|x^*\| = \|y^*\| = 1$ and $x^* \ne y^*$, then $\|\frac{1}{2}(x^* + y^*)\| < 1$.

Proof. Choose x_n, y_n in X with $\|x_n\| = 1 = \|y_n\|$, $x^*(x_n) \to \|x^*\|$, and $y^*(y_n) \to \|y^*\|$. By Theorem 6.7, $\|x^*\|\hat{x}_n \to x^*$ and $\|y^*\|\hat{y}_n \to y^*$. Thus, using Theorem 6.5 and the parallelogram law (Theorem 1.5) in X, we obtain

$$
\begin{aligned}
\|x^* + y^*\|^2 + \|x^* - y^*\|^2 &= \lim[\|(\|x^*\|\hat{x}_n + \|y^*\|\hat{y}_n)\|^2 + \|(\|x^*\|\hat{x}_n - \|y^*\|\hat{y}_n)\|^2] \\
&= \lim[\|(\|x^*\|x_n + \|y^*\|y_n)\|^2 + \|(\|x^*\|x_n - \|y^*\|y_n)\|^2] \\
&= \lim[2\|(\|x^*\|x_n)\|^2 + 2\|(\|y^*\|y_n)\|^2] \\
&= 2\|x^*\|^2 + 2\|y^*\|^2.   ∎
\end{aligned}
$$

The main geometric result of this chapter is a "strong separation" principle that implies (among other things) that (1) each point outside of a closed convex set may be strictly separated from the set by a hyperplane; (2) every closed convex set is the intersection of all the closed half-spaces that contain it; and (3) there is a distance formula from a point to a convex set in terms of distances to separating hyperplanes. It is one of the more useful theorems in the book.

6.23 Strong Separation Principle. *Let K be a closed convex set in the inner product space X and $x \in X \setminus K$. Then there exists a unique $x^* \in X^*$ such that $\|x^*\| = 1$ and*

(6.23.1) $$d(x, K) = x^*(x) - \sup x^*(K).$$

In particular,

(6.23.2) $$\sup x^*(K) < x^*(x).$$

Moreover, if x has a best approximation in K (e.g., if X or K is complete), then x^* has a representer $\left(\text{namely, } d(x, K)^{-1}[x - P_K(x)]\right)$.

Proof. Since $x \notin K$, the distance $d := d(x, K)$ is positive. To motivate the general case, we first prove the special case when $P_K(x)$ exists. In this case, set $z = d^{-1}(x - P_K(x))$. By Theorem 4.1, $\sup_{y \in K}\langle x - P_K(x), y - P_K(x)\rangle = 0$, and so $\sup_{y \in K}\langle z, y - P_K(x)\rangle = 0$. Thus $\|z\| = 1$, and

$$\langle x, z\rangle - \sup_{y \in K}\langle y, z\rangle = \langle x, z\rangle - \langle P_K(x), z\rangle = \langle x - P_K(x), z\rangle = d.$$

It follows that (6.23.1) holds with $x^* = \hat{z}$.

In general, if x has no best approximation in K, choose a sequence $\{y_n\}$ in K such that $\|x - y_n\| \to d$. Then $\{y_n\}$ is a minimizing sequence, so by the proof of Theorem 3.4(2), $\{y_n\}$ is a Cauchy sequence. Setting $z_n = d^{-1}(x - y_n)$, we see that $\{z_n\}$ is a Cauchy sequence and $\|z_n\| \to 1$. By Theorem 6.8, the functional x^* defined on X by

$$x^*(y) = \lim \langle y, z_n \rangle, \quad y \in X,$$

is in X^* and $\|x^*\| = 1$. For each $y \in K$, we have

(6.23.3)
$$
\begin{aligned}
x^*(y) &= \lim \langle y, z_n \rangle = \lim[\langle y - y_n, z_n \rangle + \langle y_n - x, z_n \rangle + \langle x, z_n \rangle] \\
&= \lim \langle y - y_n, z_n \rangle - d + x^*(x).
\end{aligned}
$$

We next prove that

(6.23.4)
$$\lim \langle y - y_n, z_n \rangle \leq 0.$$

To this end, suppose (6.23.4) fails. By passing to a subsequence, we may assume that $\langle y - y_n, z_n \rangle \geq \delta > 0$ for every n. Given any $0 < \lambda < 1$, the vector $\lambda y + (1 - \lambda)y_n$ is in K by convexity and

$$
\begin{aligned}
d^2 &\leq \|x - [\lambda y + (1 - \lambda)y_n]\|^2 = \|x - y_n + \lambda(y_n - y)\|^2 \\
&= \|x - y_n\|^2 + 2\lambda \langle x - y_n, y_n - y \rangle + \lambda^2 \|y_n - y\|^2 \\
&\leq \|x - y_n\|^2 - 2\lambda \delta d + \lambda^2 c,
\end{aligned}
$$

where $c := \sup_n \|y_n - y\|^2$. Letting $n \to \infty$, we obtain

$$d^2 \leq d^2 - \lambda[2d\delta - \lambda c].$$

For $\lambda > 0$ sufficiently small the term in brackets is positive, and we obtain the absurdity $d^2 < d^2$. This contradiction verifies (6.23.4).

From (6.23.3) and (6.23.4), we see that $x^*(y) \leq -d + x^*(x)$ for every $y \in K$. Hence $\sup x^*(K) \leq -d + x^*(x)$, and so

$$d \leq x^*(x) - \sup x^*(K) = \inf_{y \in K} x^*(x - y) \leq \inf_{y \in K} \|x - y\| = d.$$

Thus equality must hold throughout this string of inequalities, and (6.23.1) is verified.

To verify the uniqueness of x^*, suppose both x_1^* and x_2^* satisfy $\|x_i^*\| = 1$ and $d = x_i^*(x) - \sup x_i^*(K)$ $(i = 1, 2)$. Letting $x^* = \frac{1}{2}(x_1^* + x_2^*)$, we see that $\|x^*\| \leq 1$ and

$$
\begin{aligned}
x^*(x) - \sup x^*(K) &= 2^{-1}x_1^*(x) + 2^{-1}x_2^*(x) - 2^{-1}\sup_{y \in K}[x_1^*(y) + x_2^*(y)] \\
&\geq 2^{-1}x_1^*(x) + 2^{-1}x_2^*(x) - 2^{-1}\sup x_1^*(K) - 2^{-1}\sup x_2^*(K) \\
&= 2^{-1}d + 2^{-1}d = d.
\end{aligned}
$$

On the other hand, Lemma 6.21 implies $x^*(x) - \sup x^*(K) \leq d\|x^*\| \leq d$. Hence

$$x^*(x) - \sup x^*(K) = d.$$

It follows from the last inequality that $\|x^*\| = 1$. By Lemma 6.22, we deduce that $x_1^* = x_2^*$. This completes the proof. ∎

Remarks. (1) The geometric interpretation of the strong separation theorem is this. If x is any point outside a closed convex set K, there is a hyperplane H (namely, $H = \{y \in X \mid x^*(y) = \sup x^*(K)\}$) that separates K and x such that the distance from x to H is equal to the distance from x to K (see Figure 6.23.4).

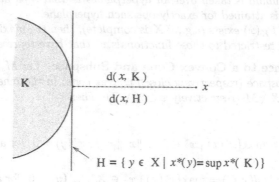

Figure 6.23.4

(2) Taking $K = \{0\}$ in Theorem 6.23, we immediately deduce that *for every* $x \in X$, *there exists* $x^* \in X^*$ *with* $\|x^*\| = 1$ *and* $x^*(x) = \|x\|$.

(3) There is a strengthening of the strong separation principle in the case where the convex set is either a convex cone or a subspace. We state it next.

6.24 Strong Separation for Subspaces and Convex Cones. *Let M (respectively C) be a closed subspace (respectively closed convex cone) in the inner product space X and $x \in X \setminus M$ (respectively $x \in X \setminus C$). Then there exists a functional $x^* \in X^*$ with $\|x^*\| = 1$, $x^*(x) > 0$, and*

$$x^*(y) = 0 \quad \text{for all } y \in M \text{ (respectively } x^*(y) \leq 0 \text{ for all } y \in C).$$

In addition, if M (respectively C) is Chebyshev (e.g., if X is complete), then x^* has a representer in $M^\perp$ (respectively in C°).

Proof. We prove the subspace case. The cone case is similar and is left as an exercise. By the separation theorem (Theorem 6.23), there exists an $x^* \in X^*$ such that $\|x^*\| = 1$ and $\sup x^*(M) < x^*(x)$. If $x^*(y) \neq 0$ for some $y \in M$, then by replacing y with $-y$ if necessary, we may assume $x^*(y) > 0$. But then the vector $y_n := ny$ is in M for every integer n, and $x^*(y_n) = nx^*(y) \to \infty$, which contradicts $\sup x^*(M) < \infty$. ∎

Using Theorems 6.23 and 6.24, Remark (1), and Lemma 6.21, we immediately obtain the following two distance formulas.

6.25 Distance to a Convex Set. *Let K be a closed convex set in the inner product space X and $x \in X \setminus K$. Then:*
(1) *(Analytic formulation)*

(6.25.1) $$d(x, K) = \max\{x^*(x) - \sup x^*(K) \mid x^* \in X^*, \|x^*\| = 1\},$$

and this maximum is attained for exactly one such functional.

(2) *(Geometric formulation)*

(6.25.2) $$d(x, K) = \max d(x, H),$$

where the maximum is taken over all hyperplanes H that separate K and x, and this maximum is attained for exactly one such hyperplane.

Moreover, if $P_K(x)$ exists (e.g., if X is complete), then the maximum in (6.25.1) may be further restricted to those functionals x^* that have representers in X.

6.26 Distance to a Convex Cone and Subspace. Let M (respectively C) be a closed subspace (respectively closed convex cone) in the inner product space X and let $x \in X \setminus M$ (respectively $x \in X \setminus C$). Then

(6.26.1)
$$d(x, M) = \max\{x^*(x) \mid x^* \in X^*, \|x^*\| = 1, \ x^*(y) = 0 \ \text{ for all } y \in M\}$$
(6.26.2)
$$(\text{respectively } d(x, C) = \max\{x^*(x) \mid x^* \in X^*, \ x^*(y) \le 0 \ \text{ for all } y \in C\}).$$

Moreover, if $P_M(x)$ (respectively $P_C(x)$) exists (e.g., if X is complete), then the maximum in (6.26.1) (respectively in (6.26.2)) may be further restricted to those functionals x^* that have representers in $M^\perp$ (respectively in C°).

In the next chapter these distance formulas will be further refined and applied.

Suppose it were somehow possible to obtain the separating hyperplane H that maximizes the expression (6.25.2), i.e., such that $d(x, H) = d(x, K)$. How would this help us to compute the best approximation $P_K(x)$? The answer is that in this case $P_K(x) = P_H(x)$, where $P_H(x)$ is given explicitly by the formula (6.17.1). This is the content of the next theorem.

6.27 Theorem. Let K be a closed convex set, $x \in X \setminus K$, and let $H = \{y \in X \mid \langle y, z \rangle = c\}$ be a hyperplane that separates K and x; that is, $z \in X \setminus \{0\}$, $c \in \mathbb{R}$, and

$$\sup_{y \in K} \langle y, z \rangle \le c \le \langle x, z \rangle.$$

Then

(6.27.1) $$P_H(x) = x - \|z\|^{-2}[\langle x, z \rangle - c]z.$$

Suppose also that $P_K(x)$ exists (e.g., if X is complete). Then the following statements are equivalent:

(1) $P_K(x) = P_H(x)$;
(2) $P_H(x) \in K$;
(3) $d(x, H) = d(x, K)$;
(4) $d(x, H) \ge d(x, K)$.

Also, if K is a convex cone, then each of these statements implies that H is a subspace (i.e., $c = 0$).

Proof. The formula for $P_H(x)$ follows by Theorem 6.17.

The implications (1) $\Longrightarrow$ (2) and (3) $\Longrightarrow$ (4) are obvious.

(2) $\Longrightarrow$ (3). If $P_H(x) \in K$, then by Lemma 6.21,

$$\|x - P_H(x)\| = d(x, H) \le d(x, K) \le \|x - P_H(x)\|.$$

Hence equality must hold throughout, and (3) holds.

(4) $\implies$ (1). If $d(x, H) \geq d(x, K)$, then by Theorem 6.16,

$$d(x, H) = \|z\|^{-1}[\langle x, z \rangle - c] \leq \|z\|^{-1}[\langle x, z \rangle - \langle P_K(x), z \rangle]$$
$$= \|z\|^{-1}\langle x - P_K(x), z \rangle \leq \|x - P_K(x)\| = d(x, K)$$
$$\leq d(x, H).$$

Hence equality must hold throughout this string of inequalities. This implies that $\|x - P_K(x)\| = d(x, H)$ and $\langle P_K(x), z \rangle = c$. Thus $P_K(x) \in H$ and $P_K(x) = P_H(x)$. This proves (1), and hence all four statements are equivalent.

To prove the last statement of the theorem, suppose K is a convex cone and all of the equivalent statements (1) through (4) hold. Now,

$$s := \sup\{\langle y, z \rangle \mid y \in K\} \leq c$$

by hypothesis. If $s < c$, then Theorem 6.16 along with Lemma 6.21 implies that $d(x, H) < d(x, H') \leq d(x, K)$, which contradicts (3). Thus $s = c$. Since $0 \in K$, $s \geq 0$. If $s > 0$, choose $y \in K$ such that $\langle y, z \rangle > s/2$. Since K is a convex cone, it follows that $y_0 = 2y \in K$, and $\langle y_0, z \rangle > s$, which is absurd. Thus $s = 0$. ∎

This theorem suggests an "algorithm" for computing best approximations to x from a convex Chebyshev set K and, simultaneously, the error $d(x, K)$. It may be briefly described as follows.

Algorithm.

(1) Obtain a hyperplane $H = \{y \in X \mid \langle y, z \rangle = c\}$ that separates K and x, and whose distance from x is maximal over all such hyperplanes; hence $d(x, H) = d(x, K)$.

(2) Then

$$d(x, K) = \|z\|^{-1}|\langle x, z \rangle - c| \quad \text{and} \quad P_K(x) = x - \|z\|^{-2}[\langle x, z \rangle - c]z.$$

Obviously, the success of this algorithm depends entirely on one's ability to determine maximal separating hyperplanes.

For the remainder of this chapter, we will study the best approximation theory from two important classes of sets: the *half-spaces* and the *polyhedra*.

Best Approximation from Half-Spaces

In this section we study the best approximation theory from half-spaces. We will see that there is a close analogy with that of hyperplanes.

6.28 Definition. *A **half-space** in an inner product space X is any set of the form*

$$\mathcal{H} = \{y \in X \mid x^*(y) \leq c\} \quad \text{or} \quad \mathcal{H} = \{y \in X \mid x^*(y) \geq c\},$$

where $x^* \in X^* \setminus \{0\}$ and $c \in \mathbb{R}$.

In other words, a half-space is the set of all points that lie either on, or to one side of, a hyperplane.

Remarks. (1) Every half-space is a nonempty closed convex set.

(2) Any half-space of the form $\mathcal{H} = \{y \in X \mid x^*(y) \geq c\}$ can obviously be written in the form $\mathcal{H} = \{y \in X \mid x_1^*(y) \leq c_1\}$, where $x_1^* = -x^*$ and $c_1 = -c$. Thus we need only consider those half-spaces where the defining inequality is $\leq$.

Half-spaces are fundamental for constructing all closed convex sets, as the following theorem attests.

6.29 Closed Convex Sets are Intersections of Half-Spaces. *Let K be a nonempty set in the inner product space X. Then K is closed and convex if and only if K is the intersection of all the half-spaces that contain K.*

Proof. The "if" part is clear, since the intersection of a collection of closed convex sets is closed and convex. For the "only if" part, suppose K is closed and convex. Let $\{\mathcal{H}_j \mid j \in J\}$ denote the indexed collection of all half-spaces that contain K. Then obviously, $\cap_{j \in J} \mathcal{H}_j \supset K$. On the other hand, if $x \in X \setminus K$, the separation theorem (Theorem 6.23) implies that there exists $x^* \in X^*$ with $\|x^*\| = 1$ and $c := \sup x^*(K) < x^*(x)$. Thus the half-space $\mathcal{H} := \{y \in X \mid x^*(y) \leq c\}$ contains K but not x and hence is in $\{\mathcal{H}_j \mid j \in J\}$. This proves that $x \notin \cap_{j \in J} \mathcal{H}_j$, and so $\cap_{j \in J} \mathcal{H}_j \subset K$. It follows that $K = \cap_{j \in J} \mathcal{H}_j$. ∎

The theories of best approximation from half-spaces and from hyperplanes are similar. Our first result gives a distance formula to half-spaces and should be compared with the distance formula to hyperplanes (Theorem 6.16).

6.30 Distance to Half-Spaces. *Let X be an inner product space, $x^* \in X^* \setminus \{0\}$, $c \in \mathbb{R}$, and $\mathcal{H} := \{y \in X \mid x^*(y) \leq c\}$. Then for each $x \in X$,*

$$(6.30.1) \qquad d(x, \mathcal{H}) = \frac{1}{\|x^*\|}[x^*(x) - c]^+.$$

Proof. If $x \in \mathcal{H}$, then $d(x, \mathcal{H}) = 0$ and $x^*(x) \leq c$, so that (6.30.1) holds. If $x \notin \mathcal{H}$, then $x^*(x) > c$, and the hyperplane $H := \{y \in X \mid x^*(y) = c\}$ separates $\mathcal{H}$ from x. By Lemma 6.21, $d(x, H) \leq d(x, \mathcal{H})$. On the other hand, $\mathcal{H} \supset H$ implies $d(x, \mathcal{H}) \leq d(x, H)$. Thus $d(x, \mathcal{H}) = d(x, H)$, and using Theorem 6.16, we obtain

$$d(x, \mathcal{H}) = \frac{1}{\|x^*\|}|x^*(x) - c| = \frac{1}{\|x^*\|}[x^*(x) - c]^+. \qquad ∎$$

6.31 Chebyschev Half-Spaces. *Let X be an inner product space, $x^* \in X^* \setminus \{0\}$, $c \in \mathbb{R}$, $H = \{y \in X \mid x^*(y) = c\}$, and $\mathcal{H} = \{y \in X \mid x^*(y) \leq c\}$. Then the following statements are equivalent:*

(1) *H is Chebyshev;*

(2) *$\mathcal{H}$ is Chebyshev;*

(3) *$\mathcal{H}$ is proximinal;*

(4) *Some $x \in X \setminus \mathcal{H}$ has a best approximation in $\mathcal{H}$;*

(5) *x^* attains its norm;*

(6) *x^* has a representer.*

Moreover, if x^* has the representer $z \in X$ (so that $\mathcal{H} = \{y \in X \mid \langle y, z \rangle \leq c\}$), then

$$(6.31.1) \qquad P_{\mathcal{H}}(x) = x - \frac{1}{\|z\|^2}[\langle x, z \rangle - c]^+ z$$

for every $x \in X$.

Proof. The equivalence of (1), (5), and (6) is from Theorem 6.17. Also, the implications (2) $\Longrightarrow$ (3) $\Longrightarrow$ (4) are obvious.

In the proof of Theorem 6.30 it was shown that for each $x \notin \mathcal{H}$, $d(x, \mathcal{H}) = d(x, H)$. Hence if (1) holds, then for all $x \notin \mathcal{H}$, Theorems 6.17 and 6.27 imply that

$$P_{\mathcal{H}}(x) = P_H(x) = x - \frac{1}{\|z\|^2}[\langle x, z \rangle - c]z,$$

where z is the representer of x^*. In case $x \in \mathcal{H}$, we have $P_{\mathcal{H}}(x) = x$, and (6.31.1) holds again. This proves that (1) $\Longrightarrow$ (2) and also verifies formula (6.31.1).

To complete the proof it suffices to show that (4) implies (5). Suppose some $x \in X \setminus \mathcal{H}$ has a best approximation $y_0 \in \mathcal{H}$. Then, using the distance formula (6.30.1), we obtain $x^*(y_0) \leq c$ and

$$\|x - y_0\| = \frac{1}{\|x^*\|}[x^*(x) - c] \leq \frac{1}{\|x^*\|}[x^*(x) - x^*(y_0)] = \frac{1}{\|x^*\|}x^*(x - y_0) \leq \|x - y_0\|.$$

Hence equality must hold throughout this string of inequalities, which implies that $\|x^*\| = x^*((x - y_0)/\|x - y_0\|)$. This proves that x^* attains its norm at $(x - y_0)/\|x - y_0\|$ and verifies (5). ∎

Best Approximation from Polyhedra

6.32 Definition. *A* **polyhedron** *or a* **polyhedral set** *in an inner product space is a nonempty set that is the intersection of a finite number of half-spaces.*

Thus Q is a polyhedron in the inner product space X if and only if

$$Q = \cap_1^n \{ y \in X \mid x_i^*(y) \leq c_i \} \neq \emptyset,$$

for some $x_i^* \in X^* \setminus \{0\}$ and $c_i \in \mathbb{R}$ $(i = 1, 2, \ldots, n)$.

We are going to characterize the polyhedra that are Chebyshev. But first we need to record a few useful facts that are interesting in their own right.

6.33 Lemma. *If a nonzero vector x is a positive linear combination of the nonzero vectors $x_1, x_2, \ldots, x_n$, then x is a positive linear combination of a linearly independent subset of $\{x_1, x_2, \ldots, x_n\}$.*

Proof. Let $x = \sum_1^n \rho_i x_i$, where $\rho_i > 0$ for all i. If $\{x_1, x_2, \ldots, x_n\}$ is linearly independent, there is nothing to do. Thus we may assume that $\{x_1, x_2, \ldots, x_n\}$ is linearly dependent. That is, $\sum_1^n \alpha_i x_i = 0$ for some scalars α_i not all zero. In fact, we may assume that $\alpha_i > 0$ for some i. Set $\lambda := \min\{\rho_i/\alpha_i \mid \alpha_i > 0\}$ and $\mu_i := \rho_i - \lambda \alpha_i$ $(i = 1, 2, \ldots, n)$. Then $\lambda > 0$, $\mu_i \geq 0$ for all i, $\mu_i = 0$ for at least one i, and $x = \sum_1^n \mu_i x_i$. This proves that x is a positive linear combination of at most $n - 1$ of the x_i's. If this subset of the x_i's is linearly independent, we're done. If not, we repeat the above argument to get x as a positive linear combination of at most $n - 2$ of the x_i's. By continuing in this way, after at most $n - 1$ steps, we must reach the conclusion of the lemma. ∎

We say that a convex cone C in the inner product space X is **finitely generated** if it is the conical hull of a finite number of vectors. Thus C is finitely generated if

$$C = \text{con}\{x_1, x_2, \ldots, x_n\} = \left\{ \sum_1^n \rho_i x_i \mid \rho_i \geq 0 \right\},$$

for some finite subset $\{x_1, x_2, \ldots, x_n\}$ of X.

6.34 Finitely Generated Cones are Chebyshev. *Every finitely generated convex cone in an inner product space is a Chebyshev set.*

Proof. Let $\{x_1, x_2, \ldots, x_n\}$ be a set of nonzero vectors in the inner product space X, and let $C = \text{con}\{x_1, x_2, \ldots, x_n\}$. Since C is a convex cone in the finite-dimensional subspace span $\{x_1, x_2, \ldots, x_n\}$ of X, it suffices by Theorem 3.8 to show that C is closed. Let $y_k \in C$ and $y_k \to y$. We must show that $y \in C$. If $y_k = 0$ for infinitely many k, it follows that $y = 0 \in C$. Thus we may assume that $y_k \neq 0$ for all k. By Lemma 6.33, each y_k is in the conical hull C_k of a linearly independent subset of $\{x_1, x_2, \ldots, x_n\}$. Since there are only finitely many different subsets C_k, by passing to a subsequence $\{k_1, k_2, \ldots\}$ of $\{1, 2, \ldots\}$, we may assume that $C_0 := C_{k_1} = C_{k_2} = \cdots$. Thus each y_{k_j} is in C_0, where C_0 is the conical hull of a linearly independent subset $\{x_{i_1}, x_{i_2}, \ldots, x_{i_m}\}$ of $\{x_1, x_2, \ldots, x_n\}$. It follows that

$$y_{k_j} = \sum_{s=1}^{m} \lambda_{k_j i_s} x_{i_s} \quad (j = 1, 2, \ldots)$$

for some $\lambda_{k_j i_s} \geq 0$. Since $y_{k_j} \to y$, it follows by Theorem 3.7(4) that

$$y = \sum_{s=1}^{m} \lambda_{i_s} x_{i_s}$$

and $\lambda_{k_j i_s} \to \lambda_{i_s}$ $(s = 1, 2, \ldots, m)$. Since $\lambda_{k_j i_s} \geq 0$ for all sets of indices, $\lambda_{i_s} \geq 0$ for all s. Thus $y \in C_0 \subset C$. ∎

6.35 Lemma. *Let $x^*, x_1^*, x_2^*, \ldots, x_n^*$ be linear functionals (but not necessarily bounded) on the inner product space X. Then the following statements are equivalent:*

(1) $x^* \in \text{span}\{x_1^*, x_2^*, \ldots, x_n^*\}$;
(2) $x^*(x) = 0$ whenever $x_i^*(x) = 0$ for $i = 1, 2, \ldots, n$;
(3) $\cap_1^n \ker x_i^* \subset \ker x^*$.

Proof. The implications (1) $\Longrightarrow$ (2) $\Longleftrightarrow$ (3) are obvious.
(2) $\Longrightarrow$ (1). Assume (2) holds and let

$$M = \{ \left(x_1^*(x), x_2^*(x), \ldots, x_n^*(x)\right) \mid x \in X \}.$$

Then M is a subspace of $l_2(n)$, and we define F on M by

$$F\left((x_1^*(x), x_2^*(x), \ldots, x_n^*(x))\right) := x^*(x), \quad x \in X.$$

[F is well-defined, since if

$$\left(x_1^*(x), \ldots, x_n^*(x)\right) = \left(x_1^*(y), \ldots, x_n^*(y)\right),$$

then $x_i^*(x - y) = 0$ for all i, so that by hypothesis (2), $x^*(x - y) = 0$, i.e., $x^*(x) = x^*(y)$.] Then F is a linear functional on M, which by Theorem 5.16 must be bounded. By Theorem 6.9, F has an extension z^* defined on all of $l_2(n)$. By Theorem 6.10, z^* has the representation

$$z^*(y) = \sum_{1}^{n} \alpha(i) y(i), \quad y \in l_2(n),$$

for some $\alpha \in l_2(n)$. In particular, for every $x \in X$,

$$x^*(x) = F\big((x_1^*(x), x_2^*(x), \ldots, x_n^*(x))\big) = z^*\big((x_1^*(x), x_2^*(x), \ldots, x_n^*(x))\big)$$
$$= \sum_1^n \alpha(i) x_i^*(x).$$

Thus, $x^* = \sum_1^n \alpha(i) x_i^* \in \operatorname{span}\{x_1^*, x_2^*, \ldots, x_n^*\}$, and (1) holds. ∎

From this result, we can readily deduce the next "interpolation theorem" for linearly independent linear functionals.

6.36 Interpolation Theorem. *Let* $\{x_1^*, x_2^*, \ldots, x_n^*\}$ *be a linearly independent set of linear functionals on the inner product space* X. *Then there exists a set* $\{x_1, x_2, \ldots, x_n\}$ *in* X *such that*

(6.36.1) $x_i^*(x_j) = \delta_{ij}$ $(i, j = 1, 2, \ldots, n)$.

In particular, for each set of n *real scalars* $\{c_1, c_2, \ldots, c_n\}$, *the element* $y = \sum_1^n c_i x_i$ *satisfies*

(6.36.2) $x_i^*(y) = c_i$ $(i = 1, 2, \ldots, n)$.

Proof. We prove (6.36.1) by induction on n. For $n = 1$, choose any $x \in X$ such that $x_1^*(x) \neq 0$ and set $x_1 = [x_1^*(x)]^{-1} x$. Then $x_1^*(x_1) = 1$. Assume now that (6.36.1) is valid for n and $\{x_1^*, x_2^*, \ldots, x_{n+1}^*\}$ is linearly independent on X. By hypothesis, there exist elements $y_1, y_2, \ldots, y_n$ in X such that

$$x_i^*(y_j) = \delta_{ij} \qquad (i, j = 1, 2, \ldots, n).$$

Next, note that there exists $x \in X$ such that $x_i^*(x) = 0$ for $i = 1, 2, \ldots, n$ and $x_{n+1}^*(x) \neq 0$. For if such an element failed to exist, then by Lemma 6.35, $x_{n+1}^* \in \operatorname{span}\{x_1^*, \ldots, x_n^*\}$, which contradicts the linear independence of $\{x_1^*, x_2^*, \ldots, x_{n+1}^*\}$. Now set

$$x_{n+1} = [x_{n+1}^*(x)]^{-1} x \quad \text{and} \quad x_i = y_i - x_{n+1}^*(y_i) x_{n+1}$$

for $i = 1, 2, \ldots, n$. Clearly, $x_j^*(x_i) = \delta_{ij}$ for $i, j = 1, 2, \ldots, n+1$. This completes the induction. ∎

Now we can characterize the polyhedra that are Chebyshev.

6.37 Chebyshev Polyhedra. *Let* X *be an inner product space, let* $x_i^* \in X^* \setminus \{0\}$ *and* $c_i \in \mathbb{R}$ *for* $i = 1, 2, \ldots, n$, *and let* $Q := \cap_1^n \{x \in X \mid x_i^*(x) \leq c_i\} \neq \emptyset$. *Then the following statements are equivalent:*

(1) *Q is Chebyshev;*
(2) *Q is proximinal;*
(3) *Each* x_i^* $(i = 1, 2, \ldots, n)$ *attains its norm;*
(4) *Each* x_i^* $(i = 1, 2, \ldots, n)$ *has a representer.*

Proof. The equivalence of (1) and (2) is just Theorem 2.4, while the equivalence of (3) and (4) is from Theorem 6.12. Although the theorem is true as stated, to simplify the proof we will hereafter *assume* that the set of functionals $\{x_1^*, x_2^*, \ldots, x_n^*\}$ is linearly independent. By Theorem 6.36, there exists a linearly independent set $\{z_1, z_2, \ldots, z_n\}$ in X such that

$$(6.37.1) \qquad x_i^*(z_j) = \delta_{ij} \quad (i, j = 1, 2, \ldots, n).$$

Then the element $z_0 = \sum_1^n c_j z_j$ satisfies $x_i^*(z_0) = c_i$ for all i, so $z_0 \in Q$ and Q is a nonempty polyhedron. Furthermore, observe that $Q = Q_0 + z_0$, where

$$(6.37.2) \qquad Q_0 := \bigcap_1^n \{x \in X \mid x_i^*(x) \leq 0\}.$$

(4) $\Longrightarrow$ (1). Suppose each x_i^* has a representer $x_i \in X$ $(i = 1, 2, \ldots, n)$. Then

$$(6.37.3) \qquad Q_0 = \bigcap_1^n \{x \in X \mid \langle x, x_i \rangle \leq 0\} = \bigcap_1^n \{x_i^\circ\}.$$

Thus, using Theorems 4.5(3) and 4.6(2), we obtain

$$Q_0 = \bigcap_1^n [\mathrm{con}\,(x_i)]^\circ = \left[\sum_1^n \mathrm{con}\,(x_i) \right]^\circ = [\mathrm{con}\,\{x_1, x_2, \ldots, x_n\}]^\circ.$$

But $\mathrm{con}\,\{x_1, x_2, \ldots, x_n\}$ is a Chebyshev convex cone by Theorem 6.34, and therefore its dual cone Q_0 is a Chebyshev convex cone by Theorem 5.6(1). Since Q is a translate of Q_0, Q must also be a Chebyshev set by Theorem 2.7(1)(v).

(1) $\Longrightarrow$ (4). Suppose that Q is a Chebyshev set. Fix any $i \in \{1, 2, \ldots, n\}$. It suffices to show that x_i^* has a representer. We may assume $\|x_j^*\| = 1$ for all j. Let $y_i = -\sum_{j \neq i} z_j$. Then

$$(6.37.4) \qquad x_i^*(y_i) = 0 > -1 = x_j^*(y_i) \quad \text{for all } j \neq i.$$

In particular, $y_i \in Q_0$.

For the remainder of the proof, we will use the notation $B_\rho(z)$ to denote the open ball in X centered at z with radius ρ. Choose any $0 < \epsilon < 1$.

Claim 1. $B_\epsilon(y_i) \cap \ker x_i^* \subset Q_0$.

For let $y \in B_\epsilon(y_i) \cap \ker x_i^*$. Then for all $j \neq i$, $x_j^*(y) = x_j^*(y - y_i) + x_j^*(y_i) < \epsilon - 1 < 0$. Thus $y \in Q_0$.

Next choose any $x \in B_{\epsilon/4}(y_i)$ such that $x_i^*(x) > 0$. Then for all $j \neq i$,

$$(6.37.5) \qquad x_j^*(x) = x_j^*(y_i) + x_j^*(x - y_i) = -1 + x_j^*(x - y_i) < -1 + \epsilon/4 < 0.$$

Claim 2. $P_{Q_0}(x) \in \ker x_i^*$.

For if not, then $x_i^*(P_{Q_0}(x)) < 0$. For each $\lambda \in [0, 1]$, define $z_\lambda := \lambda x + (1 - \lambda) P_{Q_0}(x)$. Then $x_i^*(z_0) < 0 < x_i^*(z_1)$ implies that there exists $0 < \lambda_0 < 1$ such that $x_i^*(z_{\lambda_0}) = 0$. Then $z_{\lambda_0} \in Q_0$ using (6.37.5), and

$$\|x - z_{\lambda_0}\| = (1 - \lambda_0)\|x - P_{Q_0}(x)\| < \|x - P_{Q_0}(x)\|,$$

which is impossible.

Claim 3. $P_{Q_0}(x) \in B_{\epsilon/2}(y_i)$.

For $\|x - y_i\| < \epsilon/4$ and $y_i \in Q_0$ imply that $\|x - P_{Q_0}(x)\| < \epsilon/4$. Hence

$$\|P_{Q_0}(x) - y_i\| \le \|P_{Q_0}(x) - x\| + \|x - y_i\| < \epsilon/2,$$

which proves Claim 3.

By Claim 3 we can choose $0 < \epsilon' < \epsilon/2$ such that $B_{\epsilon'}(P_{Q_0}(x)) \subset B_{\epsilon/2}(y_i)$. Set $x' := \epsilon'x + (1 - \epsilon')P_{Q_0}(x)$. Then $P_{Q_0}(x') = P_{Q_0}(x)$ by Exercise 18 at the end of Chapter 2 (or Lemma 12.1 below) and $x' \in B_{\epsilon'}(P_{Q_0}(x'))$. Also, $x_i^*(x') = \epsilon'x_i^*(x) > 0$ and, for $j \ne i$,

$$x_j^*(x') = \epsilon'x_j^*(x) + (1 - \epsilon')x_j^*(P_{Q_0}(x)) < (1 - \epsilon')[x_j^*(P_{Q_0}(x) - y_i) + x_j^*(y_i)]$$
$$< -(1 - \epsilon')^2 < 0.$$

Claim 4. $d(x', Q_0) \le d(x', \ker x_i^*)$.

For let $y \in \ker x_i^*$. If $y \in B_\epsilon(y_i)$, then Claim 1 implies $y \in Q_0$, so $d(x', Q_0) = \|x' - P_{Q_0}(x')\| \le \|x' - y\|$. If $y \notin B_\epsilon(y_i)$, then

$$\|y - x'\| = \|y - [\epsilon'x + (1 - \epsilon')P_{Q_0}(x)]\|$$
$$= \|y - y_i - [\epsilon'(x - y_i) + (1 - \epsilon')\{P_{Q_0}(x) - y_i\}]\|$$
$$\ge \|y - y_i\| - \|\epsilon'(x - y_i) + (1 - \epsilon')(P_{Q_0}(x) - y_i)\|$$
$$\ge \epsilon - [\epsilon'\|x - y_i\| + (1 - \epsilon')\|P_{Q_0}(x) - y_i\|]$$
$$\ge \epsilon - [\epsilon'(\epsilon/4) + (1 - \epsilon')(\epsilon/2)] > \epsilon - \epsilon/2$$
$$> \|x - P_{Q_0}(x)\| > \|x' - P_{Q_0}(x')\| = d(x', Q_0),$$

and the claim is proved.

Next notice that $\ker x_i^*$ is a hyperplane that separates x' and Q_0 (since $x_i^*(x') > 0 \ge x_i^*(y)$ for all $y \in Q_0$). Thus by Lemma 6.21, $d(x', \ker x_i^*) \le d(x', Q_0)$. From Claim 4, it follows that $d(x', \ker x_i^*) = d(x', Q_0)$. By Claim 2, $P_{Q_0}(x') = P_{Q_0}(x) \in \ker x_i^*$, so $P_{Q_0}(x') = P_{\ker x_i^*}(x')$. Then $x' \in X \setminus \ker x_i^*$ has a best approximation in $\ker x_i^*$. By Theorem 6.17, x_i^* has a representer. ∎

We conclude this chapter by proving a characterization theorem for best approximations from a polyhedron. It is convenient to isolate the key steps of the proof as lemmas, since they have independent interest.

6.38 Lemma. Let $\{C_1, C_2, \ldots, C_m\}$ be a collection of convex sets with $0 \in \cap_1^m C_i$. Then

$$(6.38.1) \qquad \mathrm{con}\left(\bigcap_1^m C_i\right) = \bigcap_1^m \mathrm{con}\, C_i.$$

In other words, the operation of taking the conical hull commutes with intersection whenever 0 is a common point to all the sets.

Proof. The inclusion $\mathrm{con}\,(\cap_1^m C_i) \subset \cap_1^m \mathrm{con}\, C_i$ is obvious since the right side is a convex cone that contains $\cap_1^m C_i$, and $\mathrm{con}\,(\cap_1^m C_i)$ is the smallest such convex cone.

Now let $x \in \cap_1^m \text{con}\, C_i$. By Theorem 4.4(4), we can write $x = \rho_i x_i$ for some $\rho_i > 0$ and $x_i \in C_i$ for $i = 1, 2, \ldots, m$. Then $(1/\rho_i)x = x_i \in C_i$ for each i. Letting $\rho = \sum_1^m \rho_i$ and $\lambda_i = \rho_i/\rho$, we see that $\lambda_i > 0$, $\sum_1^m \lambda_i = 1$, and

$$\frac{1}{\rho} x = \frac{\rho_i}{\rho} \left(\frac{1}{\rho_i} x \right) = \lambda_i x_i = \lambda_i x_i + (1 - \lambda_i) \cdot 0 \in C_i$$

for each i by convexity and the fact that $0 \in C_i$. Hence $\frac{1}{\rho} x \in \cap_1^m C_i$, and so

$$x \in \rho \left(\bigcap_1^m C_i \right) \subset \text{con} \left(\bigcap_1^m C_i \right)$$

by Theorem 4.4(4). ∎

Let $h \in X \setminus \{0\}, c \in \mathbb{R}$, and define the half-space H by

(6.38.4) $$H := \{x \in X \mid \langle x, h \rangle \leq c\}.$$

It is easy to verify that the interior and boundary of H are given by

$$\text{int}\, H = \{x \in X \mid \langle x, h \rangle < c\}$$

and

$$\text{bd}\, H = \{x \in X \mid \langle x, h \rangle = c\}.$$

6.39 Lemma. *Let X be an inner product space and H be the half-space defined in (6.38.4). Then for every $x_0 \in H$,*

(6.39.1) $$\text{con}\,(H - x_0) = \overline{\text{con}\,(H - x_0)} = \begin{cases} X & \text{if } x_0 \in \text{int}\, H, \\ H - x_0 & \text{if } x_0 \in \text{bd}\, H, \end{cases}$$

and

(6.39.2) $$(H - x_0)^\circ = \begin{cases} \{0\} & \text{if } x_0 \in \text{int}\, H, \\ \text{con}\,\{h\} & \text{if } x_0 \in \text{bd}\, H. \end{cases}$$

In particular, both $\text{con}\,(H - x_0)$ and $(H - x_0)^\circ$ are Chebyshev convex cones.

Proof. If $x_0 \in \text{int}\, H$, then $0 \in \text{int}\,(H - x_0)$, so $\text{con}\,(H - x_0) = X$. If $x_0 \in \text{bd}\, H$, then

(6.39.3) $$H - x_0 = \{x \in X \mid \langle x, h \rangle \leq 0\} = \{h\}^\circ$$

is a closed convex cone. This verifies (6.39.1).

Using Theorem 4.5(3), we see that if $x_0 \in \text{int}\, H$, then

$$(H - x_0)^\circ = [\text{con}\,(H - x_0)]^\circ = X^\circ = \{0\}.$$

On the other hand, let $x_0 \in \text{bd}\, H$. Since $\text{con}\,\{h\}$ is a Chebyshev convex cone by Theorem 6.34, it follows by using (3) and (7) of Theorem 4.5 and (6.39.3) that

$$\text{con}\,\{h\} = (\text{con}\,\{h\})^{\circ\circ} = \{h\}^{\circ\circ} = (H - x_0)^\circ.$$

This proves (6.39.2).

The last statement follows since H is a Chebyshev set by Theorem 6.31, so $H - x_0$ is also, and $\text{con}\,(H - x_0)$ is either X or $H - x_0$. Also, since $(H - x_0)^\circ$ is either $\{0\}$ or $\text{con}\,\{h\}$, $(H - x_0)^\circ$ is a convex Chebyshev cone by Theorem 6.34. ∎

6.40 Dual Cone to a Polyhedron. *Let X be an inner product space, and for each $i = 1, 2, \ldots, m$, let h_i be a nonzero element of X, $c_i \in \mathbb{R}$, and*

$$H_i := \{x \in X \mid \langle x, h_i \rangle \leq c_i\}.$$

Then for each $x_0 \in \cap_1^m H_i$,
(6.40.1)

$$\left(\bigcap_1^m H_i - x_0 \right)^\circ = \sum_1^m (H_i - x_0)^\circ = \sum_{i \in I(x_0)} (H_i - x_0)^\circ = \operatorname{con} \{h_i \mid i \in I(x_0)\},$$

where $I(x_0) := \{i \mid \langle x_0, h_i \rangle = c_i\} = \{i \mid x_0 \in \operatorname{bd} H_i\}$ is the **active index set** *for x_0 relative to $\cap_1^m H_i$.*

Proof. We have

$$
\begin{aligned}
\left(\bigcap_1^m H_i - x_0 \right)^\circ &= \left[\bigcap_1^m (H_i - x_0) \right]^\circ \\
&= \left[\operatorname{con} \bigcap_1^m (H_i - x_0) \right]^\circ \quad \text{(by Theorem 4.5(3))} \\
&= \left[\bigcap_1^m \operatorname{con} (H_i - x_0) \right]^\circ \quad \text{(by Theorem 6.38)} \\
&= \overline{\sum_1^m [\operatorname{con} (H_i - x_0)]^\circ} \quad \text{(by Theorem 4.6(4))} \\
&= \overline{\sum_1^m (H_i - x_0)^\circ} \quad \text{(by Theorem 4.5(3))} \\
&= \overline{\sum_{i \in I(x_0)} (H_i - x_0)^\circ} \quad \text{(by Theorem 6.39(2))} \\
&= \overline{\sum_{i \in I(x_0)} \operatorname{con} \{h_i\}} \quad \text{(by Theorem 6.39(2))} \\
&= \overline{\operatorname{con} \{h_i \mid i \in I(x_0)\}} \\
&= \operatorname{con} \{h_i \mid i \in I(x_0)\} \quad \text{(by Theorem 6.34).}
\end{aligned}
$$

But the same reasoning that established the last four equalities shows that

$$\sum_1^m (H_i - x_0)^\circ = \sum_{i \in I(x_0)} (H_i - x_0)^\circ = \operatorname{con} \{h_i \mid i \in I(x_0)\},$$

which is closed (by Theorem 6.34). ∎

Now we can characterize best approximations from a polyhedron.

6.41 Characterization of Best Approximations from a Polyhedron.
Let X be an inner product space, and for each $i = 1, 2, \ldots, m$, let $h_i \in X \setminus \{0\}$, $c_i \in \mathbb{R}$, $H_i := \{x \mid \langle x, h_i \rangle \leq c_i\}$, and $Q := \cap_1^m H_i \neq \emptyset$. If $x \in X$, then

(6.41.1)
$$P_Q(x) = x - \sum_1^m \rho_i h_i$$

for any set of scalars ρ_i that satisfy the following three conditions:

(6.41.2) $$\rho_i \geq 0 \quad (i = 1, 2, \ldots, m),$$

(6.41.3) $$\langle x, h_i \rangle - c_i - \sum_{j=1}^{m} \rho_j \langle h_j, h_i \rangle \leq 0 \quad (i = 1, 2, \ldots, m),$$

and

(6.41.4) $$\rho_i \left[\langle x, h_i \rangle - c_i - \sum_{j=1}^{m} \rho_j \langle h_j, h_i \rangle \right] = 0 \quad (i = 1, 2, \ldots, m).$$

Consequently, if $x \in X$ and $x_0 \in Q$, then $x_0 = P_Q(x)$ if and only if

(6.41.5) $$x_0 = x - \sum_{i \in I(x_0)} \rho_i h_i \text{ for some } \rho_i \geq 0.$$

Proof. Suppose the ρ_i satisfy the relations (6.41.2)–(6.41.4). Set $x_0 := x - \sum_1^m \rho_i h_i$. Then (6.41.3) and (6.41.4) imply that

(6.41.6) $$\langle x_0, h_i \rangle - c_i \leq 0 \quad \text{and} \quad \rho_i \left[\langle x_0, h_i \rangle - c_i \right] = 0 \quad \text{for} \quad i = 1, 2, \ldots, m.$$

Now, (6.41.6) implies that $x_0 \in \cap_1^m H_i$ and $\rho_i = 0$ whenever $i \notin I(x_0)$. Thus $x_0 = x - \sum_{i \in I(x_0)} \rho_i h_i$. It follows from Theorem 6.40 that

$$x - x_0 = \sum_{i \in I(x_0)} \rho_i h_i \in \text{con}\,\{h_i \mid i \in I(x_0)\} = \left(\cap_1^m H_i - x_0 \right)^\circ = (Q - x_0)^\circ.$$

By Theorem 4.3, $x_0 = P_Q(x)$.

The last statement is an easy consequence of the proof of the first part. ∎

It is worth noticing that this characterization gives rise to an *explicit formula* for best approximations in the case where the elements h_i that define the half-spaces form an *orthogonal* set. We establish this next.

6.42 Corollary. *Let the half-spaces H_i be defined as in Theorem 6.41, and let $Q = \cap_1^m H_i$. If $\{h_1, h_2, \ldots, h_m\}$ is an orthogonal set, then for each $x \in X$,*

(6.42.1) $$P_Q(x) = x - \sum_1^m \frac{1}{\|h_i\|^2} \left[\langle x, h_i \rangle - c_i \right]^+ h_i.$$

In particular,

(6.42.2) $$P_Q(x) = (1 - m)x + \sum_1^m P_{H_i}(x).$$

Proof. Using the orthogonality of the h_i, it follows from Theorem 6.41 that for each $x \in X$,

(6.42.3) $$P_Q(x) = x - \sum_1^m \rho_i h_i$$

for any set of scalars ρ_i that satisfy the following relations for each $i = 1, 2, \ldots, m$:

(i) $\rho_i \geq 0$,
(ii) $\langle x, h_i \rangle - c_i - \rho_i \|h_i\|^2 \leq 0$,
(iii) $\rho_i \left[\langle x, h_i \rangle - c_i - \rho_i \|h_i\|^2 \right] = 0$.

From (i) and (ii), we deduce

$$\rho_i \geq \frac{1}{\|h_i\|^2}\left[\langle x, h_i\rangle - c_i\right]^+ \quad \text{for each } i.$$

If $\rho_i > \|h_i\|^{-2}\left[\langle x, h_i\rangle - c_i\right]^+$ for some i, then $\rho_i > 0$ and

$$\rho_i\left[\langle x, h_i\rangle - c_i - \rho_i\|h_i\|^2\right] < \rho_i\left[\langle x, h_i\rangle - c_i - \{\langle x, h_i\rangle - c_i\}^+\right] \leq 0,$$

which contradicts (iii). Thus we must have $\rho_i = \|h_i\|^{-2}\left[\langle x, h_i\rangle - c_i\right]^+$ for each i. Substituting these values for ρ_i back into (6.42.3), we obtain (6.42.1). Finally, (6.42.2) follows by using formula (6.31.1). ∎

6.43 An Application. Let X denote the Euclidean plane $\ell_2(2)$ and let

$$Q = \{y \in X \mid y(2) - y(1) \leq 0 \text{ and } y(1) + y(2) \leq 0\}.$$

Determine the best approximation $P_Q(x)$ to x from Q when (a) $x = (1,2)$, (b) $x = (2,0)$, and (c) $x = (-3,-1)$.

Note that Q may be rewritten in the form

$$Q = \{y \in X \mid \langle y, h_1\rangle \leq 0\} \cap \{y \in X \mid \langle y, h_2\rangle \leq 0\},$$

where $h_1 = (-1,1)$ and $h_2 = (1,1)$. Then Q is a polyhedron and $\langle h_1, h_2\rangle = 0$, so Corollary 6.42 implies that for each $x \in X$,

$$P_Q(x) = x - \frac{1}{\|h_1\|^2}\left[\langle x, h_1\rangle\right]^+ h_1 - \frac{1}{\|h_2\|^2}\left[\langle x, h_2\rangle\right]^+ h_2$$

$$= x - \frac{1}{2}\left[x(2) - x(1)\right]^+ h_1 - \frac{1}{2}\left[x(1) + x(2)\right]^+ h_2.$$

From this formula it is easy to compute $P_Q(x)$ for any $x \in X$. In particular, $P_Q((1,2)) = (0,0)$, $P_Q((2,0)) = (1,-1)$, and $P_Q((-3,-1)) = (-2,-2)$.

Exercises

1. Let F be a linear functional on X. Show that

$$\ker F := \{x \in X \mid F(x) = 0\}$$

is closed if and only if F is continuous.

2. Give an alternative proof of the Fréchet–Riesz representation theorem (Theorem 6.10) by verifying the following steps. Given any $x^* \in X^* \setminus \{0\}$, let $M = \ker x^*$. Then:
   (i) M is a Chebyshev subspace.
   (ii) There exists $x_1 \in M^\perp$ with $\|x_1\| = 1$.
   [Hint: Take any $x \in X \setminus M$ and set $x_1 = d(x, M)^{-1}(x - P_M(x))$.]
   (iii) For any $y \in X$, the element $y - x^*(y)[x^*(x_1)]^{-1}x_1$ is in M.
   (iv) $x^*(x_1)x_1$ is the representer of x^*.
   [Hint: $x_1 \in M^\perp$.]

3. Show that if M and N are subspaces of X, then $X = M \oplus N$ if and only if each $x \in X$ has a unique representation in the form $x = y + z$, where $y \in M$ and $z \in N$.

4. A linear operator $U : X \to Y$ between inner product spaces is called a **linear isometry** if $\|U(x)\| = \|x\|$ for all $x \in X$.

(a) Show that every linear isometry is one-to-one.

(b) Show that U is a linear isometry if and only if U "preserves" inner products: $\langle U(x), U(y) \rangle = \langle x, y \rangle$ for all x, y in X.

[Hint: Consider the polarization identity $4\langle x, y \rangle = \|x + y\|^2 - \|x - y\|^2$.]

(c) Two inner product spaces X and Y are called **equivalent** if there exists a linear isometry from X onto Y. Show that X and Y are equivalent if and only if there exists a one-to-one mapping U from X onto Y that

(i) preserves the linear operations:

$$U(\alpha x + \beta y) = \alpha U(x) + \beta U(y),$$

(ii) preserves the inner product:

$$\langle U(x), U(y) \rangle = \langle x, y \rangle;$$

(iii) preserves the norm:

$$\|U(x)\| = \|x\|.$$

Thus two equivalent inner product spaces are essentially the same. The linear isometry amounts to nothing more than a relabeling of the elements.

(d) If two inner product spaces are equivalent and one is complete, so is the other.

(e) Every Hilbert space is equivalent to its dual space.

[Hint: Define $U : X \to X^*$ by $U(x) = \hat{x}$.]

(f) Any inner product space is equivalent to a dense subspace of its dual space.

[Hint: See hint for part (e).] In particular, by identifying an inner product space X with a dense subspace $\hat{X}$ of X^* (via an isometry U from X onto $\hat{X}$), we conclude that *every inner product space is a dense subspace of a Hilbert space.*

(g) Any two n-dimensional inner product spaces are equivalent.

[Hint: Let $\{x_1, x_2, \ldots, x_n\}$ and $\{y_1, y_2, \ldots, y_n\}$ be orthonormal bases in X and Y. Define $U : X \to Y$ by $U\left(\sum_1^n \alpha_i x_i\right) = \sum_1^n \alpha_i y_i$.]

5. Let $x^* \in X^* \setminus \{0\}$.

(a) Show that $x \in X$ is the representer for x^* if and only if x^* attains its norm at $x/\|x\|$.

(b) Show that x^* attains its norm on at most one point.

(c) Show that the set of all functionals in X^* that attain their norm is a dense subspace in X^*. What is the relationship between this set and $\hat{X}$?

6. Let X be any incomplete inner product space.

(a) Show that there exists a functional $x^* \in X^*$ that does *not* have a representer in X.

[Hint: Let $\{x_n\}$ be a Cauchy sequence in X that does not converge. Define

x^* as in (6.8.1).]

(b) Show that there exists a closed subspace in X that is not Chebyshev. [Hint: Part (a) and Theorem 6.17.]

7. Show that the dual of an inner product space is a Hilbert space. [Hint: Lemma 6.22 and Exercise 13 of Chapter 1.]

8. Let X be an inner product space. The following steps exhibit an alternative approach to Exercise 7 for verifying that X^* is a Hilbert space.

(a) Show that the sequence $\{x_n\}$ in X is Cauchy if and only if $\{\hat{x}_n\}$ is a Cauchy sequence in X^*.

(b) If $x^* \in X^*$, there exist $x_n \in X$ such that $\|x_n\| = \|x^*\|$ for every n and $\|x^* - \hat{x}_n\| \to 0$.

(c) For every pair x^*, y^* in X^*, let $\{x_n\}$ and $\{y_n\}$ be any sequences in X with $\|x^* - \hat{x}_n\| \to 0$ and $\|y^* - \hat{y}_n\| \to 0$. Define

(8.1) $$\langle x^*, y^* \rangle := \lim \langle x_n, y_n \rangle.$$

(i) Show that this limit exists and is independent of the particular sequences $\{x_n\}$ and $\{y_n\}$ chosen.

(ii) Show that (8.1) defines an inner product on X^* such that $\langle x^*, x^* \rangle = \|x^*\|^2$; that is, the norm generated by this inner product is the same as the operator norm on X^*.

(iii) Show that X^* is a Hilbert space.

9. Let $x^* \in X^*$, $\|x^*\| = 1$, and $0 < \epsilon < 1$. If $x, y \in X$, $\|x\| = \|y\| = 1$, $x^*(x) > 1 - \epsilon^2/8$, and $x^*(y) > 1 - \epsilon^2/8$, then $\|x - y\| < \epsilon$. [Hint: The parallelogram law (Theorem 1.5).]

10. Give an alternative proof of the fact that "$x^* \in X^*$ has a representer if (and only if) x^* attains its norm" along the following lines. Suppose x^* attains its norm at x_1, and let $M = \ker x^*$.

(i) For any $x \in X$, show that $P_M(x) = x - x^*(x)[\|x^*\|^{-1}]x_1$.

(ii) Deduce $x_1 \in M^{\perp}$.

(iii) Take the inner product of $P_M(x)$ with x_1.

11. Show that an inner product space X is complete if and only if each $x^* \in X^*$ attains its norm. [Hint: Theorems 6.5, 6.10, 6.12, and Exercise 7.]

12. (a) If X is finite-dimensional, then each $x^* \in X^*$ attains its norm. [Hint: The unit ball in X is compact.]

(b) Each finite-dimensional inner product space is a Hilbert space.

13. Consider the set in l_2 defined by

$$H = \left\{ y \in l_2 \ \middle| \ \sum_{n=1}^{\infty} \frac{1}{2^{n/2}} y(n) = 0 \right\}.$$

Show that H is a Chebyshev hyperplane, and compute $P_H(x)$ and $d(x, H)$ for any $x \in l_2$.

14. Which of the following hyperplanes in $C_2[-1, 1]$ are Chebyshev or antiproximinal? For those that are Chebyshev, give a formula for $P_H(x)$.

(a) $H = \{y \in C_2[-1, 1] \mid \int_{-1}^{1} y(t)\, dt = 2\}$.

(b) $H = \{y \in C_2[-1, 1] \mid \int_{-1}^{0} y(t)\, dt = \int_{0}^{1} y(t)\, dt\}$.

(c) $H = \{y \in C_2[-1,1] \mid \int_{-1}^{1} y(t)\cos t\,dt = 0\}$.

(d) $H = \{y \in C_2[-1,1] \mid \int_{-1}^{0} e^t y(t)\,dt + \int_{0}^{1}(1+t)y(t)\,dt = 0\}$.

15. (a) Find the distance from a point $x \in l_2(3)$ to the plane

$$H = \{y \in l_2(3) \mid ay(1) + by(2) + cy(3) = d\},$$

where a, b, c, and d are any prescribed real numbers with $a^2 + b^2 + c^2 \neq 0$.

(b) Find the best approximation in H to any $x \in l_2(3)$.

16. (a) Find the distance from a point $x \in l_2(3)$ to the half-space

$$H = \{y \in l_2(3) \mid ay(1) + by(2) + cy(3) \leq d\},$$

where a, b, c, and d are any prescribed real numbers with $a^2 + b^2 + c^2 \neq 0$.

(b) Find the best approximation in H to any $x \in l_2(3)$.

17. Let K be a convex Chebyshev set. Then $K = \cap_{j \in J} H_j$ for some collection of *Chebyshev* half-spaces $\{H_j \mid j \in J\}$.
[Hint: Inspect the proof of Theorem 6.29.]

18. Let $\{x^*, x_1^*, \ldots, x_n^*\} \subset X^*$ satisfy $x^*(x) \geq 0$ whenever $x_i^*(x) \geq 0$ for all i. Show that $x^* = \sum_1^n \lambda_i x_i^*$ for some $\lambda_i \geq 0$, that is, $x^* \in \text{con}\{x_1^*, x_2^*, \ldots, x_n^*\}$.
[Hint: Use Lemma 6.35 to deduce $x^* = \sum_1^n \lambda_i x_i^*$ for some $\lambda_i \in \mathbb{R}$. Then use Theorem 6.36 to deduce $\lambda_i \geq 0$ for all i.]

19. Suppose that $\{x, x_1, \ldots, x_n\} \subset X$ satisfies $\langle x, y \rangle \geq 0$ whenever $\langle x_i, y \rangle \geq 0$ for $i = 1, 2, \ldots, n$. Show that $x = \sum_1^n \lambda_i x_i$ for some $\lambda_i \geq 0$.
[Hint: Exercise 18.]

20. **(Farkas lemma)** Let A be an $m \times n$ matrix, $x \in l_2(n)$, and suppose $\langle x, y \rangle \geq 0$ whenever $y \in l_2(n)$ and $Ay \geq 0$. Show that x is a nonnegative linear combination of the rows of A.
[Hint: Exercise 19.]

21. Show that the functional, whose existence is guaranteed by the Hahn–Banach extension theorem (Theorem 6.9), is unique.

22. Let $Q = \{y \in C_2[0,1] \mid \int_0^1 y(t)\,dt \leq 0 \text{ and } \int_0^1 ty(t)\,dt \leq 1\}$. Show that if $x(t) = t^2$, then $P_Q(x)(t) = t^2 - \frac{1}{3}$. [Hint: Theorem 6.41.]

*For Exercises 23–29 below, we now assume that X is a **complex** inner product space. Let $\mathbb{C}$ denote the set of complex numbers.*

A **linear functional** on X is a function $f : X \to \mathbb{C}$ such that

$$f(\alpha x + \beta y) = \alpha f(x) + \beta f(y)$$

for all $x, y \in X$ and $\alpha, \beta \in \mathbb{C}$.

The definition of f being bounded and the norm of f is just as in the real case (see (6.1.2) and (6.1.3)). Also, just as in the real case, f is bounded if and only if it is continuous (Theorem 5.11). The **dual** of X is the set of all bounded linear functionals on X and is denoted by X^*.

23. The following results are valid as stated, just as in the real case:
(a) Theorem 6.3.
(b) Theorem 6.5.

(c) Theorem 6.7.
(d) Theorem 6.8.
(e) Theorem 6.9.
(f) Theorem 6.10.
(g) Theorem 6.12.
(h) Theorem 6.15.
(i) Theorem 6.16.
(j) Theorem 6.17.
(k) Theorem 6.22.

The complex inner product space X may be regarded as a real inner product space $X_{\mathbb{R}}$ as follows. Let $X_{\mathbb{R}} = X$ and define addition and multiplication by **real** scalars in $X_{\mathbb{R}}$ just as in X. Then, obviously, $X_{\mathbb{R}}$ is a real linear space. Define

$$\langle x, y \rangle_{\mathbb{R}} := \operatorname{Re}\langle x, y \rangle \qquad \text{for all } x, y \in X_{\mathbb{R}},$$

where $\operatorname{Re}(\alpha)$ denotes the real part of the complex scalar α.

24. Verify that $X_{\mathbb{R}}$ is a real inner product space with inner product $\langle \cdot, \cdot \rangle_{\mathbb{R}}$. Next let $x^* \in X^*$. That is, x^* is a bounded linear functional on X. Define $\mathcal{R}x^*$ on $X_{\mathbb{R}}$ by

$$\mathcal{R}x^*(x) := \operatorname{Re}[x^*(x)], \quad x \in X_{\mathbb{R}}.$$

25. For each $x^* \in X^*$, $\mathcal{R}x^*$ is a bounded linear functional on $X_{\mathbb{R}}$ (i.e., $\mathcal{R}x^* \in X_{\mathbb{R}}^*$) and $\|\mathcal{R}x^*\| = \|x^*\|$.
26. Show that the key lemma (Lemma 6.6) holds, provided that in (6.6.1) and the sentence following this equation, $x^*(x)$ is replaced by $\operatorname{Re} x^*(x)$.
27. Prove that $x^* \in X^*$ attains its norm if and only if $\mathcal{R}x^*$ attains its norm.
28. A **real hyperplane** in X, or a hyperplane in $X_{\mathbb{R}}$, is any set of the form

$$H = \{y \in X \mid \mathcal{R}x^*(y) = c\}$$
$$(= \{y \in X \mid \operatorname{Re} x^*(y) = c\}),$$

where $x^* \in X^* \setminus \{0\}$ and $c \in \mathbb{R}$.
(a) Verify that Theorem 6.15 is valid for real hyperplanes, provided that in statements (1) and (3), X is replaced by $X_{\mathbb{R}}$, and in (1), span$\{x_1\}$ is replaced by "real span$\{x_1\}$," i.e., all real multiples of x_1.
(b) Prove that the distance formula to hyperplanes (Theorem 6.16) is also valid for real hyperplanes. How must formula (6.16.1) be modified in this case?
29. A real hyperplane $H = \{y \in X \mid \mathcal{R}x^*(y) = c\}$ is said to *separate* a set K and a point x if

$$\sup \mathcal{R}x^*(K) \leq c \leq \mathcal{R}x^*(x),$$

where $\sup \mathcal{R}x^*(K) = \sup\{\operatorname{Re} x^*(y) \mid y \in K\}$. (Compare this with Definition 6.20.)
(a) What is the geometric interpretation of H separating K and x?
(b) Prove that Lemma 6.21 is valid for real hyperplanes.

(c) Prove that the strong separation principle (Theorem 6.23) is valid, provided that x^* is replaced by $\mathcal{R}x^*$. In particular, relations (6.23.1) and (6.23.2) are replaced by

$$d(x, K) = \operatorname{Re} x^*(x) - \sup \operatorname{Re} x^*(K)$$

and
$$\sup \operatorname{Re} x^*(K) < \operatorname{Re} x^*(x),$$

respectively.

(d) The distance formulas in Theorem 6.25 are valid for real hyperplanes. In particular, (6.25.1) is replaced by

$$d(x, K) = \max\{\operatorname{Re} x^*(x) - \sup \operatorname{Re} x^*(K) \mid x^* \in X^*, \ \|x^*\| = 1\},$$

and (6.25.2) is replaced by

$$d(x, K) = \max d(x, H),$$

where the maximum is taken over all real hyperplanes H that separate K and x.

(e) Prove that the analogue of Theorem 6.27 is valid for real hyperplanes. In particular, (6.27.1) is replaced by

$$P_H(x) = x - \|z\|^{-2}[\operatorname{Re}\langle x, z\rangle - c]z.$$

(f) The algorithm described following Theorem 6.27 is valid for real hyperplanes.

30. There are several results in this chapter that are valid in any (real) normed linear space X.

(a) Show that Theorems 6.3, 6.9, 6.16, and 6.21 are valid in X.

(b) Show that Theorem 6.15 holds in X, provided that the last sentence in part (4) is deleted.

(c) Statements (2), (3), and (4) of Theorem 6.17 are equivalent in X. Also, show that if $H = \{y \in X \mid x^*(y) = c\}$ is proximinal, then

$$P_H(x) = x - \left\{\left[\frac{x^*(x) - c}{\|x^*\|}\right] z \ \Big| \ z \in X, \ \|z\| = 1, \ x^*(z) = \|x^*\|\right\}.$$

In particular, H is Chebyshev if and only if x^* attains its norm at a unique point $z \in X$ with $\|z\| = 1$.

(d) Theorems 6.23 and 6.25, with the last statements concerning the representers of x^* deleted, are valid in X.

31. Exercises 1 and 18 are valid in any normed linear space.

Historical Notes

The key lemma (Lemma 6.6), which was fundamental to the main results of this chapter, was proved by Ocneanu (1982) in response to a query of mine. While the denseness of $\hat{X}$ in X^* is well-known, the quantitative version given here (Theorem 6.7) seems new. Similarly, I have not found Theorem 6.8 in the literature either.

The Hahn–Banach extension theorem (Theorem 6.9) is actually a special case of one of the major cornerstones of functional analysis, which goes by the same name, and is valid in any (real) normed linear space. It was first proved by Hahn (1927; p. 217). Banach (1929a; p. 212), (1929b; p. 226) proved not only the extension theorem in normed linear spaces, but also the linear space (or "geometric") version, which turned out to be the main tool for proving important separation theorems in locally convex spaces as well. That these theorems also have analogues in *complex* spaces as well as in real spaces is due to a clever trick of Bohnenblust and Sobczyk (1938) and, independently, Soukhomlinoff (1938). It is perhaps of some academic interest to note that, in contrast to the known proofs of the general Hahn–Banach theorem, the proof of Theorem 6.9 given here does *not* use the axiom of choice.

The Fréchet–Riesz Representation theorem (Theorem 6.10) in the space $L_2[0, 1]$ was established independently by Fréchet (1907a), (1907; p. 439) and Riesz (1907). The result in full generality was given by Riesz (1934). Unlike the proof given here, Riesz's original proof was modeled after Exercise 2 at the end of the chapter). For a linear functional, the condition that norm-attaining is equivalent to having a representer was noted in Deutsch (1982; p. 229). Formula (6.16.1) for the distance to hyperplanes (which is valid in any normed linear space) is due to Ascoli (1932).

The strong separation principle (Theorem 6.23) is a quantitative version of a separation theorem. Perhaps the most useful (qualitative) separation theorem, valid in any locally convex linear topological space, is due to Eidelheit (1936) (see also Dunford-Schwartz (1958; V. 2.8, p. 417). The distance to a convex set (6.25.1) is a particular case of a duality result that is valid in any normed linear space (see Nirenberg (1961; p. 39) or Deutsch and Maserick (1967)). Theorem 6.27 can be found in Deutsch, McCabe, and Phillips (1975). The distance formula to half-spaces (6.30.1) is an easy consequence of the distance formula to hyperplanes (6.16.1).

That finitely generated cones are Chebyshev was observed by Deutsch (1982; p. 230). Lemma 6.35, in any linear space, was established (by induction) by Dieudonné (1942; p. 109). The characterization of Chebyshev polyhedra (Theorem 6.37) was established by Deutsch (1982) in the case where all the functionals are linearly independent by the proof given here, and by Amir, Deutsch, and Li (2000) in the general case by a different proof. I have not seen Theorem 6.41 (the characterization of best approximations from a polyhedron) in the literature, although it may well be in the "folk-theorem" category.

CHAPTER 7

ERROR OF APPROXIMATION

Distance to Convex Sets

In this chapter we will be interested in determining the error $d(x, K)$ made in approximating the element x by the elements of a convex set K. We have already given an explicit formula for the distance $d(x, K)$ in the last chapter (Theorem 6.25), and a strengthening of this distance formula in the particular case where the convex set K is either a convex cone or a subspace (Theorem 6.26). Now we will extract still further refinements, improvements, and applications of these formulas.

For example, there are useful formulas for $d(x, M)$ in the cases where M is a subspace of finite dimension or finite codimension. The latter is used to solve Problem 4 of Chapter 1. The classical Weierstrass approximation theorem is established, which implies that the subspace of all polynomials is dense in $C_2[a, b]$. This result (along with the formula for the distance to a finite-dimensional subspace and a few other technical facts) implies a beautiful generalization due to Müntz (Theorem 7.26).

Many of the results of this chapter rest on the distance formula (6.25.1). We restate a slightly strengthened version here for ease of reference.

7.1 Distance to a Convex Set. *Let K be a closed convex subset of X and $x \in X \setminus K$. Then*

$$(7.1.1) \qquad d(x, K) = \max\{x^*(x) - \sup x^*(K) \mid x^* \in X^*, \ \|x^*\| = 1\},$$

and this maximum is attained for a unique functional $x_0^* \in X^*$ with $\|x_0^*\| = 1$.

Moreover, if the best approximation $P_K(x)$ exists (e.g., if X is complete), then

$$(7.1.2) \qquad d(x, K) = \max\left\{\langle x, z \rangle - \sup_{y \in K}\langle y, z \rangle \mid z \in X, \ \|z\| = 1\right\},$$

and this maximum is attained for a unique point $z_0 \in X$ with $\|z_0\| = 1$. In addition,

$$(7.1.3) \qquad P_K(x) = x - d(x, K)z_0 = x - \left[\langle x, z_0 \rangle - \sup_{y \in K}\langle y, z_0 \rangle\right]z_0.$$

Proof. Using Theorem 6.25, it remains only to verify (7.1.3). Let

$$H = \{y \in X \mid \langle y, z_0 \rangle = c\},$$

where $c = \sup_{y \in K}\langle y, z_0 \rangle$. Then H is a hyperplane such that

$$\sup_{y \in K}\langle y, z_0 \rangle = c = \langle x, z_0 \rangle - d(x, K) \leq \langle x, z_0 \rangle.$$

Hence H separates K and x. By Theorem 6.16, $d(x, H) = \langle x, z_0 \rangle - c = d(x, K)$. By Theorem 6.27,

$$P_K(x) = P_H(x) = x - [\langle x, z_0 \rangle - c]z_0 = x - d(x, K)z_0. \quad \blacksquare$$

7.2 Remarks. (1) The geometric interpretation of (7.1.1) is that the distance from x to K is the maximum of the distances from x to hyperplanes that separate x and K. The geometric interpretation of (7.1.2) is that in the case where $P_K(x)$ exists, we may further restrict the search for separating hyperplanes to those that are also Chebyshev or, equivalently, to those whose functionals have representers.

(2) Formula (7.1.3) tells us that if we can find the element $z_0 \in X$, $\|z_0\| = 1$, that maximizes the right side of (7.1.2), then we have an explicit formula for computing $P_K(x)$.

(3) Formulas (7.1.1) and (7.1.2) provide us with an easy way of obtaining *lower bounds* for the error of approximation $d(x, K)$. For if $x^* \in X^*$ and $\|x^*\| = 1$, then

$$(7.2.1) \qquad x^*(x) - \sup x^*(K) \leq d(x, K).$$

Upper bounds can, of course, always be obtained from the obvious inequality

$$d(x, K) \leq \|x - y\|, \quad y \in K.$$

(4) Theorem 7.1 may be regarded as a "max–inf" theorem. For (7.1.1) may be rewritten as

$$(7.2.2) \qquad d(x, K) = \max\left\{ \inf_{y \in K} x^*(x - y) \mid x^* \in X^*, \ \|x^*\| = 1 \right\}.$$

Moreover, the "inf" and the "max" in (7.2.2) may be interchanged:

$$(7.2.3) \qquad d(x, K) = \inf_{y \in K} \max\{x^*(x - y) \mid x^* \in X^*, \ \|x^*\| = 1\}.$$

To see this, we note that for each $y \in K$ there exists $x^* \in X^*$ with $\|x^*\| = 1$ and $x^*(x - y) = \|x - y\|$ (namely, $x^* = \|x - y\|^{-1}(\widehat{x - y})$).

Thus

$$\|x - y\| = \max\{x^*(x - y) \mid x^* \in X^*, \ \|x^*\| = 1\},$$

and hence (7.2.3) follows.

(5) Formula (7.1.1) (respectively (7.1.2)) is also valid when $x \in K$, provided that the maximum is taken over all $x^* \in X^*$ with $\|x^*\| \leq 1$ (respectively $z \in X$ with $\|z\| \leq 1$) rather than just $\|x^*\| = 1$ (respectively $\|z\| = 1$). (See Exercise 1 at the end of the chapter.)

(6) The supremum in (7.1.1) (respectively (7.1.2)) cannot be replaced in general by the maximum. (See Exercise 2 at the end of the chapter.)

(7) Clearly, the search for a maximum in (7.1.1) may be further restricted to those norm-one functionals x^* with $x^*(x) \geq \sup x^*(K)$.

In the same way that Theorem 7.1 is an improvement to Theorem 6.25, we may state the following improvement of Theorem 6.26.

7.3 Distance to a Convex Cone and a Subspace. *Let M (respectively C) be a closed subspace (respectively closed convex cone) in X and let $x \in X \setminus M$ (respectively $x \in X \setminus C$). Then*

$$(7.3.1) \quad d(x, M) = \max\{x^*(x) \mid x^* \in X^*, \ \|x^*\| = 1, \ x^*(y) = 0 \text{ for all } y \in M\}$$

(7.3.2)
(respectively $d(x, C) = \max\{x^*(x) \mid x^* \in X^*, \ \|x^*\| = 1, \ x^*(y) \leq 0 \text{ for all } y \in C\}$),

and this maxmum is attained for a unique $x_0^* \in X^*$.

Moreover, if $P_M(x)$ (respectively $P_C(x)$) exists (e.g., if X is complete), then

$$(7.3.3) \qquad\qquad d(x, M) = \max\{\langle x, z \rangle \mid z \in M^\perp, \ \|z\| = 1\}$$

$$(7.3.4) \qquad (\text{respectively } d(x, C) = \max\{\langle x, z \rangle \mid z \in C^\circ, \ \|z\| = 1\},$$

and this maximum is attained for a unique $z_0 \in M^\perp$ (respectively $z_0 \in C^\circ$) with $\|z_0\| = 1$. In addition,

$$(7.3.5) \qquad\qquad P_M(x) = x - \langle x, z_0 \rangle z_0$$

$$(7.3.6) \qquad\qquad (\text{respectively } P_C(x) = x - \langle x, z_0 \rangle z_0).$$

Clearly, to apply Theorem 7.3 for determining the distance to convex Chebyshev cones as well as determining best approximations from such sets requires knowledge of the dual cones of these sets. In Chapter 10, we will make a systematic study of dual cones and their applications in duality results such as Theorem 7.3. For the moment, we will content ourselves with the following simple but illustrative application of Theorem 7.3.

7.4 Example. Let C be the convex cone

$$C = \{\, x \in l_2(2) \mid 0 \leq x(2) \leq x(1) \,\}.$$

It is easy to verify that its dual cone is given by

$$C^\circ = \{\, y \in l_2(2) \mid y(1) \leq 0 \text{ and } y(1) + y(2) \leq 0 \,\}$$

(see Figure 7.4.1 below).

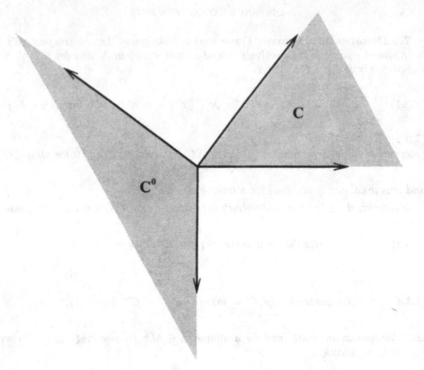

Figure 7.4.1

Let $x \in l_2(2) \setminus C$. If $x \in C^\circ$, then by Theorem 7.3

$$d(x, C) = \max\{\, \langle x, z \rangle \mid z \in C^\circ, \ \|z\| = 1 \,\} = \|x\|,$$

and the maximum is attained for $z_0 = x/\|x\|$. Moreover,

$$P_C(x) = x - \langle x, z_0 \rangle z_0 = 0.$$

(These facts are also a consequence of Theorem 5.5(4).)

Thus we may assume that $x \notin C \cup C^\circ$. We have two cases to consider.

Case 1: $x(1) > 0$ and $x(2) < 0$.

Then for each $z \in C^\circ$ with $\|z\| = 1$, we get

$$\langle x, z \rangle = x(1)z(1) + x(2)z(2) \leq x(2)z(2) \leq -x(2),$$

and equality holds if and only if $z = z_0 = (0, -1)$. Thus by Theorem 7.3,

$$d(x, C) = \max\{\, \langle x, z \rangle \mid z \in C^\circ, \ \|z\| = 1 \,\} = -x(2)$$

and

$$P_C(x) = x + x(2)z_0 = (x(1), 0).$$

Case 2: $x(1) + x(2) > 0$ and $x(2) > x(1)$.

Then for each $z \in C^\circ$ with $\|z\| = 1$, we get

$$\langle x, z \rangle = x(1)z(1) + x(2)z(2) = [x(1) - x(2)]z(1) + x(2)[z(1) + z(2)]$$
$$\leq [x(1) - x(2)]z(1),$$

and equality holds if and only if $z(1) + z(2) = 0$ and $z(1) < 0$; i.e., $z = z_0 = \frac{1}{\sqrt{2}}(-1, 1)$. Thus by Theorem 7.3,

$$d(x, C) = \max\{\, \langle x, z \rangle \mid z \in C^\circ,\ \|z\| = 1 \,\} = \frac{1}{\sqrt{2}}[x(2) - x(1)]$$

and

$$P_C(x) = x - \frac{1}{\sqrt{2}}[x(2) - x(1)]z_0 = \left(\frac{1}{2}[x(1) + x(2)],\ \frac{1}{2}[x(1) + x(2)] \right).$$

Distance to Finite-Dimensional Subspaces

We turn now to the problem of determining the distance to a finite-dimensional or finite-codimensional subspace. Recall that if $\{y_1, y_2, \ldots, y_m\}$ is a subset of X, the **Gram matrix** of this set is the $m \times m$ matrix $G(y_1, y_2, \ldots, y_m)$ whose (i, j)-entry is $\langle y_i, y_j \rangle$. The determinant of $G(y_1, y_2, \ldots, y_m)$ is called the **Gram determinant** of $\{y_1, y_2, \ldots, y_m\}$ and is denoted by $g(y_1, y_2, \ldots, y_m)$. Thus

$$g(y_1, y_2, \ldots, y_m) = \begin{vmatrix} \langle y_1, y_1 \rangle & \langle y_1, y_2 \rangle & \cdots & \langle y_1, y_m \rangle \\ \langle y_2, y_1 \rangle & \langle y_2, y_2 \rangle & \cdots & \langle y_2, y_m \rangle \\ \cdots & & & \\ \langle y_m, y_1 \rangle & \langle y_m, y_2 \rangle & \cdots & \langle y_m, y_m \rangle \end{vmatrix}.$$

7.5 Lemma. Let $\{x_1, x_2, \ldots, x_n\} \subset X$. Then $g(x_1, x_2, \ldots, x_n) \neq 0$ if and only if $\{x_1, x_2, \ldots, x_n\}$ is linearly independent.

Proof. If $\{x_1, x_2, \ldots, x_n\} \subset X$ is linearly dependent, say $x_1 = \sum_2^n \alpha_i x_i$, then by adding $-\alpha_i$ times the ith row of $g(x_1, x_2, \ldots, x_n)$ to the first row for $i = 2, 3, \ldots, n$, we obtain all zeros in the first row and hence $g(x_1, x_2, \ldots, x_n) = 0$.

Conversely, if $g(x_1, x_2, \ldots, x_n) = 0$, then matrix algebra implies that some row of $G(x_1, x_2, \ldots, x_n)$ is a linear combination of the others. Without loss of generality, we may assume that it is the first row:

$$(\langle x_1, x_1 \rangle, \ldots, \langle x_1, x_n \rangle) = \sum_{i=2}^n \alpha_i(\langle x_i, x_1 \rangle, \ldots, \langle x_i, x_n \rangle)$$

$$= \left(\left\langle \sum_{i=2}^n \alpha_i x_i, x_1 \right\rangle, \ldots, \left\langle \sum_{i=2}^n \alpha_i x_i, x_n \right\rangle \right).$$

Thus

$$\langle x_1, x_j \rangle = \left\langle \sum_{i=2}^n \alpha_i x_i, x_j \right\rangle \quad \text{for} \quad j = 1, 2, \ldots, n$$

implies that

$$\left\langle x_1 - \sum_2^n \alpha_i x_i, x_j \right\rangle = 0 \quad (j = 1, 2, \ldots, n).$$

In particular, $x_1 - \sum_2^n \alpha_i x_i \perp M := \operatorname{span}\{x_1, x_2, \ldots, x_n\}$ and also $x_1 - \sum_2^n \alpha_i x_i \in M$. That is,

$$x_1 - \sum_2^n \alpha_i x_i \in M \cap M^\perp = \{0\}$$

or $x_1 = \sum_2^n \alpha_i x_i$. Hence $\{x_1, x_2, \ldots, x_n\}$ is linearly dependent. ∎

7.6 Distance to a Finite-Dimensional Subspace. *Let*
$M = \text{span}\{x_1, x_2, \ldots, x_n\}$ *be n-dimensional in X. Then for each $x \in X$,*

(7.6.1)
$$d(x, M)^2 = \frac{g(x_1, x_2, \ldots, x_n, x)}{g(x_1, x_2, \ldots, x_n)}$$

and

(7.6.2)
$$d(x, M)^2 = \|x\|^2 - \beta^T G^{-1} \beta,$$

where $G = G(x_1, x_2, \ldots, x_n)$, G^{-1} is the inverse matrix of G, and

$$\beta^T = (\langle x, x_1 \rangle, \langle x, x_2 \rangle, \ldots, \langle x, x_n \rangle)$$

denotes the transpose of the column vector β.
    Moreover,

(7.6.3)
$$d(x, M)^2 = \begin{cases} \|x\|^2 - \dfrac{\left(\sum_1^n |\langle x, x_i \rangle|^2\right)^2}{\|\sum_1^n \langle x, x_i \rangle x_i\|^2} & \text{if } x \notin M^\perp, \\[2mm] \|x\|^2 & \text{if } x \in M^\perp. \end{cases}$$

Proof. Fix $x \in X$ and let $d = d(x, M)$. By Corollary 4.10, $P_M(x) = \sum_1^n \alpha_i x_i$,
where the α_i satisfy

(7.6.4)
$$\sum_1^n \alpha_i \langle x_i, x_j \rangle - \langle x, x_j \rangle = 0 \quad (j = 1, 2, \ldots, n).$$

Also, $x - P_M(x) \perp M$ by Theorem 4.9, so

$$d^2 = \langle x - P_M(x), x - P_M(x) \rangle = \langle x - P_M(x), x \rangle = \langle x, x \rangle - \sum_1^n \alpha_i \langle x_i, x \rangle,$$

or

(7.6.5)
$$\sum_1^n \alpha_i \langle x_i, x \rangle + [d^2 - \langle x, x \rangle] = 0.$$

By introducing the constant $\alpha_{n+1} := 1$ as the coefficient of $-\langle x, x_j \rangle$ ($j = 1, 2, \ldots, n$) and $d^2 - \langle x, x \rangle$, equations (7.6.4)–(7.6.5) represent a homogeneous linear system of $n + 1$ equations in the $n + 1$ variables $\alpha_1, \alpha_2, \ldots, \alpha_{n+1}$. Since $\alpha_{n+1} = 1$, this system has a nontrivial solution, so that the determinant of the coefficient matrix must be zero. Thus

$$\begin{vmatrix} \langle x_1, x_1 \rangle & \langle x_2, x_1 \rangle & \cdots & \langle x_n, x_1 \rangle & -\langle x, x_1 \rangle \\ \langle x_1, x_2 \rangle & \langle x_2, x_2 \rangle & \cdots & \langle x_n, x_2 \rangle & -\langle x, x_2 \rangle \\ & \cdots & & & \\ \langle x_1, x_n \rangle & \langle x_2, x_n \rangle & \cdots & \langle x_n, x_n \rangle & -\langle x, x_n \rangle \\ \langle x_1, x \rangle & \langle x_2, x \rangle & \cdots & \langle x_n, x \rangle & d^2 - \langle x, x \rangle \end{vmatrix} = 0.$$

Since the determinant is a linear function of its last column, we obtain

$$
\begin{vmatrix}
\langle x_1, x_1 \rangle & \cdots & \langle x_n, x_1 \rangle & 0 \\
\langle x_1, x_2 \rangle & \cdots & \langle x_n, x_2 \rangle & 0 \\
\cdots & & & \\
\langle x_1, x_n \rangle & \cdots & \langle x_n, x_n \rangle & 0 \\
\langle x_1, x \rangle & \cdots & \langle x_n, x \rangle & d^2
\end{vmatrix}
=
\begin{vmatrix}
\langle x_1, x_1 \rangle & \cdots & \langle x_n, x_1 \rangle & \langle x, x_1 \rangle \\
\langle x_1, x_2 \rangle & \cdots & \langle x_n, x_2 \rangle & \langle x, x_2 \rangle \\
\cdots & & & \\
\langle x_1, x_n \rangle & \cdots & \langle x_n, x_n \rangle & \langle x, x_n \rangle \\
\langle x_1, x \rangle & \cdots & \langle x_n, x \rangle & \langle x, x \rangle
\end{vmatrix}.
$$

That is,

$$
d^2 g(x_1, x_2, \ldots, x_n) = g(x_1, x_2, \ldots, x_n, x),
$$

and (7.6.1) follows.

Next observe that (7.6.4) may be rewritten in matrix notation as $\alpha^T G = \beta^T$ or $\alpha^T = \beta^T G^{-1}$. Using this and (7.6.5), we obtain

$$
d^2 = \|x\|^2 - \sum_1^n \alpha_i \langle x_i, x \rangle = \|x\|^2 - \alpha^T \beta = \|x\|^2 - \beta^T G^{-1} \beta,
$$

which establishes (7.6.2).

To verify (7.6.3), first suppose $x \in M^\perp$. Then $\beta^T = (0, 0, \ldots, 0)$, and the result follows from (7.6.2). Thus we may assume that $x \notin M^\perp$. By (7.3.3), applied to $M^\perp$ instead of M, and using the facts that $M^\perp$ is Chebyshev and $M^{\perp\perp} = M$ (see, e.g., Theorem 5.8), we obtain

$$
d(x, M^\perp) = \max \{ \langle x, z \rangle \mid z \in M, \ \|z\| = 1 \}
$$

$$
= \max \left\{ \left\langle x, \sum_1^n \alpha_i x_i \right\rangle \ \Big| \ \alpha_i \in \mathbb{R}, \ \left\| \sum_1^n \alpha_i x_i \right\| = 1 \right\}
$$

$$
= \max \left\{ \sum_1^n \alpha_i \langle x, x_i \rangle \ \Big| \ \alpha_i \in \mathbb{R}, \ \left\| \sum_1^n \alpha_i x_i \right\| = 1 \right\}.
$$

If $\|\sum_1^n \alpha_i x_i\| = 1$, then by Schwarz's inequality (in $l_2(n)$), we get

$$
\sum_1^n \alpha_i \langle x, x_i \rangle \le \left(\sum_1^n \alpha_i^2 \right)^{\frac{1}{2}} \left(\sum_1^n |\langle x, x_i \rangle|^2 \right)^{\frac{1}{2}},
$$

with equality holding if and only if there is a $\lambda > 0$ such that $\alpha_i = \lambda \langle x, x_i \rangle$ for all i. If equality holds, then since $\|\sum_1^n \alpha_i x_i\| = 1$, we obtain

$$
1 = \left\| \sum_1^n \alpha_i x_i \right\| = \left\| \sum_1^n \lambda \langle x, x_i \rangle x_i \right\| = \lambda \left\| \sum_1^n \langle x, x_i \rangle x_i \right\|,
$$

or $\lambda = \|\sum_1^n \langle x, x_i \rangle x_i\|^{-1}$. That is, $\alpha_i = \|\sum_1^n \langle x, x_j \rangle x_j\|^{-1} \langle x, x_i \rangle$ for each i. It follows that

$$
d(x, M^\perp) = \left\| \sum_1^n \langle x, x_i \rangle x_i \right\|^{-1} \sum_1^n |\langle x, x_i \rangle|^2.
$$

Using Theorem 5.8 (3), (7.6.3) is obtained. ∎

Our first application of the distance formula (Theorem 7.6) yields sharp bounds on the Gram determinant $g(x_1, x_2, \ldots, x_n)$.

7.7 Gram Determinant Inequality. *Let* $\{x_1, x_2, \ldots, x_n\}$ *be a set of nonzero vectors in* X. *Then*

(7.7.1) $0 \leq g(x_1, x_2, \ldots, x_n) \leq \|x_1\|^2 \|x_2\|^2 \cdots \|x_n\|^2.$

Moreover, equality holds on the left (respectively right) side of (7.7.1) if and only if $\{x_1, x_2, \ldots, x_n\}$ *is linearly dependent (respectively orthogonal).*

Proof. By Lemma 7.5, $g(x_1, x_2, \ldots, x_n) = 0$ if and only if $\{x_1, x_2, \ldots, x_n\}$ is linearly dependent. We next show that if $\{x_1, x_2, \ldots, x_n\}$ is linearly independent, then $g(x_1, x_2, \ldots, x_n) > 0$. The proof is by induction on n. For $n = 1$,

$$g(x_1) = \langle x_1, x_1 \rangle = \|x_1\|^2 > 0.$$

Assume that the result is true for some $n \geq 1$. By Theorem 7.6 with $x = x_{n+1}$, we see that

$$g(x_1, x_2, \ldots, x_n, x_{n+1}) = g(x_1, x_2, \ldots, x_n)d(x_{n+1}, M)^2 > 0$$

since $x_{n+1} \notin M = \text{span}\{x_1, x_2, \ldots, x_n\}$.

It remains to verify the right side of (7.7.1). Clearly, if $\{x_1, x_2, \ldots, x_n\}$ is orthogonal, then $\langle x_i, x_j \rangle = 0$ for $i \neq j$, so that

$$g(x_1, x_2, \ldots, x_n) = \begin{vmatrix} \langle x_1, x_1 \rangle & 0 & \cdots & 0 \\ 0 & \langle x_2, x_2 \rangle & \cdots & 0 \\ \cdots & & & \\ 0 & 0 & \cdots & \langle x_n, x_n \rangle \end{vmatrix} = \|x_1\|^2 \|x_2\|^2 \cdots \|x_n\|^2,$$

and equality holds on the right side of (7.7.1).

It is left to prove that

(7.7.2) $g(x_1, x_2, \ldots, x_n) \leq \|x_1\|^2 \|x_2\|^2 \cdots \|x_n\|^2,$

and if equality holds, then $\{x_1, x_2, \ldots, x_n\}$ must be orthogonal. We again proceed by induction on n. For $n = 1$,

$$g(x_1) = \langle x_1, x_1 \rangle = \|x_1\|^2,$$

and equality holds. But $\{x_1\}$ is trivially an orthogonal set. Assume that (7.7.2) is valid for some $n \geq 1$. Applying Theorem 7.6 with $x = x_{n+1}$, we obtain

$$g(x_1, x_2, \ldots, x_n, x_{n+1}) = g(x_1, x_2, \ldots, x_n)d(x_{n+1}, M)^2$$
$$\leq \|x_1\|^2 \|x_2\|^2 \cdots \|x_n\|^2 \|x_{n+1}\|^2.$$

Thus the inequality in (7.7.2) holds. Moreover, if equality holds, then we must have both

(7.7.3) $g(x_1, x_2, \ldots, x_n) = \|x_1\|^2 \|x_2\|^2 \cdots \|x_n\|^2$

and

(7.7.4) $d(x_{n+1}, M) = \|x_{n+1}\|.$

By the induction hypothesis, (7.7.3) implies that $\{x_1, x_2, \ldots, x_n\}$ is orthogonal. Further, (7.7.4) implies that $P_M(x_{n+1}) = 0$. Hence by Theorem 4.9, $x_{n+1} \in M^\perp$, and thus $\{x_1, x_2, \ldots, x_n, x_{n+1}\}$ is orthogonal. ■

If $\{x_1, x_2, \ldots, x_n\}$ is an orthonormal set, the formulas in Theorem 7.6 substantially simplify.

7.8 Corollary. Let $\{e_1, e_2, \ldots, e_n\}$ be an orthonormal basis for M. Then for each $x \in X$,

(7.8.1) $$P_M(x) = \sum_1^n \langle x, e_i \rangle e_i$$

and

(7.8.2) $$d(x, M)^2 = \|x\|^2 - \sum_1^n |\langle x, e_i \rangle|^2.$$

Finite-Codimensional Subspaces

Recall that if M and N are subspaces of the inner product space X, we write $X = M \oplus N$ if each $x \in X$ has a unique representation in the form $x = y + z$, where $y \in M$ and $z \in N$. Equivalently, $X = M \oplus N$ if and only if $X = M + N$ and $M \cap N = \{0\}$.

7.9 Definition. A closed subspace M of X is said to have **codimension** n, written codim $M = n$, if there exists an n-dimensional subspace N such that $X = M \oplus N$.

For example, by Theorem 6.15 the kernel of any nonzero bounded linear functional is a closed subspace of codimension one. The following lemma shows that this notion is well-defined.

7.10 Lemma. Let M be a closed subspace of X and let N and L be subspaces with $\dim N < \infty$ and

$$X = M \oplus N = M \oplus L.$$

Then $\dim N = \dim L$.

Proof. If the result were false, we could assume that $\dim L > n := \dim N$. Then L contains a linearly independent set $\{y_1, y_2, \ldots, y_m\}$, where $m > n$. Write $y_i = u_i + v_i$, where $u_i \in M$ and $v_i \in N$. Since N is n-dimensional and $m > n$, $\{v_1, v_2, \ldots, v_m\}$ must be linearly dependent. Thus there exist scalars α_i, not all zero, such that $\sum_1^m \alpha_i v_i = 0$. Then

$$0 \neq \sum_1^m \alpha_i y_i = \sum_1^m \alpha_i u_i + \sum_1^m \alpha_i v_i = \sum_1^m \alpha_i u_i.$$

Since $\sum_1^m \alpha_i y_i \in L$ and $\sum_1^m \alpha_i u_i \in M$, it follows that $\sum_1^m \alpha_i y_i \in M \cap L = \{0\}$, which is absurd. ∎

The following alternative description of the subspaces that have codimension n will be useful to us.

7.11 Characterization of Finite-Codimensional Subspaces. Let M be a closed subspace of X and n a positive integer. Then codim $M = n$ if and only if

$$M = \bigcap_{i=1}^n \{x \in X \mid x_i^*(x) = 0\}$$

for some linearly independent set $\{x_1^*, \ldots, x_n^*\}$ in X^*.

Proof. Let codim $M = n$. Then there exists an n-dimensional subspace $N = \text{span}\,\{x_1, x_2, \ldots, x_n\}$ such that $X = M \oplus N$. We proceed by induction on n. For $n = 1$, Theorem 7.3 allows us to choose an $x_1^* \in X^*$ with $\|x_1^*\| = 1$, $x_1^*|_M = 0$, and $d(x_1, M) = x_1^*(x_1)$. Since M is closed and $x_1 \notin M$, $d(x_1, M) > 0$. We deduce that $M = \ker x_1^*$.

Now assume that the result is valid for $n = m$ and let $X = M \oplus N$, where $N = \text{span}\,\{x_1, \ldots, x_{m+1}\}$ is $(m+1)$-dimensional. Clearly, we have $X = L \oplus N_m$, where $L = \text{span}\,\{M, x_{m+1}\}$ and $N_m = \text{span}\,\{x_1, \ldots, x_m\}$. By the induction hypothesis, $L = \cap_1^m \ker x_i^*$ for some linearly independent set $\{x_1^*, \ldots, x_m^*\}$ in X^*. But $L = M \oplus \text{span}\{x_{m+1}\}$, so by the case where $n = 1$, we have that $M = \ker y^*$ for some $y^* \in L^* \setminus \{0\}$. By the Hahn–Banach extension theorem (Theorem 6.9), there is an $x^* \in X^*$ such that $x^*|_L = y^*$ and $\|x^*\| = \|y^*\|$. Now $x^* \notin \text{span}\,\{x_1^*, \ldots, x_m^*\}$, since $x_i^*|_L = 0$ for $i = 1, \ldots, m$ and $x^*|_L = y^* \neq 0$. Thus, setting $x_{m+1}^* = x^*$, we see that $\{x_1^*, \ldots, x_{m+1}^*\}$ is linearly independent and

$$M = \ker y^* = (\ker x^*) \cap L = \cap_1^{m+1} \ker x_i^*.$$

This completes the induction.

Conversely, suppose that for some linearly independent set $\{x_1^*, \ldots, x_n^*\}$ in X^*, we have $M = \cap_1^n \ker x_i^*$. We must show that codim $M = n$. The proof is by induction. For $n = 1$, $M = \ker x_1^*$ for some $x_1^* \in X^* \setminus \{0\}$. Choose any $x_1 \in X \setminus M$ and let $N = \text{span}\{x_1\}$. By Theorem 6.15, $X = M \oplus N$ and hence M has codimension one.

Next assume that the result is valid for $n = m$ and let $M = \cap_1^{m+1} \ker x_i^*$ for some linearly independent set $\{x_1^*, \ldots, x_{m+1}^*\}$ in X^*. Letting $Y = \ker x_{m+1}^*$, we see that

$$M = Y \cap \left(\bigcap_1^m \ker x_i^* \right).$$

Next we observe that $\{x_1^*|_Y, \ldots, x_m^*|_Y\}$ is linearly independent in Y^*. For if $\sum_1^m \alpha_i x_i^*(y) = 0$ for all $y \in Y$, then by Lemma 6.35, $\sum_1^m \alpha_i x_i^* = \alpha x_{m+1}^*$ for some scalar α. By linear independence of $\{x_1^*, \ldots, x_{m+1}^*\}$, it follows that $\alpha_i = \alpha = 0$ for all i. By the induction hypothesis (applied to M in Y), there is an m-dimensional subspace N_m of Y, hence of X, such that $Y = M \oplus N_m$. Moreover, by the case where $n = 1$, we have that $X = Y \oplus N_1$, where $\dim N_1 = 1$. Setting $N = N_m + N_1$, it follows that N is an $(m+1)$-dimensional subspace. But

$$X = Y + N_1 = (M + N_m) + N_1 = M + N.$$

To complete the proof, it suffices to show that $M \cap N = \{0\}$. But if $z \in M \cap N$, then $z = y_m + y_1$, where $y_m \in N_m$ and $y_1 \in N_1$. Since

$$z - y_m = y_1 \in N_1 \cap (M + N_m) = N_1 \cap Y = \{0\}$$

implies $z = y_m \in N_m \cap M = \{0\}$, the result follows. ∎

From this result we deduce, in particular, that the subspaces of codimension one are precisely the hyperplanes through the origin.

The subspaces of finite codimension that are Chebyshev have a simple description. They are precisely those for which the defining functionals have representers. This fact, proved next, also generalizes Theorem 6.17.

7.12 Characterization of Finite-Codimensional Chebyshev Subspaces.

Let M be a closed subspace of the inner product space X with codim $M = n$. Thus

$$(7.12.1) \qquad M = \bigcap_1^n \{x \in X \mid x_i^*(x) = 0\}$$

for some linearly independent set $\{x_1^*, x_2^*, \ldots, x_n^*\}$ in X^*. Then M is Chebyshev if and only if each x_i^* has a representer in X.

In other words, a closed subspace M of codimension n is Chebyshev if and only if

$$M = \bigcap_1^n \{x \in X \mid \langle x, x_i \rangle = 0\}$$

for some linearly independent set $\{x_1, x_2, \ldots, x_n\}$ in X.

Proof. Suppose M is Chebyshev. Then $X = M \oplus M^\perp$ by Theorem 5.9. Since codim $M = n$, it follows by Lemma 7.10 that dim $M^\perp = n$. Hence $M^\perp = \operatorname{span}\{x_1, \ldots, x_n\}$ for some linearly independent set $\{x_1, \ldots, x_n\}$. Since M is Chebyshev, Theorem 5.8 (6) implies that $M = M^{\perp\perp}$. That is,

$$(7.12.2) \quad M = [\operatorname{span}\{x_1, x_2, \ldots, x_n\}]^\perp = \bigcap_1^n \{x \in X \mid \langle x, x_i \rangle = 0\} = \bigcap_1^n \ker \hat{x}_i,$$

where $\hat{x}_i$ is the linear functional on X having x_i as its representer. By (7.12.1) and (7.12.2), we have that

$$(7.12.3) \qquad \bigcap_1^n \ker x_i^* = \bigcap_1^n \ker \hat{x}_i.$$

In particular, for each $j = 1, 2, \ldots, n$, we have that $\ker x_j^* \supset \bigcap_1^n \ker \hat{x}_i$. By Lemma 6.35, $x_j^* \in \operatorname{span}\{\hat{x}_1, \hat{x}_2, \ldots, \hat{x}_n\}$, so $x_j^* = \sum_1^n \alpha_i \hat{x}_i$ for some scalars α_i. It follows that $\sum_1^n \alpha_i x_i$ is the representer for x_j^*. Hence each x_j^* has a representer in X.

Conversely, suppose each x_j^* has a representer x_j in X. Then (7.12.1) implies that

$$(7.12.4) \qquad M = \bigcap_1^n \{x \in X \mid \langle x, x_i \rangle = 0\} = \bigcap_1^n (\operatorname{span} x_i)^\perp.$$

By Theorem 4.6(2),

$$M = \left(\sum_1^n \operatorname{span} x_i\right)^\perp = (\operatorname{span}\{x_1, x_2, \ldots, x_n\})^\perp.$$

But the finite-dimensional subspace $\operatorname{span}\{x_1, x_2, \ldots, x_n\}$ is Chebyshev by Theorem 3.8, and Theorem 5.8(1) implies that the subspace $M = (\operatorname{span}\{x_1, x_2, \ldots, x_n\})^\perp$ is also Chebyshev. ∎

Next we verify a formula for the distance to a finite-codimensional subspace. It is a nice application of Theorem 5.8, where knowledge of $M^\perp$ was used to compute $P_M(x)$ and $d(x, M)$.

7.13 Distance to Finite-Codimensional Subspace. *Let* $\{x_1, x_2, \ldots, x_n\}$ *be a linearly independent set in* X *and*

$$M = \bigcap_1^n \{x \in X \mid \langle x, x_i \rangle = 0\}.$$

Then M *is a Chebyshev subspace of codimension* n, *and for each* $x \in X$,

$$(7.13.1) \qquad P_M(x) = x - \sum_1^n \alpha_i x_i,$$

where the scalars α_i *satisfy the normal equations*

$$(7.13.2) \qquad \sum_{i=1}^n \alpha_i \langle x_i, x_j \rangle = \langle x, x_j \rangle \quad (j = 1, 2, \ldots, n),$$

and

$$(7.13.3) \quad d(x, M)^2 = \sum_1^n \alpha_i \langle x, x_i \rangle = \|x\|^2 - \frac{g(x_1, x_2, \ldots, x_n, x)}{g(x_1, x_2, \ldots, x_n)} = \beta^T G^{-1} \beta,$$

where $G = G(x_1, x_2, \ldots, x_n)$ *and* $\beta^T = (\langle x, x_1 \rangle, \langle x, x_2 \rangle, \ldots, \langle x, x_n \rangle)$.
Moreover,

$$(7.13.4) \qquad d(x, M) = \begin{cases} \dfrac{\sum_1^n |\langle x, x_i \rangle|^2}{\|\sum_1^n \langle x, x_i \rangle x_i\|} & \text{if } x \notin M, \\ 0 & \text{if } x \in M. \end{cases}$$

In particular, if $\{x_1, x_2, \ldots, x_n\}$ *is an orthonormal set, then for every* $x \in X$,

$$(7.13.5) \qquad P_M(x) = x - \sum_1^n \langle x, x_i \rangle x_i,$$

and

$$(7.13.6) \qquad d(x, M) = \left[\sum_1^n |\langle x, x_i \rangle|^2 \right]^{\frac{1}{2}}.$$

Proof. By Theorem 7.12, M is a Chebyshev subspace of codimension n and $M^\perp = \text{span}\{x_1, x_2, \ldots, x_n\}$. Applying Theorem 7.6 to the subspace $M^\perp$ (instead of M), we obtain that for every $x \in X$,

$$(7.13.7) \qquad P_{M^\perp}(x) = \sum_1^n \alpha_i x_i,$$

where the scalars α_i satisfy the equations (7.13.2), and
(7.13.8)

$$d(x, M^\perp)^2 = \frac{g(x_1, x_2, \ldots, x_n, x)}{g(x_1, x_2, \ldots, x_n)} = \|x\|^2 - \sum_1^n \alpha_i \langle x, x_i \rangle = \|x\|^2 - \beta^T G^{-1} \beta.$$

In addition, if $x \notin M$, then

$$(7.13.9) \qquad d(x, M^\perp)^2 = \|x\|^2 - \frac{(\sum_1^n |\langle x, x_i \rangle|^2)^2}{\|\sum_1^n \langle x, x_i \rangle x_i\|^2}.$$

But by Theorem 5.8, $P_M(x) = x - P_{M^\perp}(x)$ and $d(x, M)^2 = \|x\|^2 - d(x, M^\perp)^2$. Combining this with (7.13.7)–(7.13.9) completes the proof. ∎

7.14 Example. Let

$$M = \left\{ y \in C_2[0,1] \,\middle|\, \int_0^1 y(t)\, dt = 0 = \int_0^1 t y(t)\, dt \right\}.$$

We will determine $P_M(x)$ and $d(x, M)$ when $x(t) = t^2$.
First observe that we can rewrite M in the form

$$M = \{\, y \in C_2[0,1] \mid \langle y, x_i \rangle = 0 \;\; (i = 1, 2) \,\},$$

where $x_i(t) = t^{i-1}$. By Theorem 7.13, M is a Chebyshev subspace of codimension 2; also,

(7.14.1) $$P_M(x) = x - (\alpha_1 x_1 + \alpha_2 x_2)$$

and

(7.14.2) $$d(x, M)^2 = \alpha_1 \langle x, x_1 \rangle + \alpha_2 \langle x, x_2 \rangle,$$

where the α_i satisfy the equations

(7.14.3)
$$\alpha_1 \langle x_1, x_1 \rangle + \alpha_2 \langle x_2, x_1 \rangle = \langle x, x_1 \rangle,$$
$$\alpha_1 \langle x_1, x_2 \rangle + \alpha_2 \langle x_2, x_2 \rangle = \langle x, x_2 \rangle.$$

Since $\langle x_1, x_1 \rangle = \int_0^1 1\, dt = 1$, $\langle x_1, x_2 \rangle = \langle x_2, x_1 \rangle = \int_0^1 t\, dt = \frac{1}{2}$, $\langle x_2, x_2 \rangle = \int_0^1 t^2\, dt = \frac{1}{3}$, $\langle x, x_1 \rangle = \int_0^1 t^2\, dt = \frac{1}{3}$, and $\langle x, x_2 \rangle = \int_0^1 t^3\, dt = \frac{1}{4}$, equations (7.14.3) become

$$\alpha_1 + \frac{1}{2}\alpha_2 = \frac{1}{3},$$
$$\frac{1}{2}\alpha_1 + \frac{1}{3}\alpha_2 = \frac{1}{4},$$

whose solution is $\alpha_1 = -\frac{1}{6}$, $\alpha_2 = 1$. Substituting this back into (7.14.1) and (7.14.2), we obtain

$$P_M(x) = x + \frac{1}{6}x_1 - x_2,$$

or

$$P_M(x)(t) = t^2 - t + \frac{1}{6},$$

and $d(x, M) = \sqrt{7}/6$.

Now we give an application of Theorem 7.13 to characterize best approximations from translates of finite-codimensional subspaces.

Recall that a translate of a subspace is called an **affine set** (or a **linear variety** or a **flat**).

7.15 Best Approximation from Finite-Codimensional Affine Sets. *Let* $\{x_1, x_2, \ldots, x_n\}$ *be linearly independent in* X, $\{c_1, c_2, \ldots, c_n\}$ *in* $\mathbb{R}$, *and*

$$V = \bigcap_{i=1}^n \{\, y \in X \mid \langle y, x_i \rangle = c_i \,\}.$$

Then V is a Chebyshev affine set, and for each $x \in X$,

$$(7.15.1) \qquad P_V(x) = x - \sum_1^n \alpha_i x_i,$$

where the scalars α_i satisfy the (modified) normal equations

$$(7.15.2) \qquad \sum_{i=1}^n \alpha_i \langle x_i, x_j \rangle = \langle x, x_j \rangle - c_j \qquad (j = 1, 2, \ldots, n),$$

and

$$(7.15.3) \qquad d(x, V) = \left[\sum_1^n \alpha_i (\langle x, x_i \rangle - c_i) \right]^{1/2}.$$

Moreover,

$$(7.15.4) \qquad d(x, V) = \begin{cases} \dfrac{\sum_1^n [\langle x, x_i \rangle - c_i]^2}{\| \sum_1^n [\langle x, x_i \rangle - c_i] x_i \|} & \text{if } x \notin V, \\ 0 & \text{if } x \in V. \end{cases}$$

In particular, if $\{x_1, x_2, \ldots, x_n\}$ is an orthonormal set, then for all $x \in X$,

$$(7.15.5) \qquad P_V(x) = x - \sum_1^n [\langle x, x_i \rangle - c_i] x_i$$

and

$$(7.15.6) \qquad d(x, V) = \left[\sum_1^n (\langle x, x_i \rangle - c_i)^2 \right]^{1/2}.$$

Proof. Since $\{x_1, x_2, \ldots, x_n\}$ is linearly independent, $g(x_1, x_2, \ldots, x_n) \neq 0$ by Lemma 7.5. This implies that there is an element $v \in \mathrm{span}\,\{x_1, x_2, \ldots, x_n\}$ that lies in V. In particular, $V \neq \emptyset$, and it is easy to verify that $V = M + v$, where

$$M = \bigcap_1^n \{y \in X \mid \langle y, x_i \rangle = 0\}.$$

By Theorem 7.13, M is a Chebyshev subspace of codimension n. The remainder of the proof follows immediately from Theorems 7.13 and 2.7. For example, V is Chebyshev and

$$P_V(x) = P_{M+v}(x) = P_M(x - v) + v = x - v - \sum_1^n \alpha_i x_i + v = x - \sum_1^n \alpha_i x_i,$$

where the α_i satisfy the equations

$$\sum_1^n \alpha_i \langle x_i, x_j \rangle = \langle x - v, x_j \rangle = \langle x, x_j \rangle - c_j \qquad (j = 1, 2, \ldots, n). \qquad \blacksquare$$

Using Theorem 7.15, we can now solve Problem 4 of Chapter 1.

Problem 4. (A control problem) The position θ of the shaft of a dc motor driven by a variable current source u is governed by the differential equation

$$(4.1) \qquad \theta''(t) + \theta'(t) = u(t), \qquad \theta(0) = \theta'(0) = 0,$$

where $u(t)$ is the field current at time t. Suppose that the boundary conditions are given by

$$(4.2) \qquad \theta(1) = 1, \qquad \theta'(1) = 0,$$

and the energy is proportional to $\int_0^1 u^2(t)dt$. Find the function u having minimum energy in the class of all real continuous functions on $[0,1]$ for which the system (4.1) and (4.2) has a solution θ.

In Chapter 2 we saw that this problem could be reformulated as follows: *If* $x_1(t) = e^{t-1}$, $x_2(t) = 1 - x_1(t)$, *and*

$$V = \{\, u \in C_2[0,1] \mid \langle u, x_1 \rangle = 0, \quad \langle u, x_2 \rangle = 1 \,\},$$

find $P_V(0)$.

By Theorem 7.15, $P_V(0) = -(\alpha_1 x_1 + \alpha_2 x_2)$, where the α_i satisfy

$$\alpha_1 \langle x_1, x_1 \rangle + \alpha_2 \langle x_2, x_1 \rangle = 0,$$
$$\alpha_1 \langle x_1, x_2 \rangle + \alpha_2 \langle x_2, x_2 \rangle = -1.$$

After a bit of algebra, we obtain

$$\alpha_1 = \frac{e-1}{3-e} \quad \text{and} \quad \alpha_2 = -\frac{1+e}{3-e}.$$

Hence

$$P_V(0) = \frac{1-e}{3-e} x_1 + \frac{(1+e)}{3-e} x_2,$$

or

$$P_V(0)(t) = \frac{1}{3-e}(1 + e - 2e^t).$$

The Weierstrass Approximation Theorem

In this section we give an elementary proof of one of the most important approximation theorems known: the classical Weierstrass (pronounced "VYE-er-shtrahss") approximation theorem. It states that any real continuous function on a finite closed interval can be uniformly approximated by polynomials. A simple corollary is that the subspace of polynomials is dense in $C_2[a,b]$.

The following simple inequality will be used a few times in the proof.

7.16 Bernoulli's Inequality. *For any real number* $h \geq -1$,

$$(7.16.1) \qquad (1+h)^n \geq 1 + nh \quad (n = 1, 2, \dots).$$

Proof. We proceed by induction on n. For $n = 1$, the result is obvious. Assuming that (7.16.1) holds for some $n \geq 1$, we see that

$$(1+h)^{n+1} = (1+h)(1+h)^n \geq (1+h)(1+nh) = 1 + (n+1)h + nh^2 \geq 1 + (n+1)h.$$

Thus the result is valid for $n + 1$. $\blacksquare$

7.17 Lemma. *For each $n \in \mathbb{N}$, the function*

$$q_n(t) = (1 - t^n)^{2^n}, \quad 0 \le t \le 1,$$

is a polynomial on $[0,1]$ with the properties that $0 \le q_n \le 1$, $q_n(t) \to 1$ uniformly on $[0,c]$ for any $0 < c < \frac{1}{2}$, and $q_n(t) \to 0$ uniformly on $[d,1]$ for any $\frac{1}{2} < d < 1$.

Proof. Clearly, q_n is a polynomial that is decreasing on $[0,1]$, $q_n(0) = 1$, and $q_n(1) = 0$. Thus $0 \le q_n \le 1$. Fix any $c \in (0, \frac{1}{2})$ and $d \in (\frac{1}{2}, 1)$.

For each $0 \le t \le c$, we use Bernoulli's inequality to obtain

$$1 \ge q_n(t) \ge q_n(c) = (1 - c^n)^{2^n} \ge 1 - 2^n c^n = 1 - (2c)^n \to 1 \quad \text{as } n \to \infty.$$

Thus $q_n \to 1$ uniformly on $[0,c]$.

For any $d \le t \le 1$, we obtain $0 \le q_n(t) \le q_n(d)$ and, using Bernoulli's inequality,

$$\frac{1}{q_n(d)} = \left[\frac{1}{1 - d^n}\right]^{2^n} = \left[1 + \frac{d^n}{1 - d^n}\right]^{2^n} \ge 1 + \frac{2^n d^n}{1 - d^n} > (2d)^n \to \infty \quad \text{as } n \to \infty.$$

This proves that $q_n \to 0$ uniformly on $[d,1]$. ∎

7.18 Lemma. *Let s be the step function defined on $[-1,1]$ by*

$$s(t) := \begin{cases} 0 & \text{if } -1 \le t < 0, \\ 1 & \text{if } \;\; 0 \le t \le 1. \end{cases}$$

Then for each $\epsilon > 0$ and $\rho > 0$, there exists a polynomial q such that $0 \le q(t) \le 1$ for all $|t| \le 1$, $|s(t) - q(t)| < \epsilon$ if $\rho \le |t| \le 1$, and $|s(t) - q(t)| \le 1$ if $|t| < \rho$.

Proof. Let q_n be the polynomials defined in Lemma 7.17. Define p_n on $[-1,1]$ by $p_n(t) = q_n(\frac{1-t}{2})$. It follows from Lemma 7.17 that $0 \le p_n \le 1$, $p_n \to 1$ uniformly on $[\rho, 1]$, and $p_n \to 0$ uniformly on $[-1, -\rho]$. Thus for n sufficiently large,

$$|s(t) - p_n(t)| < \epsilon \quad \text{on } [-1, -\rho] \cup [\rho, 1]$$

and

$$|s(t) - p_n(t)| \le 1 \quad \text{for all } t.$$

Thus $q = p_n$ works for n sufficiently large. ∎

7.19 Weierstrass Approximation Theorem. *Let f be a real continuous function on $[a,b]$. Then for each $\epsilon > 0$, there exists a polynomial p such that*

$$(7.19.1) \qquad\qquad \sup_{t \in [a,b]} |f(t) - p(t)| < \epsilon.$$

Proof. Assume first that $[a,b] = [0,1]$. By replacing f with $f - f(0)$, we may assume $f(0) = 0$. We may also assume $\epsilon < 1$. Since f is continuous on $[0,1]$, it is uniformly continuous by Theorem 5.4. Hence we may choose a partition

$$0 = t_0 < t_1 < t_2 < \cdots < t_m < t_{m+1} = 1$$

such that $|f(t) - f(t')| < \epsilon/3$ whenever $t, t' \in [t_{j-1}, t_j]$ ($j = 1, 2, \ldots, m+1$). Choose any $\rho > 0$ such that $2\rho < \min_{1 \le i \le m+1}(t_i - t_{i-1})$. Let $\epsilon_j = f(t_j) - f(t_{j-1})$

$(j = 1, 2, \ldots, m+1)$, $g(t) = \sum_1^{m+1} \epsilon_j s(t - t_j)$, and $p(t) = \sum_1^{m+1} \epsilon_j q(t - t_j)$, where q is the polynomial whose existence is guaranteed by Lemma 7.18 with ϵ replaced by $\epsilon/(3m)$.

Then p is a polynomial and $|\epsilon_j| < \epsilon/3$ for all j. If $t \in [t_{i-1}, t_i)$, then $s(t - t_j) = 1$ for all $j \leq i - 1$ and $s(t - t_j) = 0$ if $j \geq i$. Hence

$$|f(t) - g(t)| = \left| f(t) - \sum_1^{m+1} \epsilon_j s(t - t_j) \right| = \left| f(t) - \sum_1^{i-1} \epsilon_j \right| = |f(t) - f(t_{i-1})| < \frac{\epsilon}{3}.$$

Also, $|f(t_{m+1}) - g(t_{m+1})| = |f(t_{m+1}) - f(t_{m+1})| = 0$.

If $t \in (t_k - \rho, t_k + \rho)$ for some k, then, using Lemma 7.18, we obtain

$$|p(t) - g(t)| = \left| \sum_{j=1}^{m+1} \epsilon_j [q(t - t_j) - s(t - t_j)] \right|$$

$$\leq \sum_{j \neq k} |\epsilon_j| |q(t - t_j) - s(t - t_j)| + |\epsilon_k| \cdot 1$$

$$\leq \sum_{j \neq k} \frac{\epsilon}{3} \cdot \frac{\epsilon}{3m} + \frac{\epsilon}{3} = \frac{\epsilon^2}{9} + \frac{\epsilon}{3} < \frac{2\epsilon}{3}.$$

If $t \notin (t_k - \rho, t_k - \rho)$ for any k, then a slight modification of the above argument shows that

$$|p(t) - g(t)| < \frac{\epsilon}{3} \cdot \frac{\epsilon(m+1)}{3m} < \frac{\epsilon}{3}.$$

Thus, for any $t \in [0, 1]$,

$$|f(t) - p(t)| \leq |f(t) - g(t)| + |g(t) - p(t)| < \epsilon.$$

This proves (7.19.1) when $[a, b] = [0, 1]$. In the general case, if f is continuous on $[a, b]$, define F on $[0, 1]$ by $F(t) = f((b - a)t + a)$. Then F is continuous on $[0, 1]$, so by the first part of the proof, given $\epsilon > 0$, there exists a polynomial P such that $|F(t) - P(t)| < \epsilon$ for all $t \in [0, 1]$. Hence $|f((b - a)t + a) - P(t)| < \epsilon$ for $t \in [0, 1]$. Making the change of variable $u = (b - a)t + a$, we get that

$$|f(u) - p(u)| < \epsilon \quad \text{for all } u \in [a, b],$$

where $p(u) = P(\frac{u-a}{b-a})$ is a polynomial in u. ∎

7.20 Corollary. *The polynomials are dense in* $C_2[a, b]$.

Proof. Given $f \in C_2[a, b]$ and $\epsilon > 0$, Theorem 7.19 allows us to choose a polynomial p such that

$$|f(t) - p(t)| < \frac{\epsilon}{(b-a)^{1/2}} \quad \text{for all } t \in [a, b].$$

Then

$$\|f - p\|^2 = \int_a^b |f(t) - p(t)|^2 dt < \int_a^b \frac{\epsilon^2}{b - a} dt = \epsilon^2. \quad \blacksquare$$

An inner product space X is called **separable** if it contains a countable dense subset.

7.21 Proposition.

(1) *Every finite-dimensional inner product space is separable.*

(2) *$\ell_2(\Gamma)$ is separable if and only if Γ is countable.*

(3) *$C_2[a, b]$ is separable.*

In particular, $\ell_2(n)$ and ℓ_2 are separable, but $\ell_2(\mathbb{R})$ is not separable.

Proof. (1) Let X be a finite-dimensional inner product space. Choose an orthonormal basis $\{e_1, e_2, \ldots, e_n\}$ of X. Then, by (4.10.3) or Corollary 7.8, each $x \in X$ has the representation $x = \sum_1^n \langle x, e_i \rangle e_i$.

Claim: $D := \{\sum_1^n r_i e_i \mid r_i \in \mathbb{Q}\}$ is a countable dense set in X, where $\mathbb{Q}$ denotes the set of rational numbers.

First recall that $\mathbb{Q}$ is a countable dense set in $\mathbb{R}$. Also, there is a one-to-one correspondence of D with $\mathbb{Q}^n := \mathbb{Q} \times \mathbb{Q} \times \cdots \times \mathbb{Q}$ via the mapping $\sum_1^n r_i e_i \mapsto (r_1, r_2, \ldots, r_n)$. Since $\mathbb{Q}^n$ is a finite product of countable sets, it is countable. Hence D is countable.

If $x \in X$ and $\epsilon > 0$, choose rational numbers r_i $(i = 1, 2, \ldots, n)$ such that $|\langle x, e_i \rangle - r_i| < \epsilon/n$. Then

$$\left\| x - \sum_1^n r_i e_i \right\| = \left\| \sum_1^n (\langle x, e_i \rangle - r_i) e_i \right\| \leq \sum_1^n \|(\langle x, e_i \rangle - r_i) e_i\| = \sum_1^n |\langle x, e_i \rangle - r_i| < \epsilon,$$

so D is dense in X.

(2) For each $\gamma \in \Gamma$, define e_γ by $e_\gamma(i) = 0$ if $i \in \Gamma \setminus \{\gamma\}$ and $e_\gamma(\gamma) = 1$. Then $\{e_\gamma \mid \gamma \in \Gamma\}$ is an orthonormal set in $\ell_2(\Gamma)$. In particular, $\|e_\gamma - e_{\gamma'}\| = \sqrt{2}$ if $\gamma \neq \gamma'$. If D is a dense set in $\ell_2(\Gamma)$, then, for every $\gamma \in \Gamma$, there exists $y_\gamma \in D$ such that $\|e_\gamma - y_\gamma\| < \sqrt{2}/2$. If $\gamma \neq \gamma'$, we have

$$\|y_\gamma - y_{\gamma'}\| = \|y_\gamma - e_\gamma + e_\gamma - e_{\gamma'} + e_{\gamma'} - y_{\gamma'}\|$$
$$\geq \|e_\gamma - e_{\gamma'}\| - \|(y_\gamma - e_\gamma) + (e_{\gamma'} - y_{\gamma'})\|$$
$$\geq \sqrt{2} - (\|y_\gamma - e_\gamma\| + \|e_{\gamma'} - y_{\gamma'}\|) > 0.$$

That is, $y_\gamma \neq y_{\gamma'}$. It follows that D contains at least as many elements as Γ. Hence if Γ is uncountable, so is D. Thus $\ell_2(\Gamma)$ is not separable.

Note that every $x \in \ell_2(\Gamma)$ may obviously be written in the form $x = \sum_{\gamma \in \Gamma} \langle x, e_\gamma \rangle e_\gamma$, since the sum on the right clearly has the same value as x (namely, $x(\gamma) = \langle x, e_\gamma \rangle$) at each point of Γ.

If Γ is finite, then $\ell_2(\Gamma)$ is finite-dimensional and hence separable by (1). If Γ is countably infinite, say $\Gamma = \{\gamma_n \mid n = 1, 2, \ldots\}$, then for every $x \in \ell_2(\Gamma)$, we have $x = \lim_{n \to \infty} \sum_{i=1}^n \langle x, e_{\gamma_i} \rangle e_{\gamma_i}$. It follows that the set $D := \cup_1^\infty D_n$, where $D_n := \{\sum_1^n r_i e_{\gamma_i} \mid r_i \in \mathbb{Q}\}$, is dense in $\ell_2(\Gamma)$. Since each D_n is countable (by the same proof as in (1)) and D is the union of a countable number of countable sets, D must also be countable. Hence $\ell_2(\Gamma)$ is separable.

(3) To prove (3), we will just outline the steps and leave to the reader the verification of the details. First, the set of all polynomials $\mathcal{P}$ is dense in $C_2[a, b]$ by Corollary 7.20. Then observe that the set of all polynomials with rational coefficients, $\mathcal{P}_\mathbb{Q}$, is dense in $\mathcal{P}$, hence dense in $C_2[a, b]$. Finally, the set $\mathcal{P}_\mathbb{Q}$ is countable. Thus $C_2[a, b]$ is separable. ∎

Müntz's Theorem

For the remainder of the chapter we will study an interesting generalization of Corollary 7.20 due to Müntz. Recall that Corollary 7.20 implies that the space of polynomials

$$\mathcal{P} = \text{span}\{1, t, t^2, \ldots, t^n, \ldots\}$$

forms a dense subspace in $C_2[0,1]$. A natural and interesting question is whether there are *smaller* subspaces of $\mathcal{P}$ that are also dense in $C_2[0,1]$.

For example, what can we say about the subspace

$$\mathcal{D} = \text{span}\{1, t^2, t^3, t^5, t^7, t^{11}, \ldots, t^{r_n}, \ldots\},$$

where r_n denotes the nth prime number, or the subspace

$$\mathcal{B} = \text{span}\{1, t^2, t^4, t^8, \ldots, t^{2^n}, \ldots\}?$$

(Müntz's theorem will imply that the former is dense in $C_2[0,1]$, but the latter is not.)

Fix any sequence of real numbers $\{r_n\}$ (not necessarily integers) with

$$0 \le r_1 < r_2 < \cdots < r_n < \cdots$$

and $r_n \to \infty$, and let

$$\mathcal{M} := \text{span}\{t^{r_1}, t^{r_2}, \ldots, t^{r_n}, \ldots\}.$$

What are necessary and sufficient conditions on the exponents $\{r_n\}$ to ensure that $\mathcal{M}$ is dense in $C_2[0,1]$? The elegant answer to this question is the theorem of Müntz, which states that $\mathcal{M}$ is dense in $C_2[0,1]$ if and only if $\sum_2^\infty (1/r_n) = \infty$.

One intriguing aspect of this result is that it connects two seemingly unrelated properties. Namely, it connects the topological property of denseness with the number-theoretic property of the divergence of the sum of reciprocal exponents.

To simplify the proof of Müntz's theorem, we divide the steps into four lemmas each of which has independent interest.

7.22 Cauchy Determinant Formula. Fix an integer $n \ge 1$, and let $a_i, b_i, (i = 1, 2, \ldots, n)$ be $2n$ real numbers such that $a_i + b_j \ne 0$ for all i and j. Define

$$D_n := \begin{vmatrix} \dfrac{1}{a_1 + b_1} & \dfrac{1}{a_1 + b_2} & \cdots & \dfrac{1}{a_1 + b_n} \\[2mm] \dfrac{1}{a_2 + b_1} & \dfrac{1}{a_2 + b_2} & \cdots & \dfrac{1}{a_2 + b_n} \\[2mm] \cdots & & & \\[2mm] \dfrac{1}{a_n + b_1} & \dfrac{1}{a_n + b_2} & \cdots & \dfrac{1}{a_n + b_n} \end{vmatrix}.$$

Then

(7.22.1)
$$D_n = \frac{\prod_{1 \le i < j \le n}(a_j - a_i)(b_j - b_i)}{\prod_{i,j=1}^n (a_i + b_j)}.$$

Proof. The proof is by induction. For $n = 1$, $D_1 = 1/(a_1 + b_1)$. This is also the value of the the right side of (7.22.1) when $n = 1$ (since a product over the empty set of indices is 1 by definition). Next assume that (7.22.1) is valid when n is replaced by $n - 1$. Then subtract the last row of D_n from each of the other rows and remove common factors. This yields

$$D_n = \frac{\prod_{i=1}^{n-1}(a_n - a_i)}{\prod_{j=1}^{n}(a_n + b_j)} \begin{vmatrix} \dfrac{1}{a_1 + b_1} & \dfrac{1}{a_1 + b_2} & \cdots & \dfrac{1}{a_1 + b_n} \\ \cdots & & & \\ \dfrac{1}{a_{n-1} + b_1} & \dfrac{1}{a_{n-1} + b_2} & \cdots & \dfrac{1}{a_{n-1} + b_n} \\ 1 & 1 & \cdots & 1 \end{vmatrix}.$$

Now subtract the last column from each of the others and remove common factors. This yields

$$D_n = \frac{\prod_{i=1}^{n-1}(a_n - a_i)(b_n - b_i)}{\prod_{j=1}^{n}(a_n + b_j)\prod_{i=1}^{n-1}(a_i + b_n)} \begin{vmatrix} \dfrac{1}{a_1 + b_1} & \cdots & \dfrac{1}{a_1 + b_{n-1}} & 1 \\ \cdots & & & \\ \dfrac{1}{a_{n-1} + b_1} & \cdots & \dfrac{1}{a_{n-1} + b_{n-1}} & 1 \\ 0 & \cdots & 0 & 1 \end{vmatrix}$$

$$= \frac{\prod_{i=1}^{n-1}(a_n - a_i)(b_n - b_i)}{\prod_{j=1}^{n}(a_n + b_j)\prod_{i=1}^{n-1}(a_i + b_n)} D_{n-1} = \frac{\prod_{1 \le i < j \le n}(a_j - a_i)(b_j - b_i)}{\prod_{i,j=1}^{n}(a_i + b_j)},$$

which verifies (7.22.1). This completes the induction and the proof. ■

Let $\{r_n\}$ be a sequence of numbers satisfying

$$(7.22.2) \qquad 0 \le r_1 < r_2 < \cdots < r_n < \cdots, \qquad \lim_{n \to \infty} r_n = \infty,$$

and define for each $n \ge 1$,

$$(7.22.3) \qquad \mathcal{M}_n := \text{span}\{t^{r_1}, t^{r_2}, \ldots, t^{r_n}\},$$

and

$$(7.22.4) \qquad \mathcal{M} := \cup_1^\infty \mathcal{M}_n = \text{span}\{t^{r_1}, t^{r_2}, \ldots, t^{r_n}, \ldots\}.$$

7.23 Distance from t^m to $\mathcal{M}_n$. Let $\mathcal{M}_n$ be the subspace defined as in (7.22.3). Then for any integer $m \ge 0$,

$$(7.23.1) \qquad d(t^m, \mathcal{M}_n) = \frac{1}{\sqrt{2m+1}} \prod_{j=1}^{n} \frac{|r_j - m|}{r_j + m + 1}.$$

Proof. By Theorem 7.6, we have that

$$(7.23.2) \qquad d(t^m, \mathcal{M}_n)^2 = \frac{g(t^{r_1}, t^{r_2}, \ldots, t^{r_n}, t^m)}{g(t^{r_1}, t^{r_2}, \ldots, t^{r_n})}.$$

Since

$$\langle t^r, t^s \rangle = \int_0^1 t^r t^s \, dt = \frac{1}{r+s+1},$$

it follows that

$$g(t^{r_1}, t^{r_2}, \ldots, t^{r_n}) = \begin{vmatrix} \dfrac{1}{r_1+r_1+1} & \dfrac{1}{r_1+r_2+1} & \cdots & \dfrac{1}{r_1+r_n+1} \\[2mm] \dfrac{1}{r_2+r_1+1} & \dfrac{1}{r_2+r_2+1} & \cdots & \dfrac{1}{r_2+r_n+1} \\[2mm] \cdots & & & \\[2mm] \dfrac{1}{r_n+r_1+1} & \dfrac{1}{r_n+r_2+1} & \cdots & \dfrac{1}{r_n+r_n+1} \end{vmatrix}.$$

Applying Proposition 7.22 (with $a_i = b_i = r_i + \frac{1}{2}$), we obtain

$$g(t^{r_1}, t^{r_2}, \ldots, t^{r_n}) = \frac{\prod_{1 \le i < j \le n} (r_j - r_i)^2}{\prod_{i,j=1}^n (r_i + r_j + 1)}.$$

Similarly,

$$g(t^{r_1}, t^{r_2}, \ldots, t^{r_n}, t^m) = \frac{\prod_{1 \le i < j \le n} (r_j - r_i)^2 \prod_{k=1}^n (m - r_k)^2}{(2m+1) \prod_{i,j=1}^n (r_i + r_j + 1) \prod_{k=1}^n (m + r_k + 1)^2}.$$

Substituting these expressions into (7.23.2), cancelling the common factors, and taking the square root, we obtain (7.23.1). ∎

The next lemma relates the divergence of an infinite product with the divergence of an infinite series.

7.24 Lemma. *Suppose $0 < a_n \neq 1$ for every n and $a_n \to 0$. Then*

$$(7.24.1) \qquad \prod_1^\infty (1 - a_i) = 0 \qquad (\text{i.e., } \prod_1^n (1 - a_i) \to 0)$$

if and only if

$$(7.24.2) \qquad \sum_1^\infty a_i = \infty \qquad (\text{i.e., } \sum_1^n a_i \to \infty).$$

Proof. By omitting finitely many of the a_i's, we may assume that $a_n < 1$ for all n. Using L'Hospital's rule,

$$\lim_{t \to 0} \frac{\log(1-t)}{t} = \lim_{t \to 0} \frac{-1}{1-t} = -1.$$

Thus for each $0 < \epsilon < 1$, there exists an integer N such that

$$-1 - \epsilon < \frac{\log(1 - a_n)}{a_n} < -1 + \epsilon,$$

or

$$-(1 + \epsilon)a_n < \log(1 - a_n) < (-1 + \epsilon)a_n,$$

for all $n \geq N$. This implies that $\sum_1^n a_i \to \infty$ if and only if $\sum_1^n \log(1 - a_i) \to -\infty$. But $\log \prod_1^n (1 - a_i) = \sum_1^n \log(1 - a_i) \to -\infty$ if and only if $\prod_1^n (1 - a_i) \to 0$. This completes the proof. ∎

Our last lemma shows that the denseness of $\mathcal{M}$ is equivalent to being able to approximate the monomial t^m for each integer $m \geq 0$.

7.25 Lemma. *Let the exponents $\{r_n\}$ and the subspace $\mathcal{M}$ of $C_2[0, 1]$ be given as in (7.22.2) and (7.22.4), and let $\mathcal{M}_n$ be the subspace defined as in (7.22.3). Then $\mathcal{M}$ is dense in $C_2[0, 1]$ if and only if*

$$(7.25.1) \qquad\qquad \lim_{n \to \infty} d(t^m, \mathcal{M}_n) = 0$$

for each integer $m \geq 0$ with $m \notin \{r_1, r_2, \ldots, r_n, \ldots\}$.

Proof. Let $\mathcal{M}$ be dense in $C_2[0, 1]$, m a nonnegative integer, and $x(t) = t^m$. Given any $\epsilon > 0$, choose an element $y \in \mathcal{M}$ such that $\|x - y\| < \epsilon$. Since $\mathcal{M} = \cup_1^\infty \mathcal{M}_n$ and $\mathcal{M}_n \subset \mathcal{M}_{n+1}$ for all n, $y \in \mathcal{M}_n$ for n sufficiently large. Hence for n large,

$$d(x, \mathcal{M}_n) \leq \|x - y\| < \epsilon.$$

This verifies (7.25.1).

Now suppose (7.25.1) holds for each integer $m \geq 0$ with $m \neq r_i$ for every i. If $m = r_i$ for some i, then $t^m \in \mathcal{M}_n$ for all $n \geq i$, so (7.25.1) holds for this case also. That is, (7.25.1) holds for every nonnegative integer m. Next let p be any polynomial. Then $p(t) = \sum_0^k \alpha_i t^i$ for some scalars α_i, and using Theorem 2.7 (2), we see that
(7.25.2)

$$d(p, \mathcal{M}_n) = d\left(\sum_0^k \alpha_i t^i, \mathcal{M}_n\right) \leq \sum_0^k d(\alpha_i t^i, \mathcal{M}_n) = \sum_0^k |\alpha_i| d(t^i, \mathcal{M}_n) \to 0$$

as $n \to \infty$ by (7.25.1). Finally, let $x \in C_2[0, 1]$ be arbitrary and $\epsilon > 0$. By Corollary 7.20 there exists a polynomial p such that $\|x - p\| < \epsilon/2$. By (7.25.2) there exists an integer N such that $d(p, \mathcal{M}_n) < \epsilon/2$ for all $n \geq N$. Using Theorem 5.3, it follows that for $n \geq N$,

$$d(x, \mathcal{M}) \leq d(x, \mathcal{M}_n) \leq \|x - p\| + d(p, \mathcal{M}_n) < \epsilon.$$

This proves that $\mathcal{M}$ is dense in $C_2[0, 1]$. ∎

7.26 Müntz's Theorem. *Let $\{r_n\}$ be a sequence of real numbers with*

$$0 \le r_1 < r_2 < \cdots < r_n \to \infty,$$

let $\mathcal{M}_n$ be defined by

$$\mathcal{M}_n = \mathrm{span}\{t^{r_1}, t^{r_2}, \ldots, t^{r_n}\},$$

and let

$$\mathcal{M} := \bigcup_1^\infty \mathcal{M}_n = \mathrm{span}\{t^{r_1}, t^{r_2}, \ldots, t^{r_n}, \ldots\}.$$

Then the following statements are equivalent:

(1) $\mathcal{M}$ *is dense in* $C_2[0,1]$;

(2) $\lim_{n\to\infty} d(t^m, \mathcal{M}_n) = 0$ *for each nonnegative integer* $m \notin \{r_1, r_2, \ldots\}$;

(3) $\lim_{n\to\infty} d(t^m, \mathcal{M}_n) = 0$ *for some nonnegative integer* $m \notin \{r_1, r_2, \ldots\}$;

(4) $\sum_{n=2}^\infty (1/r_n) = \infty$.

Proof. The equivalence of (1) and (2) is just Lemma 7.25, while the implication (2) $\Longrightarrow$ (3) is obvious.

(3) $\Longrightarrow$ (4). Suppose $\lim_{n\to\infty} d(t^m, \mathcal{M}_n) = 0$ for some nonnegative integer m not in $\{r_1, r_2, \ldots\}$. Now, for each integer $n \ge 1$,

$$(7.26.1) \qquad \prod_1^n \frac{|r_j - m|}{r_j + m + 1} = \prod_1^n |1 - a_j|,$$

where $a_j = (2m + 1)/(r_j + m + 1)$. Note that $0 < a_j \ne 1$ for every j and $a_j \to 0$. By Lemma 7.23 and (7.26.1), we see that

$$(7.26.2) \qquad \lim_{n\to\infty} \prod |1 - a_j| = 0.$$

By Lemma 7.24, (7.26.2) is equivalent to

$$(7.26.3) \qquad \sum_{j=1}^\infty a_j = \infty.$$

That is,

$$(7.26.4) \qquad \sum_{j=1}^\infty \frac{1}{r_j + m + 1} = \infty.$$

For j sufficiently large, $r_j > m + 1$, so that

$$(7.26.5) \qquad \frac{1}{2r_j} < \frac{1}{r_j + m + 1} < \frac{1}{r_j}.$$

By the comparison test for infinite series, it follows that (7.26.4) holds if and only if (4) holds.

(4) $\Longrightarrow$ (2). Suppose that $\sum_{j=2}^\infty 1/r_j = \infty$. Then for every nonnegative integer m not in $\{r_1, r_2, \ldots\}$, (7.26.5) holds for j sufficiently large. By retracing the steps of the proof of the implication (3) $\Longrightarrow$ (4), we deduce that

$$\lim_{n\to\infty} \prod_{j=1}^n \frac{|r_j - m|}{r_j + m + 1} = 0.$$

By Lemma 7.23, this implies $\lim_{n\to\infty} d(t^m, \mathcal{M}_n) = 0$. This proves (2) and completes the proof. ∎

7.27 Remark. While the Weierstrass approximation theorem (Corollary 7.20) implies that the polynomials are dense in $C_2[a, b]$ for *any* interval $[a, b]$, Müntz's theorem is *not* generally valid in $C_2[a, b]$. For example, consider the subspace of $C_2[-1, 1]$ defined by

$$M := \text{span}\{1, t^2, t^4, \ldots, t^{2n}, \ldots\}.$$

In particular, all functions in M are *even*. Also, $\sum_1^\infty 1/(2n) = \infty$. However, taking $x(t) = t$ and any $p \in M$, we see that p is an even function so that

$$
\begin{aligned}
\|x - p\|^2 &= \int_{-1}^1 [x(t) - p(t)]^2 dt = \int_{-1}^0 [t - p(t)]^2 dt + \int_0^1 [t - p(t)]^2 dt \\
&= \int_0^1 [-t - p(-t)]^2 dt + \int_0^1 [t - p(t)]^2 dt \\
&= \int_0^1 [t + p(t)]^2 dt + \int_0^1 [t - p(t)]^2 dt \\
&= 2 \int_0^1 [t^2 + p^2(x)] dt \geq 2 \int_0^1 t^2 dt = \frac{2}{3}.
\end{aligned}
$$

Hence $d(x, M)^2 \geq \frac{2}{3}$ so that M is not dense in $C_2[-1, 1]$.

7.28 Examples. (1) The subspace of all polynomials

$$\mathcal{P} = \text{span}\{1, t, t^2, \ldots, t^n, \ldots\}$$

is dense in $C_2[0, 1]$, since $\sum_1^\infty 1/n = \infty$. Thus we recover Corollary 7.20 again. However, keep in mind that Corollary 7.20 was used in the proof of the Müntz theorem. It would be of some interest to give a direct proof of Müntz's theorem that does *not* use the Weierstrass approximation theorem.

(2) $\text{span}\{t^2, t^4, \ldots, t^{2n}, \ldots\}$ is dense in $C_2[0, 1]$, since $\sum_1^\infty 1/(2n) = \infty$.

(3) $\mathcal{M} := \text{span}\{1, t^2, t^4, t^8, \ldots, t^{2^n}, \ldots\}$ is *not* dense in $C_2[0, 1]$ since $\sum_1^\infty 1/(2^n) = 1 < \infty$. In fact, by Theorem 7.26, if m is *any* positive integer such that $m \neq 2^n$ for any integer n, then t^m *cannot* be approximated arbitrarily well by elements of $\mathcal{M}$.

Exercises

1. Verify that formula (7.1.1) (respectively (7.1.2)) is also valid when $x \in K$, provided that the maximum is taken over all $x^* \in X^*$ with $\|x^*\| \leq 1$ (respectively $z \in X$ with $\|z\| \leq 1$) rather than just $\|x^*\| = 1$ (respectively $\|z\| = 1$).

2. Let $X = l_2(2)$ and

$$K = \{y \in X \mid y(2) \geq e^{-y(1)}\}.$$

   In other words, K is the "epigraph" of the function $t \mapsto e^{-t}$.
   (a) Show that K is convex and closed.
   (b) For $z = (0, -1)$, show that $\sup_{y \in K} \langle y, z \rangle = 0$, yet $\langle y, z \rangle < 0$ for all $y \in K$.
   (c) Show that the "supremum" (respectively "infimum") in (7.1.1) or (7.1.2)

(respectively (7.2.2)) *cannot* be replaced by "maximum" (respectively "minimum"). [Hint: Part (b).]

3. Let $\{x_1, x_2, \ldots, x_n\}$ be a basis for the n-dimensional subspace M. Show that that expression

$$\frac{g(x_1, x_2, \ldots, x_n, x)}{g(x_1, x_2, \ldots, x_n)}$$

depends *only* on x and not the particular basis chosen for M.

4. Give an alternative proof of Lemma 7.10 by showing that L contains a set of n linearly independent vectors if and only if N does.

5. Let $\{x_1^*, x_2^*, \ldots, x_n^*\}$ be a linearly independent set in X^* and

$$M = \bigcap_1^n \{x \in X \mid x_i^*(x) = 0\}.$$

Verify the following distance formula:

$$d(x, M) = \max \left\{ \sum_1^n \alpha_i x_i^*(x) \,\middle|\, \alpha_i \in \mathbb{R} \ (i = 1, 2, \ldots, n), \ \left\| \sum_1^n \alpha_i x_i^* \right\| = 1 \right\}$$

for every $x \in X$. How does this generalize formula (7.13.3)?

6. Find the best approximation to
   (a) $x(t) = e^t$
   and
   (b) $x(t) = \sin t$
   from the set

$$V = \{y \in C_2[0,1] \mid \langle y, x_1 \rangle = 1, \ \langle y, x_2 \rangle = 2\},$$

   where $x_1(t) = 1$ and $x_2(t) = t$.

7. Let $\{x_1^*, x_2^*, \ldots, x_n^*\}$ be linearly independent in X^*, $\{c_1, c_2, \ldots, c_n\} \subset \mathbb{R}$, and

$$V = \bigcap_1^n \{y \in X \mid x_i^*(y) = c_i\}.$$

   Show that:
   (a) V is a nonempty closed affine set.
   (b) $V = M + v$, where

$$M = \bigcap_1^n \{y \in X \mid x_i^*(y) = 0\}$$

   and $v \in V$ is arbitrary.
   (c) V is Chebyshev if and only if each x_i^* has a representer in X.
   [Hint: Theorem 7.12.]

8. Let $\{x_1, x_2, \ldots, x_n\}$ be linearly independent in X, $\{c_1, c_2, \ldots, c_n\} \subset \mathbb{R}$, and

$$V = \bigcap_1^n \{y \in X \mid \langle y, x_i \rangle = c_i\}.$$

(a) Prove that the element of smallest norm in V is given by $v_0 = \sum_1^n \alpha_i x_i$, where the scalars α_i are chosen such that

$$\langle v_0, x_j \rangle = c_j \qquad (j = 1, 2, \ldots, n).$$

(b) Show also that

$$\min_{v \in V} \|v\| = \frac{\sum_1^n c_i^2}{\| \sum_1^n c_i x_i \|}.$$

9. Let M be a subspace of the inner product space X.

(a) Show that M is dense if and only if $x^* = 0$ whenever $x^* \in X^*$ and $x^*(y) = 0$ for all $y \in M$. Now suppose that X is a Hilbert space.

(b) Show that M is dense in X if and only if $M^\perp = \{0\}$. That is, $\langle x, y \rangle = 0$ for all $y \in M$ implies $x = 0$.

(c) Give an example to show that if X is not complete, then (b) is false. [Hint: Let $x^* \in X^*$ have no representer and set $M = \{x \in X \mid x^*(x) = 0\}$. If $M^\perp \neq \{0\}$, some point off M has a best approximation in M. Deduce that M is Chebyshev, which contradicts x^* not having a representer.]

10. Consider the subspace

$$M = \operatorname{span} \{1, t^2, t^4, \ldots, t^{2n}, \ldots\}$$

of $C_2[-1, 1]$.

(a) Prove that if x is an *even* function in $C_2[-1, 1]$, then x can be approximated arbitrarily well by elements in M. [Hint: Reduce this to a problem in $C_2[0, 1]$.]

(b) Prove that if x is an *odd* function (not identically zero), then x can *not* be approximated arbitrarily well by elements of M. [Hint: Look at Remark 7.27.]

11. Which of the following subspaces are dense in $C_2[0, 1]$? Justify your answers.

(a) $\operatorname{span} \{t, t^3, t^5, \ldots, t^{2n-1}, \ldots\}$.

(b) $\operatorname{span} \{t^\alpha, t^{\alpha+1}, t^{\alpha+2}, \ldots, t^{\alpha+n}, \ldots\}$, where $\alpha = 10^{100}$. Are there any values of $\alpha > 0$ that change your answer?

(c) $\operatorname{span} \{t, t^4, t^9, \ldots, t^{n^2}, \ldots\}$.

(d) $\operatorname{span} \{t, t^{\sqrt{2}}, t^{\sqrt{3}}, \ldots, t^{\sqrt{n}}, \ldots\}$.

(e) $\operatorname{span} \{t^{p_1}, t^{p_2}, \ldots, t^{p_n} \ldots\}$, where p_n is the nth prime number. [Hint: You may use the prime number theorem, which implies that $\lim_{n \to \infty} (n \log n)/p_n = 1$.]

12. Unlike the proof of the Weierstrass approximation theorem that we gave in Theorem 7.19, the "usual" proof uses Bernstein polynomials and is based on the following steps.

(a) Verify the "binomial theorem":

$$(t + s)^n = \sum_{k=0}^n \binom{n}{k} t^k s^{n-k}, \quad n = 1, 2, \ldots,$$

where s and t are any real numbers.

(b) For any $x \in C_2[0,1]$, the **Bernstein polynomial** of degree at most n for x is defined by

$$B_n x(t) := \sum_{k=0}^{n} \binom{n}{k} x\left(\frac{k}{n}\right) t^k (1-t)^{n-k}, \quad t \in [0,1].$$

Verify that $B_n x \in \mathcal{P}_n$, $B_n e_0 = e_0$ for $n \geq 0$, $B_n e_1 = e_1$ for $n \geq 1$, and $B_n e_2 = e_2 + (e_1/n)(e_0 - e_1)$ for $n \geq 2$, where $e_k(t) = t^k$.

(c) For any $x \in C_2[0,1]$ and $\epsilon > 0$, show that for n sufficiently large,

$$|B_n x(t) - x(t)| < \epsilon \quad \text{for all } t \in [0,1].$$

[Hint: Since x uniformly continuous, there exists $\delta = \delta(\epsilon) > 0$ such that $|x(t) - x(s)| < \epsilon/2$ whenever $|t - s| < \delta$. Choose any $n > \|x\|_\infty / \epsilon \delta^2$, where $\|x\|_\infty := \max_{0 \leq t \leq 1} |x(t)|$, and write

$$|x(t) - B_n x(t)| = \left| \sum_{k=0}^{n} \left[x(t) - x\left(\frac{k}{n}\right) \right] \binom{n}{k} t^k (1-t)^{n-k} \right|$$
$$\leq \sum_1 (t) + \sum_2 (t),$$

where

$$\sum_1 (t) := \sum_{\{k \mid |\frac{k}{n} - t| < \delta\}} \left| x(t) - x\left(\frac{k}{n}\right) \right| \binom{n}{k} t^k (1-t)^{n-k},$$

$$\sum_2 (t) := \sum_{\{k \mid |\frac{k}{n} - t| \geq \delta\}} \left| x(t) - x\left(\frac{k}{n}\right) \right| \binom{n}{k} t^k (1-t)^{n-k}.$$

Verify that $\sum_1 (t) < \epsilon/2$ and $\sum_2 (t) < \epsilon/2$ for any $t \in [0,1]$. To verify the second inequality, it may help to observe that

$$\sum_2 (t) \leq \frac{2\|x\|_\infty}{\delta^2} \sum_{k=0}^{n} \left(\frac{k}{n} - t \right)^2 \binom{n}{k} t^k (1-t)^{n-k},$$

and the latter sum may be evaluated using part (b).]

(d) For any $x \in C_2[0,1]$, $B_n x(t) \to x(t)$ uniformly on $[0,1]$.

(e) For any $x \in C_2[a,b]$, define $f(s) := x\big((b-a)s+a\big)$ for $s \in [0,1]$ and apply part (d) to f. From this we deduce that there is a sequence of polynomials that converges uniformly to x on $[a,b]$.

Historical Notes

The first statement of Theorem 7.1, in any normed linear space, can be found in Nirenberg (1961; p. 39). The distance formula (7.6.1) in terms of Gram determinants was given by Gram (1879) (see also Gram (1883)). The characterization of finite-codimensional Chebyshev subspaces (Theorem 7.12) was established by Deutsch (1982; Theorem 3.3).

The important and fundamental Weierstrass approximation theorem (Theorem 7.19, which we abbreviate WAT) was first proved by Weierstrass (1885) using a singular integral. He also deduced, as a consequence of this, that every 2π-periodic continuous function on the real line can be uniformly approximated by trigonometric polynomials. This is sometimes called the "second Weierstrass approximation theorem." Conversely, it can be shown that the second Weierstrass approximation theorem implies the WAT, so that they are logically equivalent. Runge (1885), (1885a) had all the ingredients for a proof of the WAT at his disposal. He showed that every continuous function on $[a, b]$ can be uniformly approximated arbitrarily well by rational functions (Runge (1885a)). But he had already established the tools in his 1885 paper that he could replace these rational functions by polynomials! Thus, either by overlooking this fact or not thinking it was important enough to include, Runge missed out on the opportunity to independently establish the WAT in the same year as Weierstrass.

Picard (1891) gave an alternative proof of the WAT, and a generalization to a continuous function of n variables. Lerch (1892) (see also (1903)) showed that the WAT could be deduced from Dirichlet's proof of a theorem of Fourier.

Lebesgue (1898) gave a proof of the WAT by reducing it to first uniformly approximating the function $f(t) = |t|$ on the interval $[-\sqrt{2}, \sqrt{2}]$ by polynomials. Then any piecewise linear continuous function on $[0, 1]$ can be uniformly approximated by polynomials, and finally, any continuous function on $[0, 1]$ can be uniformly approximated arbitrarily well by polynomials. Fejér (1900) gave a proof of the WAT by showing that the averages, or Cesaro sums, of the partial sums of the Fourier series of a continuous 2π-periodic function on the real line converge uniformly to the function. Mittag-Leffler (1900) gave a proof of the WAT that was similar to that of Runge and Lebesgue in the sense that the key step was reducing the problem to approximating piecewise linear functions. Landau (1908) gave a proof of the WAT based on singular integrals. Inspired by Landau's proof, de la Vallée-Poussin (1908) gave a proof of the second Weierstrass approximation theorem based on singular integrals.

Jackson (1911) showed that if f is a 2π-periodic Lipschitz continuous function, with Lipschitz constant λ, then the error

$$d(f, \mathcal{T}_n) := \inf_{T \in \mathcal{T}_n} \max_{t \in \mathbb{R}} |f(t) - T(t)|$$

is bounded above by a constant K_n that depends only on λ, and $K_n \to 0$ as $n \to \infty$. (In (1938), Achieser and Krein showed that $K_n = \pi\lambda/(2(n+1))$ is the best possible constant.)

Bernstein (1912), motivated by a probabilistic interpretation, gave a fairly elementary proof of the WAT by giving an explicit construction of the polynomials $B_n = B_n(f)$ such that B_n converges to f uniformly on $[0, 1]$. However, in practice, the Bernstein polynomials are rarely used to approximate functions, since their convergence is generally very slow. Indeed, the Bernstein polynomials $B_n(f)$ for the function $f(t) = t^2$ converge to f like $1/n$ (see, e.g., Davis (1963; pp. 116)). Korovkin (1953) (see also Korovkin (1960)) noted that the crux of Bernstein's proof was in verifying that (B_n) is a sequence of positive linear operators with the property that $(B_n f)$ converges to f uniformly on $[0, 1]$ whenever $f(t) = 1, t,$ or t^2. This led Korovkin to establish a few elegant generalizations, which, in turn, has spawned a whole area of so-called theorems of Korovkin type. Bohman (1952) had proved similar theorems a year earlier. But it was not until fairly recently that Bohman's contribution became widely known. Now most authors correctly refer to these results as "Bohman–Korovkin" theorems. One of the most elementary proofs of the WAT was given by Kuhn (1964). This is the proof given in this book (7.16–7.19) based on first approximating step functions. A nice history of the classical proofs of the WAT can be found in the (unpublished) expository masters thesis of Cakon (1987). More recent, and more accessible, is the excellent exposition on Weierstrass's work in approximation theory by Pinkus (2000).

Normally, one would expect that a proof of the WAT could be attained through Lagrange interpolating polynomials. However, Faber (1914) showed that no matter how the nodes for interpolation are prescribed, there is a continuous function f whose interpolating polynomials *fail* to converge uniformly to f on $[a, b]$. Despite this, Fejér (1930) gave a proof of the WAT using *Hermite* interpolation. Specifically, the interpolating polynomials whose derivatives are all zero at the nodes do converge uniformly to the continuous function.

Perhaps the most important and far-reaching generalization of the WAT is the result of Stone (1937) (see also Stone (1948)), which now goes by the name of the Stone–Weierstrass theorem. A fine exposition of this theorem and its extensions was given by Prolla (1993). Virtually all proofs of the Stone–Weierstrass theorem use the WAT as a key step, or at least the special case of the WAT concerned with the uniform approximation of the function $f(t) = |t|$ on the interval $[-1, 1]$ by polynomials. However, Brosowski and Deutsch (1981) have given an elementary proof of the Stone–Weierstrass theorem that does not use the WAT (nor uses the special case mentioned above) and uses only the basic facts that a closed subset of a compact set is compact, a positive continuous function on a compact set has a positive infimum, and the elementary Bernoulli inequality (Theorem 7.16). In particular, the WAT appears as a genuine corollary of this proof. Ransford (1984) used this idea to give an elementary proof of Bishop's generalization of the Stone–Weierstrass theorem.

Müntz's Theorem (Theorem 7.26) is due to Müntz (1914), with some subsequent extensions (to complex exponents) and simplifications by Szász (1916). In contrast to the proof given here, an elementary proof of Müntz's theorem that does not use the WAT was given by Burckel and Saeki (1983).

CHAPTER 8

GENERALIZED SOLUTIONS OF LINEAR EQUATIONS

Linear Operator Equations

In this chapter we consider the fundamental problem of solving the linear operator equation

$$(8.0.1) \qquad\qquad Ax = y,$$

where $A : X \to Y$ is a bounded linear operator between the inner product spaces X and Y, and y is a given element of Y. If $X = l_2(n)$ and $Y = l_2(m)$, then (8.0.1) reduces to a system of m linear equations in n unknowns. Even in this special case, however, we know that there are three possible outcomes of (8.0.1): no solution, a unique solution, or infinitely many solutions.

In the case that there is no solution to (8.0.1), it still makes sense to try to satisfy (8.0.1) "as best as possible." This suggests determining $x_0 \in X$ so as to minimize the expression $\|Ax - y\|$. That is, find $x_0 \in X$ such that

$$(8.0.2) \qquad\qquad \|Ax_0 - y\| = \inf_{x \in X} \|Ax - y\|.$$

Any $x_0 \in X$ satifying (8.0.2) is called a **generalized solution** to (8.0.1). Of course, every solution to (8.0.1) in the ordinary sense (i.e., satisfying $Ax_0 = y$) is also a generalized solution.

We will see that if X and Y are complete and the range of A is closed, there always exists a generalized solution. In general, however, there are infinitely many generalized solutions. For computational purposes, we shall be interested in the generalized solution having the smallest norm. This particular solution is unique, and it gives rise to the notion of the "generalized inverse" of A.

To obtain useful characterizations of generalized inverses, it is convenient to introduce the notion of the "adjoint" of a bounded linear map from X to Y. We begin by first recalling the definition of the the the inverse of a linear operator, and establishing some fundamental properties of the inverse.

For brevity throughout this chapter, we shall denote the set of all bounded linear operators from X into Y by $\mathcal{B}(X, Y)$. As was shown in Chapter 5 (Theorem 5.11), boundedness is equivalent to continuity for linear operators. Recall also that the norm of a bounded linear operator $A \in \mathcal{B}(X, Y)$ is given by

$$(8.0.3) \qquad \|A\| = \sup\left\{ \frac{\|Ax\|}{\|x\|} \ \Big|\ x \in X \backslash \{0\} \right\} = \sup_{\|x\| \leq 1} \|Ax\| = \sup_{\|x\| = 1} \|Ax\|,$$

and

$$(8.0.4) \qquad\qquad \|Ax\| \leq \|A\| \, \|x\| \quad \text{for all } x \in X.$$

Here, and in the sequel, we will often write Ax instead of $A(x)$ whenever A is a linear operator.

There is a useful alternative expression of the norm of an operator $A \in \mathcal{B}(X, Y)$. Namely,

$$\|A\| = \sup\left\{ \frac{|\langle Ax, y \rangle|}{\|x\| \|y\|} \,\middle|\, x \in X \backslash \{0\}, \ y \in Y \backslash \{0\} \right\}$$

(8.0.5)
$$= \sup\{ |\langle Ax, y \rangle| \mid x \in X, \ \|x\| \le 1, \ y \in Y, \ \|y\| \le 1 \}$$

$$= \sup\{ |\langle Ax, y \rangle| \mid x \in X, \ \|x\| = 1, \ y \in Y, \ \|y\| = 1 \}.$$

To verify this, note that by Schwarz's inequality,

$$\frac{|\langle Ax, y \rangle|}{\|x\| \|y\|} \le \frac{\|Ax\| \|y\|}{\|x\| \|y\|} \le \|A\|.$$

Hence

$$\sup\left\{ \frac{|\langle Ax, y \rangle|}{\|x\| \|y\|} \,\middle|\, x \in X \backslash \{0\}, \ Y \backslash \{0\} \right\} \le \|A\|.$$

On the other hand, choose $x_n \in X$, $\|x_n\| = 1$, such that $\|Ax_n\| \to \|A\|$. Then

$$\left\langle Ax_n, \frac{Ax_n}{\|Ax_n\|} \right\rangle = \|Ax_n\| \to \|A\|$$

implies that

$$\|A\| \le \sup\{ |\langle Ax, y \rangle| \mid x \in X, \ \|x\| = 1, \ y \in Y, \ \|y\| = 1 \}$$

$$\le \sup\{ |\langle Ax, y \rangle| \mid x \in X, \ \|x\| \le 1, \ y \in Y, \ \|y\| \le 1 \}$$

$$\le \sup\left\{ \frac{|\langle Ax, y \rangle|}{\|x\| \|y\|} \,\middle|\, x \in X \backslash \{0\}, \ y \in Y \backslash \{0\} \right\} \le \|A\|.$$

This verifies (8.0.5).

Also, if $A \in \mathcal{B}(X, Y)$ and $B \in \mathcal{B}(Y, Z)$, then the composition $B \circ A : X \to Z$ will often be written as juxtaposition: $BA := B \circ A$, so that $BAx = B(Ax)$ for all $x \in X$.

8.1 Lemma. If $A \in \mathcal{B}(X, Y)$ and $B \in \mathcal{B}(Y, Z)$, then $BA \in \mathcal{B}(X, Z)$ and

(8.1.1)
$$\|BA\| \le \|B\| \|A\|.$$

Proof. To verify the linearity of BA is a simple exercise using the linearity of A and B. For boundedness, note that for each $x \in X$,

$$\|BAx\| = \|B(Ax)\| \le \|B\| \|Ax\| \le \|B\| \|A\| \|x\|.$$

Hence BA is bounded and (8.1.1) holds. ∎

8.2 Theorem. *With the usual pointwise operations of addition and scalar multiplication, $\mathcal{B}(X, Y)$ is a normed linear space. Moreover, $\mathcal{B}(X, Y)$ is complete if Y is complete.*

Proof. If A, B in $\mathcal{B}(X, Y)$, x_1, x_2 in X, and α_1, α_2 in $\mathbb{R}$, we see that

$$
\begin{aligned}
(A + B)(\alpha_1 x_1 + \alpha_2 x_2) &= A(\alpha_1 x_1 + \alpha_2 x_2) + B(\alpha_1 x_1 + \alpha_2 x_2) \\
&= \alpha_1 A x_1 + \alpha_2 A x_2 + \alpha_1 B x_1 + \alpha_2 B x_2 \\
&= \alpha_1 (A x_1 + B x_1) + \alpha_2 (A x_2 + B x_2) \\
&= \alpha_1 (A + B) x_1 + \alpha_2 (A + B) x_2.
\end{aligned}
$$

Thus $A + B$ is linear. Moreover, for any $x \in X$,

$$
\|(A + B)x\| = \|Ax + Bx\| \le \|Ax\| + \|Bx\| \le \|A\|\,\|x\| + \|B\|\,\|x\| = (\|A\| + \|B\|)\|x\|.
$$

This proves that $A + B$ is bounded and $\|A + B\| \le \|A\| + \|B\|$.

Similarly, it is easy to verify that αA is a bounded linear operator if A is, and

$$
\|\alpha A\| = |\alpha|\|A\|.
$$

Clearly, $\|A\| \ge 0$ and $\|A\| = 0$ if and only if $Ax = 0$ for all x, i.e., $A = 0$. This proves that $\mathcal{B}(X, Y)$ is a normed linear space.

Finally, suppose that Y is complete. Let $\{A_n\}$ be a Cauchy sequence in $\mathcal{B}(X, Y)$. Then for each $\epsilon > 0$ there exists $N = N(\epsilon) \in \mathbb{N}$ such that

$$(8.2.1) \qquad \|A_n - A_m\| < \epsilon \quad \text{for all} \quad n, m \ge N.$$

For each $x \in X$, it follows that

$$(8.2.2) \qquad \|A_n x - A_m x\| = \|(A_n - A_m)x\| \le \|A_n - A_m\|\,\|x\| \le \epsilon\|x\|$$

for all $n, m \ge N$. Thus $\{A_n x\}$ is a Cauchy sequence in Y for each $x \in X$. Since Y is complete, there exists a point $A(x) \in Y$ such that $A_n(x) \to A(x)$ for each $x \in X$. To complete the proof, it suffices to show that $A \in \mathcal{B}(X, Y)$ and $\|A_n - A\| \to 0$.

Now, for any $x_1, x_2 \in X$, $\alpha_1, \alpha_2 \in \mathbb{R}$, we have (using Theorem 1.12),

$$
\begin{aligned}
A(\alpha_1 x_1 + \alpha_2 x_2) &= \lim A_n(\alpha_1 x_1 + \alpha_2 x_2) = \lim[\alpha_1 A_n x_1 + \alpha_2 A_n x_2] \\
&= \alpha_1 A(x_1) + \alpha_2 A(x_2).
\end{aligned}
$$

Thus $A : X \to Y$ is linear.

Passing to the limit in (8.2.2) as $m \to \infty$, we see that

$$(8.2.3) \qquad \|A_n x - Ax\| \le \epsilon\|x\|$$

for all $n \ge N$, $x \in X$. This implies that $A - A_N \in \mathcal{B}(X, Y)$, and hence $A = (A - A_N) + A_N \in \mathcal{B}(X, Y)$, and

$$(8.2.4) \qquad \|A_n - A\| \le \epsilon \quad \text{for all } n \ge N. \quad \blacksquare$$

8.3 Dual Spaces are Complete. *If X is any inner product space (complete or not), then X^* is complete.*

Proof. Note that $X^* = \mathcal{B}(X, \mathbb{R})$ and $\mathbb{R}$ is complete, so we can apply Theorem 8.2. ■

When X and Y are both finite-dimensional, there is a natural way of representing linear operators in $\mathcal{B}(X, Y)$ by *matrices*.

8.4 Definition. *Let $\{x_1, x_2, \ldots, x_n\}$ (respectively $\{y_1, y_2, \ldots, y_m\}$) be an orthonormal basis for X (respectively Y) and $A \in \mathcal{B}(X, Y)$. The* **matrix** *of A relative to these bases is the $m \times n$ matrix*

$$(8.4.1) \qquad \begin{bmatrix} a_{11} & a_{12} & \cdots & a_{1n} \\ a_{21} & a_{22} & \cdots & a_{2n} \\ \cdots & & & \\ a_{m1} & a_{m2} & \cdots & a_{mn} \end{bmatrix},$$

where

$$(8.4.2) \qquad a_{ij} = \langle Ax_j, y_i \rangle \quad (i = 1, 2, \ldots, m; \; j = 1, 2, \ldots, n).$$

Now, each vector $x \in X$ may be written uniquely in the form $x = \sum_1^n \langle x, x_i \rangle x_i$ by Corollary 4.10. Let us "identify" x with its (column) vector of coefficients

$$(8.4.3) \qquad (\langle x, x_1 \rangle, \langle x, x_2 \rangle, \ldots, \langle x, x_n \rangle)^T$$

in $l_2(n)$. More precisely, the mapping $x \mapsto (\langle x, x_1 \rangle, \langle x, x_2 \rangle, \ldots, \langle x, x_n \rangle)^T$ from X to $l_2(n)$ is linear, one-to-one, onto, and preserves both the inner product and norm. In the language of Exercise 4 at the end of Chapter 6, the spaces X and $l_2(n)$ are *equivalent*.

Similarly, each $y \in Y$ may be identified with the vector

$$(\langle y, y_1 \rangle, \langle y, y_2 \rangle, \ldots, \langle y, y_m \rangle)^T$$

in $l_2(m)$. In particular, Ax may be identified with

$$(\langle Ax, y_1 \rangle, \langle Ax, y_2 \rangle, \ldots, \langle Ax, y_m \rangle)^T.$$

But $x = \sum_1^n \langle x, x_j \rangle x_j$ implies that $Ax = \sum_1^n \langle x, x_j \rangle Ax_j$, and hence

$$\langle Ax, y_i \rangle = \sum_{j=1}^n \langle x, x_j \rangle \langle Ax_j, y_i \rangle = \sum_{j=1}^n a_{ij} \langle x, x_j \rangle.$$

Thus Ax can be identified with the m-tuple

$$(\langle Ax, y_1 \rangle, \langle Ax, y_2 \rangle, \ldots, \langle Ax, y_m \rangle)^T$$

$$= \left(\sum_1^n a_{1j} \langle x, x_j \rangle, \sum_1^n a_{2j} \langle x, x_j \rangle, \ldots, \sum_1^n a_{mj} \langle x, x_j \rangle \right)^T.$$

But this m-tuple is just the formal matrix product

(8.4.4)
$$\begin{bmatrix} a_{11} & a_{12} & \cdots & a_{1n} \\ a_{21} & a_{22} & \cdots & a_{2n} \\ \cdots & & & \\ a_{m1} & a_{m2} & \cdots & a_{mn} \end{bmatrix} \begin{bmatrix} \langle x, x_1 \rangle \\ \langle x, x_2 \rangle \\ \cdots \\ \langle x, x_n \rangle \end{bmatrix}.$$

For this reason, we often identify the operator A with its matrix (8.4.1), x with the n-tuple (8.4.3), and Ax with (8.4.4). In short, we are identifying X (respectively Y) with $l_2(n)$ (respectively $l_2(m)$) and the basis $\{x_1, x_2, \ldots, x_n\}$ (respectively $\{y_1, x_2, \ldots, y_m\}$) with the canonical orthonormal basis $\{e_1, e_2, \ldots, e_n\}$ in $l_2(n)$ (respectively $\{e_1, e_2, \ldots, e_m\}$ in $l_2(m)$), where $e_i(j) = \delta_{ij}$ for all i, j.

To determine the matrix of a given operator A, it is helpful to notice that Ax_i can be identified with the ith *column* of the matrix for A:

$$Ax_i = \begin{bmatrix} a_{1i} \\ a_{2i} \\ \cdots \\ a_{mi} \end{bmatrix} = (a_{1i}, a_{2i}, \ldots, a_{mi})^T.$$

This follows by substituting x_i for x in (8.4.4).

8.5 Example. Let $\{x_1, x_2, x_3\}$ (respectively $\{y_1, y_2\}$) be an orthonormal basis for a 3-dimensional (respectively 2-dimensional) inner product space X (respectively Y). The mapping $A : X \to Y$ is defined by

$$A(\alpha_1 x_1 + \alpha_2 x_2 + \alpha_3 x_3) = (\alpha_1 - \alpha_2)y_1 + (3\alpha_1 + 2\alpha_2 + \alpha_3)y_2.$$

Determine the matrix of A relative to these bases.

We have $Ax_1 = y_1 + 3y_2 = (1, 3)^T$, $Ax_2 = -y_1 + 2y_2 = (-1, 2)^T$, and $Ax_3 = y_2 = (0, 1)^T$. Thus the matrix for A is

$$A = \begin{bmatrix} 1 & -1 & 0 \\ 3 & 2 & 1 \end{bmatrix}.$$

8.6 Definition. *The **identity** operator on X is the mapping $I = I_X : X \to X$ defined by*

$$Ix = x, \quad x \in X.$$

When there is not likely to be any confusion as to the space to which we are referring, the space subscript on the identity operator will usually be omitted.

Note that $I \in \mathcal{B}(X, X)$ and $\|I\| = 1$. Moreover, if X is n-dimensional and $\{x_1, x_2, \ldots, x_n\}$ is any orthonormal basis in X, then the matrix of I relative to this basis is the $n \times n$ matrix with 1's on the diagonal and 0's elsewhere:

$$\begin{bmatrix} 1 & 0 & \cdots & 0 \\ 0 & 1 & \cdots & 0 \\ & \cdots & & \\ 0 & 0 & \cdots & 1 \end{bmatrix}.$$

That is, the (i, j) entry is δ_{ij}.

8.7 Definitions. *If F is any mapping from X into Y, then the **range** of F is the set*

$$\mathcal{R}(F) = F(X) := \{\, F(x) \mid x \in X \,\},$$

*and the **null space** of F is the set*

$$\mathcal{N}(F) := \{\, x \in X \mid F(x) = 0 \,\}.$$

If $A \in \mathcal{B}(X,Y)$, then its range and null space are linear subspaces of X and Y, respectively. Moreover, $\mathcal{N}(A)$ is closed, since A is continuous. However, the range of A need not be closed.

8.8 Example. (A bounded linear operator with nonclosed range) Define A : $l_2 \to l_2$ by

$$Ax = \left(x(1), \frac{1}{2}x(2), \frac{1}{3}x(3), \ldots, \frac{1}{n}x(n), \ldots \right).$$

Then $A \in \mathcal{B}(l_2, l_2)$ and $\|A\| = 1$. Clearly, $\mathcal{R}(A)$ contains *all* sequences with finitely many nonzero components. Hence $\overline{\mathcal{R}(A)} = l_2$. However, the vector

$$y = \left(1, \frac{1}{2}, \frac{1}{3}, \ldots, \frac{1}{n}, \ldots \right)$$

is in $l_2 \backslash \mathcal{R}(A)$, and thus $\mathcal{R}(A)$ is not closed.

Recall that a mapping $F : X \to Y$ is called **injective** (or an **injection** or **one-to-one**) if $F(x_1) = F(x_2)$ implies $x_1 = x_2$. F is called **surjective** (or a **surjection** or **onto**) if $\mathcal{R}(F) = Y$. Finally, F is called **bijective** (or a **bijection**) if it is both injective and surjective.

It is easy to see that a *linear* mapping F is injective if and only if $\mathcal{N}(F) = \{0\}$. (This follows from the fact that $F(x_1) = F(x_2) \Leftrightarrow F(x_1 - x_2) = 0$.)

8.9 Lemma. *Let F be any function from X into Y. The following statements are equivalent:*

(1) F is bijective;

(2) There exists a (unique) map $G : Y \to X$ such that

$$(8.9.1) \qquad\qquad F \circ G = I_Y \quad and \quad G \circ F = I_X.$$

Moreover, if F linear, then so is G.

Proof. $(1) \Rightarrow (2)$. Suppose F is bijective. Then for each $y \in Y$ there is an $x \in X$ such that $F(x) = y$. Since F is injective, x is unique. Thus we can define a function $G : Y \to X$ by setting

$$G(y) = x \quad \Leftrightarrow \quad F(x) = y.$$

Clearly,

$$x = G(y) = G(F(x)) = G \circ F(x)$$

for all $x \in X$ implies that $I_X = G \circ F$.

Similarly,

$$y = F(x) = F(G(y)) = F \circ G(y)$$

for all $y \in Y$ implies that $F \circ G = I_Y$. This proves (8.9.1).

Next we show that G is unique. If $H : Y \to X$ is any function that satisfies

$$H \circ F = I_X \quad \text{and} \quad F \circ H = I_Y,$$

then

$$H = H \circ I_Y = H \circ (F \circ G) = (H \circ F) \circ G = I_X \circ G = G,$$

which proves uniqueness.

(2) $\Rightarrow$ (1). Suppose there is a function $G : Y \to X$ satisfying (8.9.1). Let $F(x_1) = F(x_2)$. Then

$$x_1 = I_X(x_1) = G \circ F(x_1) = G(F(x_1)) = G(F(x_2)) = G \circ F(x_2) = I_X(x_2) = x_2.$$

Thus F is injective.

Now let $y \in Y$. Then

$$y = I_Y(y) = F \circ G(y) = F(G(y)).$$

Taking $x = G(y) \in X$ shows that $F(x) = y$. That is, F is surjective, hence bijective.

Finally, suppose F is linear. We will show that G is also linear. Let $y_i \in Y$, $\alpha_i \in \mathbb{R}$, $x_i = G(y_i)$ $(i = 1, 2)$, and $x = G(\alpha_1 y_1 + \alpha_2 y_2)$. Then

$$\begin{aligned}
F(x) &= F[G(\alpha_1 y_1 + \alpha_2 y_2)] = \alpha_1 y_1 + \alpha_2 y_2 \\
&= \alpha_1 F(G(y_1)) + \alpha_2 F(G(y_2)) \\
&= \alpha_1 F(x_1) + \alpha_2 F(x_2) = F(\alpha_1 x_1 + \alpha_2 x_2).
\end{aligned}$$

Since F is one-to-one, $x = \alpha_1 x_1 + \alpha_2 x_2$. That is,

$$G(\alpha_1 y_1 + \alpha_2 y_2) = \alpha_1 G(y_1) + \alpha_2 G(y_2).$$

Hence G is linear. ∎

For any function $F : X \to Y$ that is bijective, the unique function G from Y to X satifying (8.5.1) is called the **inverse** of F and is denoted by F^{-1}. Thus

$$F \circ F^{-1} = I_Y \quad \text{and} \quad F^{-1} \circ F = I_X.$$

Also, by Lemma 8.9 (applied to F^{-1} instead of F), we see that $F^{-1} : Y \to X$ is bijective, and

$$(F^{-1})^{-1} = F.$$

Moreover, if F is also linear, then so is F^{-1}.

8.10 Example. Let X be an n-dimensional Hilbert space, let $\{x_1, x_2, \ldots, x_n\}$ be an orthonormal basis for X, let $\lambda_1, \lambda_2, \ldots, \lambda_n$ be n real numbers, and let the mapping $A : X \to X$ be defined by

$$A(x) = \sum_1^n \lambda_i \langle x, x_i \rangle x_i, \quad x \in X.$$

Then:

(a) $A \in \mathcal{B}(X, X)$ and the matrix of A relative to $\{x_1, x_2, \ldots, x_n\}$ is

$$
\begin{bmatrix}
\lambda_1 & 0 & \cdots & 0 \\
0 & \lambda_2 & \cdots & 0 \\
\cdots & & & \\
0 & 0 & \cdots & \lambda_n
\end{bmatrix}.
$$

(b) $\|A\| = \max_i |\lambda_i|$.
(c) A has an inverse if and only if $\lambda_i \neq 0$ for all i.
(d) If $\lambda_i \neq 0$ for all i, then $A^{-1} : X \to X$ is given by

$$
A^{-1}(x) = \sum_1^n \lambda_i^{-1} \langle x, x_i \rangle x_i, \quad x \in X,
$$

and the matrix for A^{-1} relative to $\{x_1, x_2, \ldots, x_n\}$ is

$$
\begin{bmatrix}
1/\lambda_1 & 0 & \cdots & 0 \\
0 & 1/\lambda_2 & \cdots & 0 \\
\cdots & & & \\
0 & 0 & \cdots & 1/\lambda_n
\end{bmatrix}.
$$

(e) $\|A^{-1}\| = \max_i |1/\lambda_i|$.

It is clear that A is linear, and since X is finite-dimensional, A must be bounded by Theorem 5.16. Also,

$$
a_{ij} = \langle Ax_j, x_i \rangle = \left\langle \sum_{k=1}^n \lambda_k \langle x_j, x_k \rangle x_k, x_i \right\rangle = \langle \lambda_j x_j, x_i \rangle = \lambda_j \delta_{ij}.
$$

This verifies part (a).

To verify (b), note that

$$
\|Ax\|^2 = \left\| \sum_1^n \lambda_i \langle x, x_i \rangle x_i \right\|^2 = \sum_1^n |\lambda_i|^2 |\langle x, x_i \rangle|^2
$$

$$
\leq \max_i |\lambda_i|^2 \sum_1^n |\langle x, x_i \rangle|^2 = \max_i |\lambda_i|^2 \|x\|^2
$$

using the orthogonality of the x_i's. Thus $\|Ax\| \leq \max_i |\lambda_i| \|x\|$ for all x, and hence $\|A\| \leq \max_i |\lambda_i|$. To see that equality holds, choose an index i_0 such that $|\lambda_{i_0}| = \max_i |\lambda_i|$. Then with $x = x_{i_0}$, we see that $\|x\| = 1$ and $\|Ax\| = \|\lambda_{i_0} x_{i_0}\| = |\lambda_{i_0}| = \max_i |\lambda_i|$. This proves (b).

If $\lambda_i = 0$ for some i, say $\lambda_1 = 0$, then

$$
Ax_1 = \sum_1^n \lambda_i \langle x_1, x_i \rangle x_i = \sum_2^n \lambda_i \delta_{1i} x_i = 0
$$

implies that $\mathcal{N}(A) \neq \{0\}$ and A is not injective. Hence A has no inverse.

Conversely, if $\lambda_i \neq 0$ for all i, consider the mapping $B : X \to X$ defined by

$$B(x) = \sum_1^n \lambda_i^{-1} \langle x, x_i \rangle x_i, \quad x \in X.$$

By parts (a) and (b), B is a bounded linear map with $\|B\| = \max_i |\lambda_i^{-1}|$. Also, for each $x \in X$,

$$ABx = A(Bx) = A\left(\sum_1^n \lambda_i^{-1} \langle x, x_i \rangle x_i\right)$$

$$= \sum_{i=1}^n \lambda_i^{-1} \langle x, x_i \rangle A x_i = \sum_{i=1}^n \lambda_i^{-1} \langle x, x_i \rangle \sum_{j=1}^n \lambda_j \langle x_i, x_j \rangle x_j$$

$$= \sum_{i=1}^n \lambda_i^{-1} \langle x, x_i \rangle \lambda_i x_i = \sum_1^n \langle x, x_i \rangle x_i = x.$$

Thus $AB = I$. Similarly, $BA = I$. By Lemma 8.9, A has an inverse and $B = A^{-1}$. This proves (c), and parts (d) and (e) follow immediately.

A natural question suggested by Lemma 8.9 is the following. If A is a *bounded* linear operator from X to Y that is bijective, must A^{-1} also be bounded?

We will see that the answer is affirmative when both X and Y are complete, but not in general.

8.11 Definition. *An operator $A \in \mathcal{B}(X, Y)$ is* **bounded below** *if there is a constant $\rho > 0$ such that*

$$(8.11.1) \qquad \|Ax\| \geq \rho \|x\| \quad \text{for all } x \in X,$$

equivalently, if $\rho := \inf\{\|Ax\| \mid x \in X,\ \|x\| = 1\} > 0$. In this case, we also say that A is bounded below by ρ.

8.12 Lemma. *Let $A \in \mathcal{B}(X, Y)$.*

(1) If A is bounded below, then A is injective.

(2) A has an inverse that is bounded if and only if A is both surjective and bounded below.

Proof. (1) Suppose A is bounded below by $\rho > 0$. If $Ax_1 = Ax_2$, then

$$0 = \|Ax_1 - Ax_2\| = \|A(x_1 - x_2)\| \geq \rho\|x_1 - x_2\|$$

implies that $x_1 = x_2$. Thus A is injective.

(2) Suppose A has an inverse that is bounded. Then A is bijective, and for each $x \in X$,

$$\|x\| = \|A^{-1}Ax\| \leq \|A^{-1}\|\,\|Ax\|.$$

If $X \neq \{0\}$, then $A^{-1} \neq 0$ implies

$$\|Ax\| \geq \frac{1}{\|A^{-1}\|} \|x\| \quad \text{for all } x \in X.$$

Thus A is bounded below (by $\|A^{-1}\|^{-1}$). If $X = \{0\}$, then A is trivially bounded below (by any $\rho > 0$).

Conversely, suppose A is both surjective and bounded below by $\rho > 0$. By (1), A is injective. By Lemma 8.9, A^{-1} is also linear.

It remains to verify that A^{-1} is bounded. For any $y \in Y$, there is an $x \in X$ such that $Ax = y$. Then

$$\|A^{-1}y\| = \|A^{-1}(Ax)\| = \|x\| \le \frac{1}{\rho}\|Ax\| = \frac{1}{\rho}\|y\|.$$

This proves that A^{-1} is bounded and, in fact, $\|A^{-1}\| \le \rho^{-1}$. ∎

The Uniform Boundedness and Open Mapping Theorems

Thus to determine whether A has a bounded inverse requires, in particular, checking whether A is bounded below. This may be difficult to verify in practice. However, in the case where both X and Y are *complete*, there is an easily verifiable criterion; namely, A is bijective. To prove this, we need a few facts that are of independent interest.

Recall that a subset D of X is called **dense** in X if for each $x \in X$ and $\epsilon > 0$, there exists $y \in D$ such that $\|x - y\| < \epsilon$. In other words, $B(x, \epsilon) \cap D \ne \emptyset$. Equivalently, $D \cap U \ne \emptyset$ for every nonempty open set U in X.

8.13 Baire's Theorem. *Let X be a complete subset of an inner product space.*

(1) *If $\{D_n\}$ is a sequence of dense open sets in X, then $\cap_1^\infty D_n$ is dense in X.*

(2) *If $\{S_n\}$ is a sequence of sets in X with $X = \cup_1^\infty S_n$, then $\overline{S_n}$ contains an interior point for some n.*

Proof. (1) Let $D = \cap_1^\infty D_n$ and let U be a nonempty open set in X. We are to show that $U \cap D \ne \emptyset$. Since D_1 is dense, $U \cap D_1 \ne \emptyset$. Since $U \cap D_1$ is open, it follows that there is an open ball $B(x_1, \epsilon_1) \subset U \cap D_1$, where $0 < \epsilon_1 \le 1$. Since $B[x_1, \epsilon_1/2] \subset B(x_1, \epsilon_1)$, it follows that $U \cap D_1$ contains the closed ball $B_1 := B[x_1, \epsilon_1/2]$. Similarly, since $B(x_1, \epsilon_1/2) \cap D_2$ is nonempty and open, it must contain a closed ball $B_2 := B[x_2, \epsilon_2]$, where $0 < \epsilon_2 \le \frac{1}{2}$. Continuing in this way, we obtain for each integer n a closed ball $B_n := B[x_n, \epsilon_n]$ contained in $B_{n-1} \cap D_n$, where $0 < \epsilon_n \le \frac{1}{n}$. For every $n \ge m$, we see that $x_n \in B_m$, and hence

$$\|x_n - x_m\| \le \epsilon_m \le \frac{1}{m}.$$

It follows that $\{x_n\}$ is a Cauchy sequence. Since X is complete, there is $x \in X$ such that $x_n \to x$. For each m, the set B_m is closed and contains x_n for all $n \ge m$. Hence $x \in B_m \subset D_m$ for every m. It follows that $x \in U \cap D$.

(2) Since X is complete, it is closed, and hence $X = \cup_1^\infty \overline{S_n}$. If the result were false, then $\overline{S_n}$ would have no interior point for each n. Hence $D_n := X \backslash \overline{S_n}$ would be a dense open set for each n. But

$$\emptyset = X \backslash \cup_1^\infty \overline{S_n} = \cap_1^\infty (X \backslash \overline{S_n}) = \cap_1^\infty D_n,$$

which contradicts part (1). ∎

One important consequence of Baire's theorem is the following theorem, which shows that if a collection of operators is pointwise bounded, it must actually be uniformly bounded.

8.14 Uniform Boundedness Theorem. *Suppose X is a Hilbert space, Y is an inner product space, and $\mathcal{A} \subset \mathcal{B}(X, Y)$. If*

$$(8.14.1) \qquad \sup\{\|Ax\| \mid A \in \mathcal{A}\} < \infty \quad \text{for each } x \in X,$$

then

$$(8.14.2) \qquad \sup\{\|A\| \mid A \in \mathcal{A}\} < \infty.$$

Proof. For each $n \in \mathbb{N}$, set

$$S_n := \{x \in X \mid \|Ax\| \leq n \text{ for each } A \in \mathcal{A}\}.$$

Then S_n is closed, and $X = \cup_1^\infty S_n$ by (8.14.1). By Baire's theorem (Theorem 8.13), some S_n has an interior point. Thus there exist $N \in \mathbb{N}$, $x_0 \in S_N$, and $\epsilon > 0$ such that $B(x_0, \epsilon) \subset S_N$. Thus for all $A \in \mathcal{A}$, $\|Ax\| \leq N$ for all $x \in B(x_0, \epsilon)$. In particular, $\|Ax_0\| \leq N$, and if $\|x - x_0\| < \epsilon$, then

$$\|A(x - x_0)\| \leq \|Ax\| + \|Ax_0\| \leq 2N.$$

That is, $\|Az\| \leq 2N$ whenever $\|z\| < \epsilon$. Thus, $\|Aw\| \leq 2N/\epsilon$ whenever $\|w\| < 1$. This proves that $\|A\| \leq 2N/\epsilon$ for all $A \in \mathcal{A}$, and hence (8.14.2) holds. $\blacksquare$

8.15 Corollary. *Let X be a Hilbert space, Y an inner product space, and $A_n \in \mathcal{B}(X, Y)$ for $n = 1, 2, \ldots$. If*

$$(8.15.1) \qquad A(x) := \lim A_n x$$

exists for each $x \in X$, then $A \in \mathcal{B}(X, Y)$,

$$(8.15.2) \qquad \sup_n \|A_n\| < \infty,$$

and

$$(8.15.3) \qquad \|A\| \leq \liminf_n \|A_n\|.$$

Proof. Since $\{A_n x\}$ converges for each x, the collection $\mathcal{A} = \{A_n \mid n \in \mathbb{N}\}$ is pointwise bounded: $\sup_n \|A_n x\| < \infty$ for each $x \in X$. By the uniform boundedness theorem (Theorem 8.14), (8.15.2) holds. Further, the mapping A defined by (8.15.1) is clearly linear, since all the A_n are, and

$$\|A(x)\| = \lim_n \|A_n x\| = \liminf_n \|A_n x\| \leq \liminf_n \|A_n\| \, \|x\|$$

for all $x \in X$. Thus A is bounded and (8.15.3) holds. $\blacksquare$

8.16 Lemma. *Let X and Y be inner product spaces and $A \in \mathcal{B}(X, Y)$. Then the following statements are equivalent:*

(1) *$A(U)$ is open in Y for each open set U in X;*

(2) *There exists $\delta > 0$ such that*

$$(8.16.1) \qquad\qquad A(B(X)) \supset \delta B(Y),$$

where $B(X)$ (respectively $B(Y)$) denotes the open unit ball in X (respectively Y); and

(3) *There exists $\rho > 0$ such that for each $y \in Y$, there is an $x \in X$ with $Ax = y$ and $\|x\| \le \rho \|y\|$.*

Proof. (1) $\Rightarrow$ (2). Suppose (1) holds. Then $A(B(X))$ is open. Since $0 = A(0) \in A(B(X))$, there exists $\delta > 0$ such that

$$\delta B(Y) = B(0, \delta) \subset A(B(X)).$$

(2) $\Rightarrow$ (3). If (2) holds, then

$$B(Y) \subset \delta^{-1} A(B(X)) = A(\delta^{-1} B(X)).$$

Fix any $\lambda \in (0, 1)$. Then for every $y \in Y \setminus \{0\}$, $\lambda \|y\|^{-1} y \in B(Y)$. Thus there exists $x_0 \in \delta^{-1} B(X)$ such that $Ax_0 = \lambda \|y\|^{-1} y$. Then $y = A(\|y\| \lambda^{-1} x_0)$, and setting $x = \|y\| \lambda^{-1} x_0$, we see that $Ax = y$ and $\|x\| = \|y\| \lambda^{-1} \|x_0\| \le \delta^{-1} \lambda^{-1} \|y\|$. Taking $\rho = (\delta \lambda)^{-1}$, (3) follows.

(3) $\Rightarrow$ (1). Assume that (3) holds and let U be open in X. We must show that $A(U)$ is open in Y. Let $y_0 \in A(U)$ and choose $x_0 \in U$ such that $y_0 = Ax_0$. Since U is open, there exists $\epsilon > 0$ such that

$$x_0 + \epsilon B(X) = B(x_0, \epsilon) \subset U.$$

Then

$$y_0 + \epsilon A(B(X)) = A(x_0 + \epsilon B(X)) \subset A(U).$$

But by our hypothesis, for each $y \in B(Y)$, there exists $x \in X$ such that $Ax = y$ and $\|x\| < \rho$. That is, $B(Y) \subset A(\rho B(X)) = \rho A(B(X))$. Thus

$$B(y_0, \epsilon \rho^{-1}) = y_0 + \epsilon \rho^{-1} B(Y)) \subset y_0 + \epsilon A(B(X)) \subset A(U),$$

and hence y_0 is an interior point of $A(U)$. Since y_0 was arbitrary, $A(U)$ is open. ∎

8.17 The Open Mapping Theorem. *Let X and Y be Hilbert spaces. If A is a bounded linear operator from X onto Y, then $A(U)$ is open in Y for every open set U in X. Hence all the statements of Lemma 8.16 hold.*

Proof. Let $U = B(X)$ and $V = B(Y)$. By Lemma 8.16, it suffices to prove that there exists $\delta > 0$ such that

$$(8.17.1) \qquad\qquad A(U) \supset \delta V.$$

Given $y \in Y$, there exists $x \in X$ such that $Ax = y$. Let k be any integer such that $\|x\| < k$. It follows that $y \in A(kU)$. This proves that

$$Y = \cup_{k=1}^{\infty} A(kU).$$

Since Y is complete, it follows from Baire's theorem that $\overline{A(kU)}$ contains an interior point y_0 for some $k \in \mathbb{N}$. Hence there is an $\eta > 0$ such that

$$(8.17.2) \qquad W := B(y_0, \eta) \subset \overline{A(kU)}.$$

For any $y \in Y$ with $\|y\| < \eta$, we have that $y + y_0 \in W$. From (8.17.2), we can choose sequences $\{x'_n\}, \{x''_n\}$ in kU such that

$$Ax'_n \to y_0 \quad \text{and} \quad Ax''_n \to y + y_0.$$

Setting $x_n = x''_n - x'_n$, we see that $\|x_n\| < 2k$ and $Ax_n \to y$. This proves that for each $y \in Y$ with $\|y\| < \eta$ and for each $\epsilon > 0$, there exists $x \in X$ with

$$(8.17.3) \qquad \|x\| < 2k \quad \text{and} \quad \|Ax - y\| < \epsilon.$$

Claim: If $\delta = \eta/(4k)$, then for each $y \in Y$ and $\epsilon > 0$, there exists $x \in X$ with

$$(8.17.4) \qquad \|x\| \le \frac{1}{\delta}\|y\| \quad \text{and} \quad \|Ax - y\| < \epsilon.$$

To prove the claim, let $y \in Y \backslash \{0\}$. Then $y' = \eta(2\|y\|)^{-1}y$ has the property that $\|y'\| < \eta$. By (8.17.3), there exists $x' \in X$ with $\|x'\| < 2k$ and $\|Ax' - y'\| < \epsilon'$, where $\epsilon' = \eta(2\|y\|)^{-1}\epsilon$. Then $x = 2\|y\|\eta^{-1}x'$ satisfies $\|x\| < 4k\eta^{-1}\|y\| = \delta^{-1}\|y\|$ and

$$\|Ax - y\| = 2\|y\|\eta^{-1}\|Ax' - y'\| < 2\|y\|\eta^{-1}\epsilon' = \epsilon.$$

This verifies (8.17.4). If $y = 0$, (8.17.4) is trivially true with $x = 0$.

It remains to prove (8.17.1). Fix any $y \in \delta V$ and $\epsilon > 0$. By (8.17.4) there exists $x_1 \in X$ with $\|x_1\| < 1$ and

$$(8.17.5) \qquad \|y - Ax_1\| < \frac{1}{2}\delta\epsilon.$$

Applying the claim with y replaced by $y - Ax_1$, we obtain $x_2 \in X$ with

$$(8.17.6) \qquad \|x_2\| < \frac{1}{2}\epsilon \quad \text{and} \quad \|y - Ax_1 - Ax_2\| < \frac{1}{2^2}\delta\epsilon.$$

Continuing in this fashion, we obtain a sequence $\{x_n\}$ in X with

$$(8.17.7) \qquad \|x_{n+1}\| < \frac{1}{2^n}\epsilon$$

and

$$(8.17.8) \qquad \|y - Ax_1 - Ax_2 - \cdots - Ax_n\| < \frac{1}{2^n}\delta\epsilon \quad (n = 1, 2, \ldots).$$

Setting $s_n = \sum_1^n x_i$, we see from (8.17.7) that for $n > m$,

$$\|s_n - s_m\| = \|x_{m+1} + x_{m+2} + \cdots + x_n\| \le \frac{1}{2^m}\epsilon + \frac{1}{2^{m+1}}\epsilon + \cdots + \frac{1}{2^{n-1}}\epsilon < \frac{1}{2^{m-1}}\epsilon.$$

It follows that $\{s_n\}$ is a Cauchy sequence. Since X is complete, there exists an $x \in X$ such that $s_n \to x$. Using (8.17.7), we get

$$\|s_n\| \le \sum_1^n \|x_i\| < \|x_1\| + \frac{1}{2}\epsilon + \frac{1}{2^2}\epsilon + \cdots + \frac{1}{2^{n-1}}\epsilon.$$

Thus

$$\|x\| = \lim \|s_n\| \le \|x_1\| + \left(\sum_1^\infty \frac{1}{2^n}\right)\epsilon < 1 + \epsilon.$$

Since A is continuous, $As_n \to Ax$. Passing to the limit in (8.17.8), we deduce that $Ax = y$.

We have shown that $A((1+\epsilon)U) \supset \delta V$, and hence $A(U) \supset (1+\epsilon)^{-1}\delta V$, for every $\epsilon > 0$. But $\delta V = \cup_{\epsilon>0}(1+\epsilon)^{-1}\delta V$, so that (8.17.1) holds. ∎

The Closed Range and Bounded Inverse Theorems

8.18 Closed Range Theorem. *Suppose X and Y are Hilbert spaces and $A \in \mathcal{B}(X,Y) \setminus \{0\}$. Then the following statements are equivalent:*

(1) *A has closed range;*
(2) *There exists $\rho > 0$ such that $\|Ax\| \ge \rho\|x\|$ for all $x \in \mathcal{N}(A)^\perp$;*
(3) *$\rho := \inf\{\|Ax\| \mid x \in \mathcal{N}(A)^\perp, \|x\| = 1\} > 0$.*

Proof. The equivalence of (2) and (3) is clear.

(1) $\Rightarrow$ (2). Suppose A has closed range. Let $X_0 := \mathcal{N}(A)^\perp$, $Y_0 := \mathcal{R}(A)$, and $A_0 := A|_{X_0}$. Note that X_0 and Y_0 are closed subspaces of Hilbert spaces and hence are themselves Hilbert spaces. Also, $A_0 : X_0 \to Y_0$ is clearly linear and surjective. Finally, A_0 is injective, since if $A_0 x_1 = A_0 x_2$ for x_1, $x_2 \in X_0$, then $A(x_1 - x_2) = 0$ implies $x_1 - x_2 \in \mathcal{N}(A) \cap X_0 = \{0\}$. That is, $x_1 = x_2$.

Applying the open mapping theorem (Theorem 8.17) to $A_0 \in \mathcal{B}(X_0, Y_0)$, we obtain a $\delta > 0$ such that

$$(8.18.1) \qquad A(B(X_0)) \supset \delta B(Y_0).$$

Now, $A_0^{-1} : Y_0 \to X_0$ exists and is linear by Lemma 8.9. We deduce from (8.18.1) that

$$B(X_0) = A_0^{-1}[A(B(X_0))] \supset A_0^{-1}[\delta B(Y_0)] = \delta A_0^{-1}(B(Y_0)),$$

or

$$A_0^{-1}[B(Y_0)] \subset \frac{1}{\delta}B(X_0).$$

It follows that

$$\|A_0^{-1}\| = \sup_{y_0 \in B(Y_0)} \|A_0^{-1}(y_0)\| \le \frac{1}{\delta} \sup_{y_0 \in B(Y_0)} \|y_0\| = \frac{1}{\delta},$$

and hence A_0^{-1} is bounded.

By Lemma 8.12, A_0 is bounded below: there exists $\rho > 0$ such that $\|A_0 x_0\| \ge \rho\|x_0\|$ for all $x_0 \in X_0$. In other words, (2) holds.

(2) $\Rightarrow$(1). Suppose (2) holds. To show that $\mathcal{R}(A)$ is closed, let $y_n \in \mathcal{R}(A)$ and $y_n \to y$. Then $y_n = Ax_n$ for some $x_n \in X$. Write $x_n = u_n + v_n$, where $u_n \in \mathcal{N}(A)$ and $v_n \in \mathcal{N}(A)^\perp$. Then

$$\rho\|v_n - v_m\| \le \|A(v_n - v_m)\| = \|A(x_n - x_m)\| = \|y_n - y_m\| \to 0$$

as n, $m \to \infty$. Thus $\{v_n\}$ is a Cauchy sequence in $\mathcal{N}(A)^\perp$. By completeness, there exists $v \in \mathcal{N}(A)^\perp$ such that $v_n \to v$. Hence $y_n = Ax_n = Av_n \to Av$, which implies that $y = Av \in \mathcal{R}(A)$, and thus $\mathcal{R}(A)$ is closed. ∎

8.19 Bounded Inverse Theorem. *Let X and Y be Hilbert spaces and $A \in \mathcal{B}(X, Y)$. Then A has a bounded inverse if and only if A is bijective.*

Proof. The necessity is immediate. For the sufficiency, assume that A is bijective. Then $\mathcal{N}(A) = \{0\}$ implies $\mathcal{N}(A)^\perp = X$. Also, A is surjective implies that $\mathcal{R}(A) = Y$ is closed. By the closed range theorem (Theorem 8.18), A is bounded below. By Lemma 8.12, A has a bounded inverse. ∎

The Closed Graph Theorem

As a final application of the open mapping theorem, we prove the "closed graph" theorem and an important consequence of it that will be used in the next two chapters.

If X and Y are inner product spaces, let $X \times Y$ denote the set of all ordered pairs (x, y) with $x \in X$ and $y \in Y$. Addition and scalar multiplication are defined "componentwise" in $X \times Y$; that is,

(8.19.1) $$(x_1, y_1) + (x_2, y_2) := (x_1 + x_2, y_1 + y_2),$$

and

(8.19.2) $$\alpha(x, y) := (\alpha x, \alpha y).$$

With these operations, it is not difficult to verify that $X \times Y$ is a linear space. In fact, one can define an inner product in $X \times Y$ by

(8.19.3) $$\langle (x_1, y_1), (x_2, y_2) \rangle := \langle x_1, x_2 \rangle + \langle y_1, y_2 \rangle.$$

Of course, the first (respectively second) inner product on the right denotes the inner product in the space X (respectively Y).

8.20 Proposition. *$X \times Y$ is an inner product space. It is complete if and only if both X and Y are complete.*

We leave the simple verification of this proposition as an exercise. Note that the norm in $X \times Y$ is given by

$$\|(x, y)\| := \sqrt{\langle (x, y), (x, y) \rangle} = \sqrt{\|x\|^2 + \|y\|^2}.$$

In particular, convergence in $X \times Y$ is componentwise:

(8.20.1) $\|(x_n, y_n) - (x, y)\| \to 0 \quad \Leftrightarrow \quad \|x_n - x\| \to 0 \text{ and } \|y_n - y\| \to 0.$

8.21 Definition. *The **graph** of a mapping $F : X \to Y$ is the set*

$$\mathcal{G}(F) := \{(x, F(x)) \in X \times Y \mid x \in X\}.$$

*F is said to have a **closed graph** if its graph is closed as a subset of $X \times Y$.*

Equivalently, using (8.20.1), it is easy to see that F has a closed graph if $x_n \to x$ and $F(x_n) \to y$ implies that $y = F(x)$.

It is obvious that every continuous mapping has a closed graph. The converse is true when both X and Y are complete, but not in general.

8.22 Example (A discontinuous linear map with a closed graph). Let X denote the inner product space

$$X = \{x \in C_2[0,1] \mid x \text{ is a polynomial and } x(0) = 0\}.$$

That is, X is the space of all polynomials on $[0, 1]$ without constant term. Define $F : X \to C_2[0,1]$ by

$$F(x) = x', \quad x \in X,$$

where the prime denotes differentiation. Clearly, F is linear. We will show that F *has a closed graph but is not continuous.*

To see that F is not continuous, let $x_n(t) = \sqrt{2n+1}\, t^n$ $(n = 1, 2, \dots)$. Then $\|x_n\| = 1$ and $\|F(x_n)\| > n$. Thus F is unbounded, hence discontinuous by Theorem 5.11.

To see that F has a closed graph, let $x_n \to x$ and $F(x_n) \to y$. We must show that $y = F(x)$. For any $t \in [0, 1]$, the fundamental theorem of calculus and Schwarz's inequality imply

$$\left| x_n(t) - \int_0^t y(s)ds \right| = \left| \int_0^t [x_n'(s) - y(s)]ds \right| \leq \int_0^1 |x_n'(s) - y(s)|ds \leq \|x_n' - y\|.$$

It follows that with $h(t) = \int_0^t y(s)ds$, we have $h' = y$ and

$$\|x_n - h\|^2 = \int_0^1 |x_n(t) - h(t)|^2 dt \leq \|x_n' - y\|^2.$$

Since $x_n' = F(x_n) \to y$, it follows that $x_n \to h$. But $x_n \to x$. Hence $x = h$, and so

$$F(x) = x' = h' = y.$$

However, if both X and Y are complete, then a linear map with a closed graph must be continuous. This is the content of our next theorem.

8.23 Closed Graph Theorem. *Let X and Y be Hilbert spaces and let $A : X \to Y$ be a linear map that has a closed graph. Then A is continuous.*

Proof. First note that $\mathcal{G}(A)$, the graph of A, is a closed linear subspace of $X \times Y$, hence must be complete, since $X \times Y$ is. Define Q_1 and Q_2 on $\mathcal{G}(A)$ by

$$Q_1((x, Ax)) = x$$

and

$$Q_2((x, Ax)) = Ax, \quad (x, Ax) \in \mathcal{G}(A).$$

Then Q_1 and Q_2 are linear on $\mathcal{G}(A)$,

$$\|Q_1((x, Ax))\| = \|x\| \leq \|(x, Ax)\|,$$

and

$$\|Q_2((x, Ax))\| = \|Ax\| \leq \|(x, Ax)\|.$$

This implies that both Q_1 and Q_2 are bounded and $\|Q_i\| \leq 1$ ($i = 1, 2$). That is, $Q_1 \in \mathcal{B}(\mathcal{G}(A), X)$ and $Q_2 \in \mathcal{B}(\mathcal{G}(A), Y)$.

Note that Q_1 is bijective onto X. By the bounded inverse theorem (Theorem 8.19), Q_1 has a bounded inverse $Q_1^{-1} \in \mathcal{B}(X, \mathcal{G}(A))$. Clearly,

$$Ax = Q_2((x, Ax)) = Q_2(Q_1^{-1}(x)) = Q_2 \circ Q_1^{-1}(x)$$

for each $x \in X$. By Lemma 8.1, $A = Q_2 \circ Q_1^{-1}$ is continuous. ∎

The following application of the closed graph theorem will be used in the next two chapters.

8.24 Proposition. *Let U and V be closed subspaces in the Hilbert space X such that $U + V$ is closed and $U \cap V = \{0\}$. Then the mapping $Q : U + V \to U$ defined by*

$$Q(u + v) = u, \quad \text{for every } u \in U \text{ and } v \in V,$$

is a continuous linear mapping.

Proof. Let $X_0 = U + V$. Since $U \cap V = \{0\}$, this is a direct sum: $X_0 = U \oplus V$. Moreover, since X_0 and U are closed subspaces in X, they are also Hilbert spaces. It is easy to see that $Q : X_0 \to U$ is well-defined, linear, $Q(u) = u$ for all $u \in U$, and $Q(v) = 0$ for all $v \in V$.

Next we verify that Q has a closed graph. To see this, let $u_n + v_n \to x$ and $Q(u_n + v_n) \to y$. We must show that $y = Q(x)$. Since $U + V$ is closed, $x = u + v$ for some $u \in U$ and $v \in V$. Also, $u_n = Q(u_n + v_n) \to y \in U$ since U is closed. Hence $v_n = (u_n + v_n) - u_n \to x - y = u + v - y$. Since $v_n \in V$ for all n and V is closed, $u + v - y \in V$. Thus

$$Q(x) = Q(u + v) = Q(u + v - y) + Q(y) = Q(y) = y.$$

This proves that Q has a closed graph and hence, by the closed graph theorem (Theorem 8.23), Q is continuous. ∎

Adjoint of a Linear Operator

Now we turn to the study of the "adjoint" of a bounded linear operator.

8.25 The Adjoint Operator. *Let X and Y be inner product spaces and $A \in \mathcal{B}(X, Y)$.*

(1) If there exists a mapping $B : Y \to X$ that satisfies

$$(8.25.1) \qquad\qquad \langle Ax, y \rangle = \langle x, B(y) \rangle$$

for all $x \in X$ and $y \in Y$, then there is only one such mapping. It is called the **adjoint** *of A and is denoted by A^*.*

(2) If X is complete, then A^* always exists.

(3) If A^* exists, then A^* is a bounded linear operator from Y into X, i.e., $A^* \in \mathcal{B}(Y, X)$,

$$(8.25.2) \qquad\qquad \|A^*\| = \|A\|,$$

and

$$(8.25.3) \qquad\qquad \|A\|^2 = \|A^*A\| = \|AA^*\|.$$

In particular, if X is a Hilbert space, the adjoint of A is a bounded linear mapping $A^* : Y \to X$ that is defined by

$$(8.25.4) \qquad \langle Ax, y \rangle = \langle x, A^*y \rangle \quad \text{for all } x \in X, \ y \in Y.$$

Proof. (1) Suppose $C : Y \to X$ satisfies (8.25.1) as B does. Then

$$\langle x, C(y) \rangle = \langle x, B(y) \rangle$$

for all $x \in X$ and $y \in Y$. Hence, for each $y \in Y$,

$$\langle x, C(y) - B(y) \rangle = 0$$

for all $x \in X$. It follows that $C(y) - B(y) = 0$, or $C(y) = B(y)$. Since y was arbitrary, $C = B$.

(2) Now suppose X is complete. Fix any $y \in Y$. Define f_y on X by

$$(8.25.5) \qquad\qquad f_y(x) = \langle Ax, y \rangle, \quad x \in X.$$

It is obvious that f_y is linear, since A is. Moreover,

$$|f_y(x)| \leq \|Ax\| \, \|y\| \leq \|A\| \, \|x\| \, \|y\|.$$

Hence f_y is bounded and $\|f_y\| \leq \|A\| \, \|y\|$. That is, $f_y \in X^*$. Since X is complete, Theorem 6.10 implies that f_y has a unique representer $z_y \in X$. That is,

$$(8.25.6) \qquad\qquad f_y(x) = \langle x, z_y \rangle, \quad x \in X.$$

Define $B : Y \to X$ by $B(y) = z_y$. Then (8.25.5) and (8.25.6) imply

$$(8.25.7) \qquad\qquad \langle Ax, y \rangle = \langle x, B(y) \rangle.$$

By part (1), $B = A^*$.

(3) For all $y_1, y_2 \in Y$ and $\alpha_1, \alpha_2 \in \mathbb{R}$, (8.25.1) implies

$$\langle x, A^*(\alpha_1 y_1 + \alpha_2 y_2) \rangle = \langle Ax, \alpha_1 y_1 + \alpha_2 y_2 \rangle = \alpha_1 \langle Ax, y_1 \rangle + \alpha_2 \langle Ax, y_2 \rangle$$
$$= \alpha_1 \langle x, A^*(y_1) \rangle + \alpha_2 \langle x, A^*(y_2) \rangle = \langle x, \alpha_1 A^*(y_1) + \alpha_2 A^*(y_2) \rangle$$

for all $x \in X$. Thus

$$A^*(\alpha_1 y_1 + \alpha_2 y_2) = \alpha_1 A^*(y_1) + \alpha_2 A^*(y_2),$$

and hence A^* is linear.

Furthermore, for all $y \in Y$, we have that $x = A^*y \in X$, so that

$$\|A^*y\|^2 = \langle A^*y, A^*y \rangle = \langle x, A^*y \rangle = \langle Ax, y \rangle$$
$$\leq \|Ax\| \, \|y\| \leq \|A\| \, \|x\| \, \|y\| = \|A\| \, \|A^*y\| \, \|y\|.$$

Cancelling $\|A^*y\|$, we deduce that

$$\|A^*y\| \leq \|A\| \, \|y\|, \quad y \in Y.$$

Thus A^* is bounded and $\|A^*\| \leq \|A\|$. Then, for all $x \in X$,

$$\|Ax\|^2 = \langle Ax, Ax \rangle = \langle x, A^*(Ax) \rangle \leq \|x\| \, \|A^*(Ax)\| \leq \|x\| \, \|A^*\| \, \|Ax\|.$$

This implies $\|Ax\| \leq \|A^*\| \, \|x\|$ for all x, so that $\|A\| \leq \|A^*\|$. Hence $\|A^*\| = \|A\|$. Using Lemma 8.1, we obtain

$$\|A^*A\| \leq \|A^*\| \, \|A\| = \|A\|^2.$$

Conversely, for any $x \in X$ with $\|x\| = 1$,

$$\|Ax\|^2 = \langle Ax, Ax \rangle = \langle A^*Ax, x \rangle \leq \|A^*Ax\| \leq \|A^*A\|.$$

Hence

$$\|A\|^2 = \sup_{\|x\|=1} \|Ax\|^2 \leq \|A^*A\| \leq \|A\|^2,$$

and this proves the first equality of (8.25.3). The second equality is proved similarly. The last statement of the theorem follows from (1) and (2). ∎

A mapping $A \in \mathcal{B}(X, X)$ is called **self-adjoint** if A^* exists and $A^* = A$.

8.26 Examples. (1) (**The zero mapping**) Let $0 : X \to Y$ denote the zero mapping: $0(x) = 0$ for all x. Clearly, 0^* is the zero mapping from Y to X. If $X = Y$, then 0 is self-adjoint.

(2) (**The identity mapping**) If I is the identity map on X, then for all $x, y \in X$,

$$\langle Ix, y \rangle = \langle x, y \rangle = \langle x, Iy \rangle.$$

Thus $I^* = I$ and I is self-adjoint.

(3) (**The metric projection**) Let M be a Chebyshev subspace of X. Then by Theorem 5.13, $\langle P_M(x), y \rangle = \langle x, P_M(y) \rangle$ for all $x, y \in X$. Hence $P_M^* = P_M$ and P_M is self-adjoint.

(4) (**Adjoint of a matrix**) Let $A \in \mathcal{B}(l_2(n), l_2(m))$ have the matrix

$$A = \begin{bmatrix} a_{11} & a_{12} & \cdots & a_{1n} \\ a_{21} & a_{22} & \cdots & a_{2n} \\ \cdots & & & \\ a_{m1} & a_{m2} & \cdots & a_{mn} \end{bmatrix}$$

relative to a given orthonormal basis $\{x_1, x_2, \ldots, x_n\}$ in $l_2(n)$ and $\{y_1, y_2, \ldots, y_m\}$ in $l_2(m)$. Thus

(8.26.1) $a_{ij} = \langle Ax_j, y_i \rangle \quad (i = 1, 2, \ldots, m; \; j = 1, 2, \ldots, n).$

Since $l_2(n)$ is finite-dimensional, it is complete (Theorem 3.7), so that A^* exists by Lemma 8.25. Also, using (8.26.1), the matrix of $A^* : l_2(m) \to l_2(n)$ is given by

$$(8.26.2) \qquad a_{ij}^* = \langle A^* y_j, x_i \rangle = \langle y_j, Ax_i \rangle = \langle Ax_i, y_j \rangle = a_{ji}$$

for $i = 1, 2, \ldots, n$; $j = 1, 2, \ldots, m$. That is, the matrix of A^* is just the *transpose* of the matrix of A. For example, if $A = \begin{bmatrix} 1 & 2 & 3 \\ 4 & 5 & 6 \end{bmatrix}$, then $A^* = \begin{bmatrix} 1 & 4 \\ 2 & 5 \\ 3 & 6 \end{bmatrix}$.

It follows that if $l_2(n) = l_2(m)$ (i.e., if $n = m$), then A is self-adjoint if and only if the matrix of A is **symmetric**, i.e., the matrix of A is equal to its transpose.

(5) **(Integral operator)** Consider the operator defined in Proposition 5.17. Thus $A : C_2[a, b] \to C_2[a, b]$ is given by

$$(8.26.3) \qquad (Ax)(s) = \int_a^b k(s, t)x(t)\, dt, \quad s \in [a, b],$$

where k is a continuous function on $[a, b] \times [a, b]$. We saw there that A is in $\mathcal{B}(C_2[a, b], C_2[a, b])$. Now by interchanging the order of integration, we get

$$\langle Ax, y \rangle = \int_a^b \left[\int_a^b k(s, t)x(t)\, dt \right] y(s)\, ds = \int_a^b x(t) \left[\int_a^b k(s, t)y(s)\, ds \right] dt$$
$$= \langle x, B(y) \rangle,$$

where

$$(8.26.4) \qquad B(y)(t) := \int_a^b k(s, t)y(s)\, ds, \quad t \in [a, b].$$

It follows from Theorem 8.25 that A^* exists and $A^* = B$. Comparing (8.26.3) and (8.26.4), we see that A is self-adjoint if (and actually only if) k is symmetric; that is,

$$(8.26.5) \qquad k(s, t) = k(t, s) \quad \text{for all } s,\ t \in [a, b].$$

Next we show that without the completeness hypothesis, the adjoint of an operator may *not* exist.

8.27 Theorem. *Let X and Y be inner product spaces with $Y \neq \{0\}$. Then A^* exists for each $A \in \mathcal{B}(X, Y)$ if and only if X is complete.*

Proof. The sufficiency was already established in Theorem 8.25.

For the necessity, suppose A^* exists for every $A \in \mathcal{B}(X, Y)$. If X were not complete, then by the remark following Theorem 6.10 (see also Exercise 6 at the end of Chapter 6), there would exist $x^* \in X^*$ that has no representer. Fixing any $y_1 \in Y \setminus \{0\}$, define A on $\dot{X}$ by

$$Ax = x^*(x)y_1, \quad x \in X.$$

Then $A \in \mathcal{B}(X, Y)$ and for each $x \in X$ and $y \in Y$,

$$\langle x, A^* y \rangle = \langle Ax, y \rangle = \langle x^*(x)y_1, y \rangle = x^*(x)\langle y_1, y \rangle.$$

Putting $y = y_1$ yields $\langle x, A^* y_1 \rangle = x^*(x)\|y_1\|^2$, or

$$x^*(x) = \langle x, A^*(\|y_1\|^{-2} y_1) \rangle, \quad x \in X.$$

Thus x^* has the representer $A^*(\|y_1\|^{-2} y_1)$, which is a contradiction. ∎

Remark. During the course of the proof we essentially proved that *if* $A = x^* \in X^* := \mathcal{B}(X, \mathbb{R})$, *then* A^* *exists if and only if* A *has a representer. Moreover, if* A *has the representer* x_1, *then* $A^* : \mathbb{R} \to X$ *is given by*

$$A^* y = y x_1, \quad y \in \mathbb{R}.$$

More generally, we can determine precisely *which* bounded linear operators from X into a finite-dimensional inner product space have adjoints, and we leave the proof as an exercise (see Exercise 15 at the end of the chapter).

8.28 Representation Theorem for Linear Operators.
Let X *be an inner product space,* Y *a finite-dimensional inner product space with an orthonormal basis* $\{y_1, y_2, \ldots, y_n\}$, *and* $A \in \mathcal{B}(X, Y)$. *Then:*
(1) *There exist* $x_i^* \in X^*$ $(i = 1, 2, \ldots, n)$ *such that*

$$(8.28.1) \qquad Ax = \sum_1^n x_i^*(x) y_i \quad \text{for all } x \in X.$$

(2) A^* *exists if and only if each* x_i^* *has a representer in* X.
(3) *If* x_i^* *has the representer* $x_i \in X$ $(i = 1, 2, \ldots, n)$, *then* $A^* \in \mathcal{B}(Y, X)$ *is given by*

$$(8.28.2) \qquad A^* y = \sum_1^n \langle y, y_i \rangle x_i \quad \text{for all } y \in Y.$$

In particular, if X is a Hilbert space, then there exist $x_1, x_2, \ldots, x_n$ in X such that

$$(8.28.3) \qquad Ax = \sum_1^n \langle x, x_i \rangle y_i \quad \text{for all } x \in X,$$

and (8.28.2) holds.

The adjoint has other useful properties that are listed in the next few results.

8.29 Lemma. Let X *and* Y *be inner product spaces,* A, $B \in \mathcal{B}(X, Y)$, *and* $\alpha \in \mathbb{R}$. *If* A^* *and* B^* *exist, so do* $(A + B)^*$, $(\alpha A)^*$, *and* $A^{**} := (A^*)^*$. *Moreover,*
(1) $(A + B)^* = A^* + B^*$,
(2) $(\alpha A)^* = \alpha A^*$,
(3) $A^{**} = A$.
This result is an easy consequence of the definitions involved and is left as an exercise.

8.30 Lemma. Let X, Y, *and* Z *be inner product spaces,* $A \in \mathcal{B}(Y, Z)$, *and* $B \in \mathcal{B}(X, Y)$. *If* A^* *and* B^* *exist, so does* $(AB)^*$, *and*

$$(8.30.1) \qquad (AB)^* = B^* A^*.$$

Proof. We already saw in Lemma 8.1 that $AB \in \mathcal{B}(X, Z)$. Hence

$$\langle ABx, z \rangle = \langle A(Bx), z \rangle = \langle Bx, A^* z \rangle = \langle x, B^*(A^* z) \rangle = \langle x, B^* A^* z \rangle$$

for all $x \in X$, $z \in Z$. Thus $(AB)^* = B^* A^*$. ∎

8.31 Theorem. *Let X and Y be Hilbert spaces and $A \in \mathcal{B}(X,Y)$. Then A has a bounded inverse if and only if A^* does. In this case,*

$$(8.31.1) \qquad (A^*)^{-1} = (A^{-1})^*.$$

Proof. If A has a bounded inverse, then, using Lemma 8.30,

$$I = I^* = (AA^{-1})^* = (A^{-1})^* A^*.$$

Similarly, $I = A^*(A^{-1})^*$. It follows from Lemma 8.9 that $(A^*)^{-1} = (A^{-1})^*$. Thus A^* has a bounded inverse given by (8.31.1).

Conversely, suppose A^* has a bounded inverse. By the first part of the proof, A^{**} has a bounded inverse. But by Lemma 8.29, $A^{**} = A$. ∎

If $A \in \mathcal{B}(X,X)$, it is not true in general that $\|A^n\| = \|A\|^n$ for each $n \in \mathbb{N}$. But this will be the case when A is self-adjoint or, more generally, whenever A is **normal**, i.e., whenever A commutes with its adjoint: $AA^* = A^*A$.

8.32 Lemma. *Let $A \in \mathcal{B}(X,X)$ and suppose A^* exists (e.g., if X is complete). Then:*

(1) $(A^n)^*$ exists for each $n \in \mathbb{N}$ and

$$(8.32.1) \qquad (A^n)^* = (A^*)^n.$$

In particular, if A is self-adjoint, so is A^n.

(2) If A is normal, then

$$(8.32.2) \qquad \|A^n\| = \|A\|^n \quad \text{for each } n \in \mathbb{N}.$$

In particular, (8.32.2) holds if A is self-adjoint.

Proof. (1) When $n = 1$, the result is trivially true. Assume that the result holds for some $n \geq 1$. Then for all $x, y \in X$,

$$\langle A^{n+1}x, y \rangle = \langle A(A^n x), y \rangle = \langle A^n x, A^* y \rangle = \langle x, (A^n)^* A^* y \rangle$$
$$= \langle x, (A^*)^n A^* y \rangle = \langle x, (A^*)^{n+1} y \rangle.$$

This shows that $(A^{n+1})^*$ exists and is equal to $(A^*)^{n+1}$. Thus, by induction, the result is valid for all n.

(2) Assume first that A is self-adjoint. Then by Theorem 8.25, $\|A^2\| = \|A\|^2$. Since A^2 is self-adjoint by part (1), it follows that $\|A^4\| = \|A^2\|^2 = \|A\|^4$. Continuing inductively in this fashion, we deduce that

$$(8.32.3) \qquad \|A^n\| = \|A\|^n$$

whenever $n = 2^m$, $m \in \mathbb{N}$.

Now suppose that A is normal, but not necessarily self-adjoint. A simple induction shows that A (respectively A^*) commutes with any integer power of A^* (respectively A). It follows that any power of A (respectively A^*) commutes with any power of A^* (respectively A). Then, using (1), we obtain

$$(A^n)^* A^n = (A^*)^n A^n = (A^* A)^n = (AA^*)^n = A^n (A^*)^n = A^n (A^n)^*.$$

That is, A^n is normal. By Theorem 8.25,

$$(8.32.4) \qquad \|A^n\|^2 = \|A^n(A^n)^*\| = \|A^n(A^*)^n\| = \|(AA^*)^n\|$$

for any $n \in \mathbb{N}$. But AA^* is self-adjoint, so that by (8.32.3) (applied to AA^* instead of A), we deduce

$$(8.32.5) \qquad \|(AA^*)^n\| = \|AA^*\|^n = \|A\|^{2n}$$

whenever $n = 2^m$, $m \in \mathbb{N}$. Combining (8.32.4) and (8.32.5), we see that

$$(8.32.6) \qquad \|A^n\| = \|A\|^n$$

whenever $n = 2^m$, $m \in \mathbb{N}$. For a general n, choose $m \in \mathbb{N}$ such that $r := 2^m - n \geq 0$. Since (8.32.6) is valid with n replaced by $n + r = 2^m$, we have, using Lemma 8.1, that

$$\|A\|^{n+r} = \|A^{n+r}\| = \|A^n A^r\| \leq \|A^n\| \|A^r\| \leq \|A^n\| \|A\|^r.$$

Hence $\|A\|^n \leq \|A^n\|$. Since the reverse inequality is always valid by Lemma 8.1, it follows that $\|A^n\| = \|A\|^n$. ∎

There are interesting duality relationships that hold between the null spaces and ranges of A and A^*.

8.33 Lemma. *Let X and Y be inner product spaces, $A \in \mathcal{B}(X,Y)$, and suppose A^* exists (e.g., if X is complete). Then:*
(1) $\mathcal{N}(A) = \mathcal{R}(A^*)^\perp$ *and* $\mathcal{N}(A^*) = \mathcal{R}(A)^\perp$.
(2) *If X (respectively Y) is complete, then* $\mathcal{N}(A)^\perp = \overline{\mathcal{R}(A^*)}$ *(respectively $\mathcal{N}(A^*)^\perp = \overline{\mathcal{R}(A)}$).*
(3) $\mathcal{N}(A^*A) = \mathcal{N}(A)$.

Proof. (1) $x \in \mathcal{N}(A) \Leftrightarrow Ax = 0 \Leftrightarrow \langle Ax, y \rangle = 0$ for all $y \in Y \Leftrightarrow \langle x, A^*y \rangle = 0$ for all $y \in Y \Leftrightarrow x \in \mathcal{R}(A^*)^\perp$. Thus $\mathcal{N}(A) = \mathcal{R}(A^*)^\perp$.

Since A^* exists, Lemma 8.29 implies $A^{**} = A$. Substituting A^* for A in the first part yields $\mathcal{N}(A^*) = \mathcal{R}(A)^\perp$.

(2) Assume that X is complete. By (1), $\mathcal{N}(A) = \mathcal{R}(A^*)^\perp$. Since $\overline{\mathcal{R}(A^*)}$ is a closed subspace of X, it is Chebyshev by Theorem 3.5 and

$$\mathcal{N}(A)^\perp = \mathcal{R}(A^*)^{\perp\perp} = \overline{\mathcal{R}(A^*)}^{\perp\perp} = \overline{\mathcal{R}(A^*)}$$

using Theorem 4.5. Similarly, $\mathcal{N}(A^*)^\perp = \overline{\mathcal{R}(A)}$ if Y is complete.

(3) Clearly, $\mathcal{N}(A) \subset \mathcal{N}(A^*A)$. If $x \in \mathcal{N}(A^*A)$, then $A^*Ax = 0$ implies

$$0 = \langle A^*Ax, x \rangle = \langle Ax, Ax \rangle = \|Ax\|^2.$$

Thus $Ax = 0$ or $x \in \mathcal{N}(A)$. ∎

Generalized Solutions to Operator Equations

Now we can characterize the generalized solutions to the equation

$$(8.33.1) \qquad Ax = y.$$

8.34 Characterization of Generalized Solutions. *Let X and Y be inner product spaces, $A \in \mathcal{B}(X, Y)$, and suppose A^* exists (e.g., if X is complete). For given elements $x_0 \in X$ and $y \in Y$, the following statements are equivalent:*

(1) *x_0 is a generalized solution to $Ax = y$; that is, $\|Ax_0 - y\| \le \|Ax - y\|$ for all $x \in X$;*

(2) *$Ax_0 = P_{\mathcal{R}(A)}(y)$;*

(3) *$A^*Ax_0 = A^*y$.*

Proof. (1) and (2) are obviously equivalent.

(2) $\Leftrightarrow$ (3). Using the characterization theorem (Theorem 4.9) with $M = \mathcal{R}(A)$, we see that $Ax_0 = P_{\mathcal{R}(A)}(y) \Leftrightarrow y - Ax_0 \in \mathcal{R}(A)^\perp \Leftrightarrow y - Ax_0 \in \mathcal{N}(A^*)$ (using Lemma 8.33) $\Leftrightarrow A^*(y - Ax_0) = 0 \Leftrightarrow$ (3) holds. ∎

Note that this theorem characterizes the generalized solutions of (8.33.1) as the ordinary (or exact) solutions to the linear equation

$$(8.34.1) \qquad\qquad A^*Ax = A^*y.$$

One consequence of this equivalence is that there are available methods for solving equations such as (8.34.1). In particular, if the range of either A or A^* is finite-dimensional, then (8.34.1) reduces to solving a finite linear system of equations, where the number of unknowns is equal to the number of equations.

However, this does not answer the question of whether generalized solutions even exist, and if they do, whether they are unique.

For each $y \in Y$, let $G(y)$ denote the set of all generalized solutions to (8.33.1). By Theorem 8.34,

$$G(y) = \{ x \in X \mid A^*Ax = A^*y \}.$$

Using Theorem 8.34 again, we see that generalized solutions exist, i.e., $G(y) \ne \emptyset$, if the range of A is Chebyshev in Y. This will be the case, for example, when $\mathcal{R}(A)$ is closed and Y is complete. But even when $\mathcal{R}(A)$ is Chebyshev, $G(y)$ may contain more than one point. For example, if $x_0 \in G(y)$, then $Ax_0 = P_{\mathcal{R}(A)}(y)$. But $A(x_0 + x) = P_{\mathcal{R}(A)}(y)$ is also valid for every $x \in \mathcal{N}(A)$. That is, $x_0 + x \in G(y)$ for each $x \in \mathcal{N}(A)$. Hence if $\mathcal{N}(A) \ne \{0\}$, that is, if A is not injective, then $G(y)$ contains infinitely many points whenever it contains one. In fact, $G(y)$ contains at most one point if and only if $\mathcal{N}(A) = \{0\}$, i.e., A is injective.

We summarize these remarks in the following theorem.

8.35 Set of Generalized Solutions. *Let X and Y be inner product spaces, $A \in \mathcal{B}(X, Y)$, and suppose A^* exists (e.g., if X is complete). For each $y \in Y$, let $G(y)$ denote the set of all generalized solutions to*

$$(8.35.1) \qquad\qquad Ax = y.$$

Then:

(1) *$G(y) = \{ x \in X \mid A^*Ax = A^*y \}$.*

(2) *$G(y) = \mathcal{N}(A) + x(y)$ for any $x(y) \in G(y)$. In particular, $G(y)$ is a closed affine set in X.*

(3) *$G(y) \ne \emptyset$ for all $y \in Y$ if and only if $\mathcal{R}(A)$ is a Chebyshev subspace in Y. This is the case, for example, when Y is complete and $\mathcal{R}(A)$ is closed.*

(4) *The following statements are equivalent:*

(a) $G(y)$ contains at most one point for every $y \in Y$;
(b) $G(y)$ is a singleton for some $y \in Y$;
(c) A is injective.

(5) $G(y)$ is a single point for all $y \in Y$ if and only if $\mathcal{R}(A)$ is Chebyshev and A is injective.

(6) If X and Y are complete and $\mathcal{R}(A)$ is closed, then $G(y)$ is a (nonempty) Chebyshev set for each $y \in Y$.

Proof. Everything has essentially already been verified with the exception of (2) and (6). From (1), we deduce that if $x(y) \in G(y)$, then

$$G(y) = \mathcal{N}(A^*A) + x(y).$$

But $\mathcal{N}(A^*A) = \mathcal{N}(A)$ from Lemma 8.33, and this verifies (2).

To verify (6), note that $G(y)$ is a closed convex set in X by (2), which by (3) is nonempty for each $y \in Y$ if Y is complete and $\mathcal{R}(A)$ is closed. Since X is complete, $G(y)$ is Chebyshev by Theorem 3.5. ∎

Generalized Inverse

In practical applications it is often necessary to compute *some* generalized solution. One natural candidate that has been the object of wide study is the minimal norm generalized solution.

8.36 Definition. *Let X and Y be Hilbert spaces, $A \in \mathcal{B}(X,Y)$ and suppose $\mathcal{R}(A)$ is closed. Then (by Theorem 8.35), for each $y \in Y$, the set $G(y)$ of all generalized solutions to*

$$(8.36.1) \qquad\qquad\qquad Ax = y$$

has a unique element $A^-(y) := P_{G(y)}(0)$ of minimal norm. This element is called the **minimal norm generalized solution** *to (8.36.1), and the mapping $A^- : Y \to X$ thus defined is called the* **generalized inverse** *of A.*

We will see later that the generalized inverse is a bounded linear mapping. In certain simple cases, the generalized inverse is related to ordinary inverses as follows.

8.37 Lemma. (1) *If A^{-1} exists, then $A^- = A^{-1}$.*
(2) *If $(A^*A)^{-1}$ exists, then $A^- = (A^*A)^{-1}A^*$.*

We leave the simple proof as an exercise.

The following characterization of the minimal norm generalized solution is fundamental for its practical computation, for the computation of the generalized inverse, and for showing that the generalized inverse is a bounded linear mapping.

8.38 Characterizing the Minimal Norm Generalized Solution. *Let X and Y be Hilbert spaces and suppose $A \in \mathcal{B}(X,Y)$ has closed range. Then, for every $y \in Y$,*

$$A^-(y) = G(y) \cap \mathcal{N}(A)^\perp = \{x \in \mathcal{N}(A)^\perp \mid A^*Ax = A^*y\}$$
$$= \{x \in \mathcal{N}(A)^\perp \mid Ax = P_{\mathcal{R}(A)}y\}.$$

Proof. From Theorem 8.35, we have that $G(y) = \mathcal{N}(A) + x(y)$ for any $x(y) \in G(y)$. Since $A^-(y) = P_{G(y)}(0)$, it follows from Exercise 2 at the end of Chapter 4 that $A^-(y) \in \mathcal{N}(A)^\perp$. Thus $x = A^-(y)$ if and only if $x \in G(y)$ and $x \in \mathcal{N}(A)^\perp$. But $x \in G(y)$ if and only if $A^*Ax = A^*y$ (by Theorem 8.35) if and only if $Ax = P_{\mathcal{R}(A)}y$ (by Theorem 8.34). ∎

Just as in the above theorem, it will be seen that a critical assumption for several important facts concerning the generalized inverse is that the range of a certain operator be closed. The next result shows that this is equivalent to the range of its adjoint being closed.

8.39 Lemma. Let X and Y be Hilbert spaces and $A \in \mathcal{B}(X, Y)$. Then $\mathcal{R}(A)$ is closed if and only if $\mathcal{R}(A^*)$ is closed.

Proof. If $\mathcal{R}(A)$ is closed, the closed range theorem (Theorem 8.18) implies that there exists $\delta > 0$ such that

$$(8.39.1) \qquad \|Ax\| \geq \delta \|x\| \quad \text{for all } x \in \mathcal{N}(A)^\perp.$$

Since $\mathcal{N}(A)^\perp$ is a closed subspace in X, it is Chebyshev by Theorem 3.5, and hence

$$\|Ax\| = \|A(P_{\mathcal{N}(A)^\perp}x + P_{\mathcal{N}(A)}x)\| = \|AP_{\mathcal{N}(A)^\perp}x\| \geq \delta \|P_{\mathcal{N}(A)^\perp}x\|$$

for all $x \in X$ by (8.39.1).

If $y \in \overline{\mathcal{R}(A^*)}$, then Theorem 4.5 and Lemma 8.33 imply $\overline{\mathcal{R}(A^*)} = \overline{\mathcal{R}(A^*)}^{\perp\perp} = \mathcal{R}(A^*)^{\perp\perp} = \mathcal{N}(A)^\perp$, so that $y \in \mathcal{N}(A)^\perp$. Define f on $\mathcal{R}(A)$ by

$$(8.39.2) \qquad f(Ax) := \langle x, y \rangle \quad \text{for all } x \in X.$$

Then f is well-defined, since if $Ax_1 = Ax_2$, then $x_1 - x_2 \in \mathcal{N}(A)$, and hence

$$0 = \langle x_1 - x_2, y \rangle = \langle x_1, y \rangle - \langle x_2, y \rangle.$$

Moreover, f is linear and bounded; the latter fact follows because

$$|f(Ax)| = |\langle x, y \rangle| = |\langle P_{\mathcal{N}(A)^\perp}x + P_{\mathcal{N}(A)}x, y \rangle| = |\langle P_{\mathcal{N}(A)^\perp}x, y \rangle|$$

$$\leq \|P_{\mathcal{N}(A)^\perp}(x)\| \|y\| \leq \frac{1}{\delta} \|Ax\| \|y\|$$

implies $\|f\| \leq \|y\|/\delta$.

By the Fréchet–Riesz representation theorem (Theorem 6.10), f has a representer $Ax_0 \in \mathcal{R}(A)$. That is,

$$(8.39.3) \qquad f(Ax) = \langle Ax, Ax_0 \rangle \quad \text{for all } x \in X.$$

From (8.39.2)–(8.39.3), we obtain

$$\langle x, y \rangle = \langle Ax, Ax_0 \rangle = \langle x, A^*Ax_0 \rangle$$

for all $x \in X$. This implies that $y = A^*Ax_0 \in \mathcal{R}(A^*)$. Thus $\overline{\mathcal{R}(A^*)} \subset \mathcal{R}(A^*)$ and $\mathcal{R}(A^*)$ is closed.

Conversely, if $\mathcal{R}(A^*)$ is closed, the above argument shows that $\mathcal{R}(A^{**})$ is closed. But $A^{**} = A$ by Lemma 8.29. ∎

For the next several results we will be working under the same hypotheses. Namely, the hypothesis

$(8.39.4)$ **X and Y are Hilbert spaces and $A \in \mathcal{B}(X, Y)$ has closed range.**

8.40 Lemma. *If hypothesis* (8.39.4) *holds, then*

(8.40.1) $$\mathcal{R}(A^-A) = \mathcal{R}(A^-) = \mathcal{R}(A^*A) = \mathcal{R}(A^*) = \mathcal{N}(A)^\perp.$$

In particular, the ranges of A^-, A^-A, *and* A^*A *are all closed.*
Proof. By Lemma 8.39, $\mathcal{R}(A^*)$ is closed. By Lemma 8.33,

$$\mathcal{N}(A)^\perp = \overline{\mathcal{R}(A^*)} = \mathcal{R}(A^*),$$

which proves the last equality of (8.40.1).

Clearly, $\mathcal{R}(A^*A) \subset \mathcal{R}(A^*)$. Conversely, if $x \in \mathcal{R}(A^*)$, then $x = A^*y$ for some $y \in Y$. Thus, using Lemma 8.33,

$$x = A^*(P_{\mathcal{N}(A^*)}y + P_{\mathcal{N}(A^*)^\perp}y) = A^*P_{\mathcal{N}(A^*)^\perp}y = A^*P_{\mathcal{R}(A)}y \in \mathcal{R}(A^*A).$$

This proves $\mathcal{R}(A^*) = \mathcal{R}(A^*A)$.

For each $y \in Y$, using Theorem 8.38 and Lemmas 8.33 and 8.39, $A^-y \in \mathcal{N}(A)^\perp = \mathcal{R}(A^*)$. Thus $\mathcal{R}(A^-) \subset \mathcal{R}(A^*)$. For the reverse inclusion, let $x \in \mathcal{R}(A^*) = \mathcal{N}(A)^\perp$ and set $y = Ax$. Then $A^*Ax = A^*y$. By Theorem 8.38, $x = A^-y \in \mathcal{R}(A^-)$. Thus $\mathcal{R}(A^*) = \mathcal{R}(A^-)$.

Finally, observe that $\mathcal{R}(A^-A) \subset \mathcal{R}(A^-)$ is obvious. If $x \in \mathcal{R}(A^-)$, then $x = A^-y$ for some $y \in Y$. But by Theorem 8.38, $A^-y = A^-P_{\mathcal{R}(A)}y \in \mathcal{R}(A^-A)$. Thus $\mathcal{R}(A^-A) = \mathcal{R}(A^-)$. ∎

Now we are in a position to prove that the generalized inverse is a bounded linear operator.

8.41 Generalized Inverse is Linear and Bounded. *If* $A \in \mathcal{B}(X,Y)$ *and hypothesis* (8.39.4) *holds, then* $A^- \in \mathcal{B}(Y,X)$.
Proof. Let $y_i \in Y$, $\alpha_i \in \mathbb{R}$ $(i = 1,2)$. By Theorem 8.38, $A^-(y_i) \in \mathcal{N}(A)^\perp \cap G(y_i)$ and $x_0 := \alpha_1 A^-(y_1) + \alpha_2 A^-(y_2) \in \mathcal{N}(A)^\perp$, since $\mathcal{N}(A)^\perp$ is a subspace. Further, since $A^-(y_i) \in G(y_i)$, we have

$$A^*Ax_0 = A^*A[\alpha_1 A^-(y_1) + \alpha_2 A^-(y_2)] = \alpha_1 A^*A[A^-(y_1)] + \alpha_2 A^*A[A^-(y_2)]$$
$$= \alpha_1 A^*y_1 + \alpha_2 A^*y_2 = A^*(\alpha_1 y_1 + \alpha_2 y_2).$$

Thus $x_0 \in G(\alpha_1 y_1 + \alpha_2 y_2)$. By Theorem 8.38, $x_0 = A^-(\alpha_1 y_1 + \alpha_2 y_2)$. That is,

$$A^-(\alpha_1 y_1 + \alpha_2 y_2) = \alpha_1 A^-(y_1) + \alpha_2 A^-(y_2),$$

and hence A^- is linear.

By Lemma 8.40, $\mathcal{R}(A^-) = \mathcal{R}(A^*) = \mathcal{N}(A)^\perp$. It follows from the closed range theorem (Theorem 8.18) that $\|A(A^-y)\| \geq \rho\|A^-y\|$ for all $y \in Y$, and some $\rho > 0$. Since $AA^-y = P_{\mathcal{R}(A)}y$ by Theorem 8.38, we obtain

$$\rho\|A^-y\| \leq \|P_{\mathcal{R}(A)}y\| \leq \|y\| \quad \text{for all } y \in Y.$$

Hence A^- is bounded and $\|A^-\| \leq 1/\rho$. ∎

There are alternative ways of describing the generalized inverse. An explicit formula, which is often used later, is the following.

8.42 Theorem. *If hypothesis (8.39.4) holds, then*

$$(8.42.1) \qquad\qquad A^- = (A|_{\mathcal{N}(A)^\perp})^{-1} P_{\mathcal{R}(A)}.$$

Proof. We first show that the expression on the right side of (8.42.1) is well-defined, i.e., that the restriction of A to $\mathcal{N}(A)^\perp$ has a bounded inverse on $\mathcal{R}(A)$. Let $A_0 := A|_{\mathcal{N}(A)^\perp}$. Then A_0 is a bounded linear operator from the Hilbert space $\mathcal{N}(A)^\perp$ (in X) to the Hilbert space $\mathcal{R}(A)$ (in Y). Moreover, A_0 is bijective. To see that A is injective, suppose $A_0 x_1 = A_0 x_2$ for some $x_i \in \mathcal{N}(A)^\perp$. Then $x_1 - x_2 \in \mathcal{N}(A)^\perp$ and $A(x_1 - x_2) = A_0(x_1 - x_2) = 0$, so $x_1 - x_2 \in \mathcal{N}(A)^\perp \cap \mathcal{N}(A) = \{0\}$, or $x_1 = x_2$. To see that A_0 is surjective, let $y \in \mathcal{R}(A)$. Then there exists $x \in X$ such that $y = Ax$. Then

$$y = A(P_{\mathcal{N}(A)} x + P_{\mathcal{N}(A)^\perp} x) = A P_{\mathcal{N}(A)^\perp} x = A_0 P_{\mathcal{N}(A)^\perp} x \in \mathcal{R}(A_0).$$

Thus A_0 is surjective. By the bounded inverse theorem (Theorem 8.19), we see that A_0 has a bounded inverse, i.e., $A_0^{-1} \in \mathcal{B}(\mathcal{R}(A), \mathcal{N}(A)^\perp)$. Let

$$B := A_0^{-1} P_{\mathcal{R}(A)}.$$

We will show that $B = A^-$. To this end, first note that $B \in \mathcal{B}(Y, \mathcal{N}(A)^\perp) \subset \mathcal{B}(Y, X)$. Fix any $y \in Y$. Since the range of B is contained in $\mathcal{N}(A)^\perp$, we have $By \in \mathcal{N}(A)^\perp$. Also,

$$ABy = A_0 By = A_0 A_0^{-1} P_{\mathcal{R}(A)} y = P_{\mathcal{R}(A)} y.$$

By Theorem 8.38, $By = A^- y$. Since $y \in Y$ was arbitrary, $B = A^-$. ∎

Less explicit descriptions of the generalized inverse of an operator, which are nevertheless useful in the applications, are collected in the next theorem.

8.43 Characterization of Generalized Inverse. *Let hypothesis (8.39.4) hold and let $B \in \mathcal{B}(Y, X)$. Then the following four statements are equivalent:*
- (1) $B = A^-$;
- (2) (i) $BAx = x$ for all $x \in \mathcal{N}(A)^\perp$,
    (ii) $By = 0$ for all $y \in \mathcal{R}(A)^\perp$;
- (3) (i) $AB = P_{\mathcal{R}(A)}$,
    (ii) $BA = P_{\mathcal{R}(B)}$;
- (4) (i) $(AB)^* = AB$,
    (ii) $(BA)^* = BA$,
    (iii) $ABA = A$,
    (iv) $BAB = B$.

Proof. (1) $\Rightarrow$ (2). Assume (1) holds. Using Theorem 8.42, we can write $B = (A|_{\mathcal{N}(A)^\perp})^{-1} P_{\mathcal{R}(A)}$. Thus for each $x \in \mathcal{N}(A)^\perp$, we have

$$BAx = (A|_{\mathcal{N}(A)^\perp})^{-1} P_{\mathcal{R}(A)} Ax = (A|_{\mathcal{N}(A)^\perp})^{-1} Ax = x.$$

If $y \in \mathcal{R}(A)^\perp$, then $P_{\mathcal{R}(A)} y = 0$, and so $By = 0$. Thus (2) holds.

(2) $\Rightarrow$ (3). Assume (2) holds. Then, for each $y \in Y$, we have

$$ABy = AB(P_{\mathcal{R}(A)}y + P_{\mathcal{R}(A)^\perp}y) = ABP_{\mathcal{R}(A)}y.$$

Choose $x \in X$ such that $Ax = P_{\mathcal{R}(A)}y$. Since $x = P_{\mathcal{N}(A)}x + P_{\mathcal{N}(A)^\perp}x$, we see that we can write $P_{\mathcal{R}(A)}y = Ax_1$, where $x_1 = P_{\mathcal{N}(A)^\perp}x \in \mathcal{N}(A)^\perp$. Hence

$$ABy = ABP_{\mathcal{R}(A)}y = ABAx_1 = Ax_1 = P_{\mathcal{R}(A)}y,$$

where (2)(i) was used for the third equality. Since y was arbitrary, it follows that $AB = P_{\mathcal{R}(A)}$; that is, (3)(i) holds.

Similarly, for each $x \in X$,

$$BAx = BA(P_{\mathcal{N}(A)}x + P_{\mathcal{N}(A)^\perp}x) = BAP_{\mathcal{N}(A)^\perp}x = P_{\mathcal{N}(A)^\perp}x = P_{\mathcal{R}(A^*)}x,$$

where we used Lemmas 8.33 and 8.39 for the last equality. Hence $BA = P_{\mathcal{R}(A^*)}$. It follows that $\mathcal{R}(A^*) \subset \mathcal{R}(B)$. On the other hand, if $x \in \mathcal{R}(B)$, then $x = By$ for some $y \in Y$ and

$$x = B(P_{\mathcal{R}(A)}y + P_{\mathcal{R}(A)^\perp}y) = BP_{\mathcal{R}(A)}y.$$

We then have $P_{\mathcal{R}(A)}y = Az$ for some $z \in X$. Thus,

$$x = BAz = BA(P_{\mathcal{N}(A)}z + P_{\mathcal{N}(A)^\perp}z) = BAP_{\mathcal{N}(A)^\perp}z$$
$$= P_{\mathcal{N}(A)^\perp}z \in \mathcal{N}(A)^\perp = \mathcal{R}(A^*)$$

using Lemmas 8.33 and 8.39. This proves $\mathcal{R}(B) \subset \mathcal{R}(A^*)$ and hence $\mathcal{R}(B) = \mathcal{R}(A)$. If follows that $BA = P_{\mathcal{R}(B)}$.

(3) $\Rightarrow$ (4). Assume (3) holds. Then since projections are self-adjoint by Example 8.26(3), it follows that AB and BA are self-adjoint. Also, $ABA = P_{\mathcal{R}(A)}A = A$ and $BAB = P_{\mathcal{R}(B)}B = B$. Thus (4) holds.

(4) $\Rightarrow$ (1). Assume (4) holds. Then AB is self-adjoint and $(AB)^2 = (ABA)B = AB$, so AB is idempotent. It follows (see Exercise 13 at the end of the chapter) that AB is the orthogonal projection onto its range:

(8.43.1) $$AB = P_{\mathcal{R}(AB)}.$$

We claim that $\mathcal{R}(AB) = \mathcal{R}(A)$. To see this, first note that $\mathcal{R}(AB) \subset \mathcal{R}(A)$ is obvious. Conversely, since $A = ABA$, we see that $\mathcal{R}(A) = \mathcal{R}(ABA) \subset \mathcal{R}(AB)$, which proves the claim.

If follows from this claim and (8.43.1) that $AB = P_{\mathcal{R}(A)}$. By symmetry, $BA = P_{\mathcal{R}(B)}$. Next we observe that $\mathcal{R}(B) = \mathcal{R}(A^*)$. To see this, note that $B = BAB = (BA)^*B = A^*B^*B$, and so $\mathcal{R}(B) \subset \mathcal{R}(A^*)$. On the other hand, $A = ABA$ implies that $A^* = (BA)^*A^* = BAA^*$ and hence $\mathcal{R}(A^*) \subset \mathcal{R}(B)$. This proves $\mathcal{R}(B) = \mathcal{R}(A^*)$. Combining these results we obtain

$$AB = P_{\mathcal{R}(A)} \quad \text{and} \quad BA = P_{\mathcal{R}(A^*)} = P_{\mathcal{N}(A)^\perp}.$$

If $y \in \mathcal{R}(A)$, then $y = Ax$ for some $x \in X$, so $y = A(P_{\mathcal{N}(A)}x + P_{\mathcal{N}(A)^\perp}x) = AP_{\mathcal{N}(A)^\perp}x$, and

$$A^-y = (A|_{\mathcal{N}(A)^\perp})^{-1}P_{\mathcal{R}(A)}y = (A|_{\mathcal{N}(A)^\perp})^{-1}AP_{\mathcal{N}(A)^\perp}x = P_{\mathcal{N}(A)^\perp}x = BAx = By.$$

On the other hand, if $y \in \mathcal{R}(A)^\perp$, then $P_{\mathcal{R}(A)}y = 0$ and

$$A^-y = (A|_{\mathcal{N}(A)^\perp})^{-1}P_{\mathcal{R}(A)}y = 0 = BP_{\mathcal{R}(A)}y = BABy = By.$$

This shows that A^- and B are equal on $\mathcal{R}(A) \cup \mathcal{R}(A)^\perp$. Since $Y = \mathcal{R}(A) \oplus \mathcal{R}(A)^\perp$, it follows that $B = A^-$, and thus (1) holds. ∎

It is now fairly simple to compute some examples of generalized inverses based on Theorem 8.43.

8.44 Examples.

(1) The generalized inverse of $0 \in \mathcal{B}(X, Y)$ is $0 \in \mathcal{B}(Y, X)$.

(2) If M is a closed subspace of the Hilbert space X, then $P_M^- = P_M$. In other words, the generalized inverse of an othogonal projection is the orthogonal projection itself.

(3) If $A \in \mathcal{B}(\ell_2(n), \ell_2(n))$ is the "diagonal" operator defined by

$$A\left(\sum_1^n \langle x, e_i \rangle e_i\right) = \sum_1^n \alpha_i \langle x, e_i \rangle e_i, \quad x \in \ell_2(n)$$

for some scalars $\alpha_i \in \mathbb{R}$, then $A^- \in \mathcal{B}(\ell_2(n), \ell_2(n))$ is given by

$$A^-\left(\sum_1^n \langle x, e_i \rangle e_i\right) = \sum_{\{i \mid \alpha_i \neq 0\}} \frac{1}{\alpha_i} \langle x, e_i \rangle e_i, \quad x \in \ell_2(n).$$

In other words, if the matrix for A is the $n \times n$ diagonal matrix

$$\begin{bmatrix} \alpha_1 & 0 & \cdots & 0 \\ 0 & \alpha_2 & \cdots & 0 \\ \vdots & \vdots & \ddots & \vdots \\ 0 & 0 & \cdots & \alpha_n \end{bmatrix},$$

then the matrix for A^- is the $n \times n$ diagonal matrix

$$\begin{bmatrix} \tilde{\alpha}_1 & 0 & \cdots & 0 \\ 0 & \tilde{\alpha}_2 & \cdots & 0 \\ \vdots & \vdots & \ddots & \vdots \\ 0 & 0 & \cdots & \tilde{\alpha}_n \end{bmatrix},$$

where

$$\tilde{\alpha}_i = \begin{cases} 1/\alpha_i & \text{if } \alpha_i \neq 0, \\ 0 & \text{if } \alpha_i = 0. \end{cases}$$

(4) If A is the linear functional defined on $\ell_2(n)$ by

$$Ax = \langle x, a \rangle = \sum_1^n x(i)a(i), \quad x \in \ell_2(n),$$

for some $a \in \ell_2(n) \setminus \{0\}$, then $A^- \in \mathcal{B}(\mathbb{R}, \ell_2(n))$ is given by

$$A^- y = \frac{1}{\|a\|^2} A^* y = \frac{1}{\|a\|^2} ay, \quad y \in \mathbb{R}.$$

We note that using Theorem 8.43, (1) is clear and (2) follows, since $\mathcal{R}(P_M) = M$ and $P_M P_M = P_M^2 = P_M$. Similarly, examples (3) and (4) can be easily verified using Theorem 8.43.

For the actual computation of the generalized inverse, the next result is often useful, since when either X or Y is finite-dimensional, it reduces the computation of the generalized inverse of A to that of the generalized inverse of a *matrix*.

8.45 Theorem. If hypothesis (8.39.4) holds, then

(8.45.1) $$A^- = (A^*A)^- A^* = A^*(AA^*)^-.$$

Proof. We first observe that by Lemma 8.40,

(8.45.2) $$\mathcal{R}(A^*) = \mathcal{R}(A^*A).$$

For each $y \in Y$, we deduce from Theorem 8.43(1) (applied to A^*A rather than A) and (8.45.2) that

$$A^*A(A^*A)^- A^*y = P_{\mathcal{R}(A^*A)}(A^*y) = P_{\mathcal{R}(A^*)}(A^*y) = A^*y.$$

From Theorem 8.34, it follows that $x_0 = (A^*A)^- A^*y$ is a generalized solution to the equation $Ax = y$. Moreover, using Lemma 8.40, the fact that A^*A is self-adjoint, and (8.45.2), we see that

$$x_0 \in \mathcal{R}((A^*A)^-) = \mathcal{R}((A^*A)^*) = \mathcal{R}(A^*A) = \mathcal{R}(A^*) = \mathcal{N}(A)^\perp.$$

By Theorem 8.38, $x_0 = A^-y$. Thus

$$(A^*A)^- A^*y = A^-y.$$

Since this holds for every $y \in Y$, the first equality in (8.45.1) holds. The second equality is proved similarly (see Exercise 23 at the end of the chapter). ∎

8.46 Application. Let X be a Hilbert space, $\{x_1, x_2, \ldots, x_n\} \subset X$, and define $A : X \to l_2(n)$ by

(8.46.1) $$Ax = (\langle x, x_1 \rangle, \langle x, x_2 \rangle, \ldots, \langle x, x_n \rangle), \quad x \in X.$$

By Theorem 8.28, $A^* : l_2(n) \to X$ is given by

(8.46.2) $$A^*y = \sum_1^n y(i)x_i, \quad y \in l_2(n).$$

We want to determine the generalized inverse of A.

By Theorem 8.45, $A^- = A^*(AA^*)^-$. Letting $B = AA^*$, the problem reduces to determining B^-. But $B : l_2(n) \to l_2(n)$ is given by

(8.46.3)
$$By = AA^*y = A\left(\sum_1^n y(i)x_i\right) = \sum_1^n y(i)Ax_i$$
$$= \sum_{i=1}^n y(i)(\langle x_i, x_1 \rangle, \langle x_i, x_2 \rangle, \ldots, \langle x_i, x_n \rangle).$$

Then the matrix of B relative to the canonical orthonormal basis $\{e_i, e_2, \ldots, e_n\}$ in $l_2(n)$, where $e_i(j) = \delta_{ij}$, is given by

$$b_{ij} = \langle Be_j, e_i \rangle = \left\langle \sum_{k=1}^n e_j(k)(\langle x_k, x_1 \rangle, \langle x_k, x_2 \rangle, \ldots, \langle x_k, x_n \rangle), e_i \right\rangle$$
$$= \sum_{k=1}^n e_j(k)\langle (\langle x_k, x_1 \rangle, \langle x_k, x_2 \rangle, \ldots, \langle x_k, x_n \rangle), e_i \rangle$$
$$= \sum_{k=1}^n e_j(k)\langle x_k, x_i \rangle = \langle x_j, x_i \rangle = \langle x_i, x_j \rangle.$$

That is, the matrix of B is the Gram matrix $B = G(x_1, x_2, \ldots, x_n)$. Thus

(8.46.4) $$A^- = A^* B^-,$$

and the problem reduces to finding the generalized inverse of the $n \times n$ symmetric Gram matrix B.

In particular, if $\{x_1, x_2, \ldots, x_n\}$ is linearly independent, then B has an inverse by Theorem 7.7, so that

(8.46.5) $$A^- = A^* B^{-1}.$$

Moreover, if $\{x_1, x_2, \ldots, x_n\}$ is actually orthonormal, then $B = I$, so that

(8.46.6) $$A^- = A^*.$$

We conclude this chapter by showing how Theorem 8.38 can be used to compute the generalized inverse of a matrix.

8.47 Example. Let us compute the generalized inverse of the matrix

$$A = \begin{bmatrix} -1 & 1 \\ 1 & -1 \end{bmatrix}.$$

(That is, A is the operator in $\mathcal{B}(l_2(2), l_2(2))$ given by

$$Ax = [-x(1) + x(2)]e_1 + [x(1) - x(2)]e_2 = [-x(1) + x(2)](e_1 - e_2),$$

where $e_i(j) = \delta_{ij}$ for $i, j = 1, 2$.)

Fix any $y \in l_2(2)$. We want to determine $x = A^-(y)$ using Theorem 8.38. According to that theorem, we seek the element x in $\mathcal{N}(A)^\perp$ that also satisfies

(8.47.1) $$A^* A x = A^* y.$$

By Lemma 8.40, $\mathcal{N}(A)^\perp = \mathcal{R}(A^*) = \mathcal{R}(A)$, since $A^* = A$. Thus we must determine the $x \in \mathcal{R}(A)$ satisfying (8.47.1). But

$$\mathcal{R}(A) = \{ \alpha(e_1 - e_2) \mid \alpha \in \mathbb{R} \} = \text{span}\{e_1 - e_2\}.$$

Thus $x \in \mathcal{R}(A) \Leftrightarrow x = \alpha(e_1 - e_2)$ for some $\alpha \in \mathbb{R}$. Further, if x also satisfies (8.47.1), then $A^2 x = Ay$. Now,

$$A^2 x = A(Ax) = A[-2\alpha(e_1 - e_2)] = -2\alpha A(e_1 - e_2) = 4\alpha(e_1 - e_2)$$

and

$$Ay = [-y(1) + y(2)](e_1 - e_2).$$

Hence $A^2 x = Ay \Leftrightarrow 4\alpha = -y(1) + y(2)$, or $\alpha = \frac{1}{4}[-y(1) + y(2)]$. That is, $x = A^-(y) \Leftrightarrow x = \frac{1}{4}[-y(1) + y(2)](e_1 - e_2)$. This proves that the generalized inverse of A is given by

(8.47.2) $$A^-(y) = \frac{1}{4}[-y(1) + y(2)](e_1 - e_2) \quad \text{for all } y \in \ell_2(2).$$

It follows that the matrix of A^- is given by

$$A^- = \begin{bmatrix} -\frac{1}{4} & \frac{1}{4} \\ \frac{1}{4} & -\frac{1}{4} \end{bmatrix}.$$

Exercises

1. Assume that X and Y are inner product spaces with X n-dimensional, and let $A : X \to Y$ be linear.
   (a) If A is injective, then $\mathcal{R}(A)$ is n-dimensional.
   (b) If A has an inverse, then Y is n-dimensional.

2. Suppose A, A_n are in $\mathcal{B}(X, Y)$, x, x_n in X, $A_n \to A$, and $x_n \to x$. Show that $A_n x_n \to Ax$.

3. Let $\{e_1, e_2, \ldots, e_n\}$ be the canonical orthonormal basis in $\ell_2(n)$, i.e., $e_i(j) = \delta_{ij}$ for $i, j = 1, 2, \ldots, n$. Given any subset $\{x_1, x_2, \ldots, x_n\}$ of $\ell_2(n)$, consider the operator A on $\ell_2(n)$, defined by

$$Ax = \sum_{1}^{n} \langle x, x_i \rangle e_i \quad \text{for all } x \in \ell_2(n).$$

   (a) Show that A is a bounded linear operator from $\ell_2(n)$ into itself.
   (b) Show that A is injective if and only if $\{x_1, x_2, \ldots, x_n\}$ is linearly independent.
   (c) Show that A is surjective if and only if $\{x_1, x_2, \ldots, x_n\}$ is linearly independent. [Hint: The Gram determinant $g(x_1, x_2, \ldots, x_n) \neq 0$ if and only if $\{x_1, x_2, \ldots, x_n\}$ is linearly independent.]
   (d) Show that A is bijective if and only if $\{x_1, x_2, \ldots, x_n\}$ is linearly independent.
   (e) Show that the matrix of A relative to the basis $\{e_1, e_2, \ldots, e_n\}$ is given by $[a_{ij}]_{i,j=1}^{n}$, where $a_{ij} = \langle e_j, x_i \rangle = x_i(j)$.
   (f) If $x_i = \sum_{j=1}^{i} e_j$ $(i = 1, 2, \ldots, n)$, what is the matrix of A?

4. If $A \in \mathcal{B}(X, Y)$ has a bounded inverse, show that

$$\|A^{-1}\| = \left(\inf_{\|x\|=1} \|Ax\| \right)^{-1}.$$

5. Suppose X is an inner product space, $x \in X$, and $\{x_n\}$ in X satisfies

$$x^*(x_n) \to x^*(x) \quad \text{for all } x^* \in X^*.$$

   Show that $\{x_n\}$ is bounded: $\sup_n \|x_n\| < \infty$, and $\|x\| \leq \liminf_n \|x_n\|$. [Hint: For $n \in \mathbb{N}$, define $A_n : X^* \mapsto \mathbb{R}$ by $A_n(x^*) = x^*(x_n)$, $x^* \in X^*$. By Corollary 8.15, $\sup_n \|A_n\| < \infty$. But $\|x_n\| = \|A_n\|$ using Theorem 6.25 with $K = \{0\}$.]

6. Let X be a Hilbert space and let $\{e_n \mid n = 1, 2, \ldots\}$ be an orthonormal sequence in X. If $\sum_{1}^{\infty} \langle x, e_n \rangle$ converges for every $x \in X$, show that $\sum_{1}^{\infty} e_n$ converges. [Hint: Consider the sequence of functionals $\{x_n^*\}$ defined on X by $x_n^*(x) = \sum_{k=1}^{n} \langle x, e_k \rangle$ for every $x \in X$. Use the uniform boundedness theorem (Theorem 8.14).]

7. Define A on l_2 by

$$Ax = (x(1), x(2) - x(1), x(3) - x(2), \ldots, x(n) - x(n-1), \ldots).$$

   Show that
   (a) $A \in \mathcal{B}(l_2, l_2)$ and $\|A\| \leq 2$.

(b) A is injective.

(c) A is not surjective.

(d) $y \in \mathcal{R}(A)$ if and only if

$$\sum_{n=1}^{\infty} \left[\sum_{i=1}^{n} y(i) \right]^2 < \infty.$$

8. Let X be a Hilbert space, $A \in \mathcal{B}(X, X)$, and $\|A\| < 1$. Show that $I - A$ has a bounded inverse given by

$$(I - A)^{-1} = \lim_n (I + A + A^2 + \cdots + A^n),$$

where the limit is in the norm of the space $\mathcal{B}(X, X)$. [Hint: Let $B_n = I + A + A^2 + \cdots + A^n$, $n = 1, 2, \ldots$. Show that $\lim_n A^n = 0$, $B := \lim B_n$ exists, and $(I - A)B_n = I - A^{n+1} = B_n(I - A)$.]

9. Let X be a Hilbert space, $A \in \mathcal{B}(X, X)$, and $\|I - A\| < 1$. Show that A has a bounded inverse given by

$$A^{-1} = \lim_n [I + (I - A) + (I - A)^2 + \cdots + (I - A)^n].$$

[Hint: Problem 8.]

10. Verify Proposition 8.20. That is, prove that $X \times Y$ is an inner product space; moreover, $X \times Y$ is complete if and only if both X and Y are complete. [Hint: $\{(x_n, y_n)\}$ is Cauchy in $X \times Y$ if and only if $\{x_n\}$ is Cauchy in X and $\{y_n\}$ is Cauchy in Y.]

11. Prove Lemma 8.29. That is, if both A, $B \in \mathcal{B}(X, Y)$ have adjoints, then $A + B$, αA, and A^* also have adjoints, which are given by

(a) $(A + B)^* = A^* + B^*$,

(b) $(\alpha A)^* = \bar{\alpha} A^*$, and

(c) $A^{**} = A$.

12. If $A \in \mathcal{B}(X, Y)$ has an adjoint, then

(a) AA^* and A^*A are both self-adjoint,

(b) AA^* is "nonnegative" on Y, i.e., $\langle AA^*y, y \rangle \geq 0$ for all $y \in Y$,

(c) A^*A is nonnegative on X.

13. Let X be a Hilbert space and $P \in \mathcal{B}(X, X)$. Show that P is idempotent and self-adjoint (i.e., $P^2 = P$, and $P^* = P$) if and only if $P = P_M$ for some closed subspace M in X.

14. Let X be a Hilbert space and M and N closed subspaces. With the help of Exercise 13, prove:

(a) $P_M P_N$ is an orthogonal projection $\Leftrightarrow P_M$ and P_N commute: $P_M P_N = P_N P_M$. In this case, $P_M P_N = P_{M \cap N}$.

(b) $P_M + P_N$ is an orthogonal projection $\Leftrightarrow P_N P_M = 0$. In this case, $P_M + P_N = P_{M+N}$.

(c) If P_M and P_N commute, then $P_M + P_N - P_M P_N$ is an orthogonal projection.

(d) If P_M and P_N commute, then $P_M + P_N - 2P_M P_N$ is an orthogonal projection.

15. Let $A \in \mathcal{B}(X, Y)$ and suppose A^* exists.

(a) Show that $\mathcal{R}(A)$ is dense in Y if and only if A^* is injective. [Hint:

Lemma 8.33.]

(b) Show that $\mathcal{R}(A^*)$ is dense in X if and only if A is injective.

(c) If $\mathcal{R}(A)$ is closed, then A (respectively A^*) is injective if and only if A^* (respectively A) is surjective.

16. Prove Proposition 8.28. That is, let X be an inner product space, Y an n-dimensional inner product space with orthonormal basis $\{y_1, y_2, \ldots, y_n\}$, and $A \in \mathcal{B}(X, Y)$. Show that

(a) There exist $x_i^* \in X^*$ $(i = 1, 2, \ldots, n)$ such that

$$Ax = \sum_1^n x_i^*(x) y_i \quad \text{for all } x \in X.$$

(b) A^* exists if and only if each x_i^* has a representer in X.

(c) If x_i^* has the representer $x_i \in X$ $(i = 1, 2, \ldots, n)$, then A^* is given by

$$A^* y = \sum_{i=1}^n \langle y, y_i \rangle x_i \quad \text{for all } y \in Y.$$

17. Give an example of $A \in \mathcal{B}(X, X)$ such that $\|A^2\| < \|A\|^2$. Hence show that Lemma 8.32(2) is false if A is not normal.

18. If P and Q are metric projections onto closed subspaces of a Hilbert space, prove that $\overline{\mathcal{R}(PQP)} = \overline{\mathcal{R}(PQ)}$. [Hint: Lemma 8.33.]

19. Prove Lemma 8.37. That is, verify

(a) If A^{-1} exists, then $A^- = A^{-1}$.

(b) If $(A^*A)^{-1}$ exists, then $A^- = (A^*A)^{-1}A^*$.

20. If $A \in \mathcal{B}(X, Y)$ and $B \in \mathcal{B}(Y, Z)$ each have a bounded inverse, then $BA \in \mathcal{B}(X, Z)$ has a bounded inverse given by

$$(BA)^{-1} = A^{-1}B^{-1}.$$

21. Show that if $A \in \mathcal{B}(X, Y)$ has closed range, then $(A^*)^- = (A^-)^*$. In other words, the generalized inverse of the adjoint is the adjoint of the generalized inverse.

22. Compute the generalized inverse of each of the two matrices

$$\begin{bmatrix} 1 & 3 & 5 \\ 2 & 4 & 6 \end{bmatrix} \quad \text{and} \quad \begin{bmatrix} 1 & 2 \\ 3 & 4 \\ 5 & 6 \end{bmatrix}.$$

[Hint: Exercise 21 will reduce the amount of work.]

23. Prove the second equality in Theorem 8.45. That is, if X and Y are Hilbert spaces and $A \in \mathcal{B}(X, Y)$ has closed range, then $A^- = A^*(AA^*)^-$. [Hint: Inspect the proof of the first equality in (8.45.1) and use Theorem 8.43(3) applied to AA^* rather than A.]

24. Use Theorem 8.43 to verify Example 8.44(3). That is, if $A \in \mathcal{B}(\ell_2(n), \ell_2(n))$ is defined by

$$Ax = A\left(\sum_1^n \langle x, e_i \rangle\right) = \sum_1^n \alpha_i \langle x, e_i \rangle e_i, \quad x \in \ell_2(n),$$

for some scalars $\alpha_i \in \mathbb{R}$, then $A^- \in \mathcal{B}(\ell_2(n), \ell_2(n))$ is given by

$$A^- x = A^- \left(\sum_1^n \langle x, e_i \rangle e_i \right) = \sum_1^n \tilde{\alpha}_i \langle x, e_i \rangle e_i, \quad x \in \ell_2(n),$$

where

$$\tilde{\alpha}_i = \begin{cases} 1/\alpha_i & \text{if } \alpha_i \neq 0, \\ 0 & \text{if } \alpha_i = 0. \end{cases}$$

25. Use Theorem 8.43 to verify Example 8.44(4). That is, if A is the bounded linear functional on $\ell_2(n)$ defined by

$$Ax = \langle x, a \rangle = \sum_1^n x(i) a(i), \quad x \in \ell_2(n),$$

for some $a \in \ell_2(n) \setminus \{0\}$, then $A^- \in \mathcal{B}(\mathbb{R}, \ell_2(n))$ is given by

$$A^- y = \frac{1}{\|a\|^2} A^* y = \frac{a}{\|a\|^2} y, \quad y \in \mathbb{R}.$$

26. If $A \in \mathcal{B}(X, Y)$ and $\alpha \in \mathbb{R}$, show that $(\alpha A)^- = \frac{1}{\alpha} A^-$ if $\alpha \neq 0$, and $(\alpha A)^- = 0$ if $\alpha = 0$.

The next three exercises show that it is sometimes convenient to work in a "product space" setting for certain problems that do not initially appear to be in the form of a usual problem of best approximation in Hilbert space.

27. Let X be a Hilbert space and $x_i \in X$ for $i = 1, 2, \ldots, n$. Show that x minimizes $\sum_1^n \|x - x_i\|^2$ if and only if $x = \frac{1}{n} \sum_1^n x_i$. In other words, the vector that minimizes the sum of the squares of its distances from a finite number of points is just the average of these points. [Hint: Consider the product space $X^n := X \times X \times \cdots \times X$ (with n factors X) consisting of all n-tuples of vectors $(x_1, x_2, \ldots, x_n)$, where $x_i \in X$ for all i, with the inner product and norm defined by $\langle (x_1, x_2, \ldots, x_n), (y_1, y_2, \ldots, y_n) \rangle := \sum_1^n \langle x_i, y_i \rangle$ and $\|(x_1, x_2, \ldots, x_n)\|^2 = \sum_1^n \|x_i\|^2$. Show that X^n is a Hilbert space (see Exercise 10). Let $D := \{(x, x, \ldots, x) \mid x \in X\}$ denote the "diagonal" subspace in X^n. Observe that $(x_0, x_0, \ldots, x_0) \in D$ is the best approximation to $(x_1, x_2, \ldots, x_n)$ from D if and only if x_0 minimizes $\sum_1^n \|x - x_i\|^2$. Finally, apply the characterization of best approximations (Theorem 4.9 in the space X^n) to deduce that $(x_0, x_0, \ldots, x_0)$ is the best approximation to $(x_1, x_2, \ldots, x_n)$ from D if and only if $x_0 = \frac{1}{n} \sum_1^n x_i$.]

28. More generally than Exercise 27, let $A \in \mathcal{B}(X, Y)$ and $y_i \in Y$ ($i = 1, \ldots, n$). Show that x minimizes $\sum_1^n \|Ax - y_i\|^2$ if and only if x is a generalized solution to $Ax = \frac{1}{n} \sum_1^n y_i$. [Hint: Consider the map $\hat{A} : X \to Y^n$ defined by $\hat{A}(x) := (Ax, Ax, \ldots, Ax)$. Apply known results to $\hat{A}$ instead of A.]

29. More generally than Exercise 28, let $A_i \in \mathcal{B}(X, Y)$ and $y_i \in Y$ ($i = 1, 2, \ldots, n$). Show that $x \in X$ minimizes $\sum_1^n \|A_i x - y_i\|^2$ if and only if x is a solution of the equation $(\sum_1^n A_i^* A_i) x = \sum_1^n A_i^* y_i$.

Historical Notes

Baire (1899) proved (part (1) of) Baire's theorem 8.13 in the particular case where $X = \ell_2(n)$. (The case $n = 1$ had been proved two years earlier by Osgood (1897).) But the result and proof immediately generalize when $\ell_2(n)$ is replaced by any *complete metric space*. This was done explicitly by Kuratowski (1930) and Banach (1930).

The uniform boundedness theorem (Theorem 8.14) was proved when X is a Banach space and Y is the scalar field by Hildebrandt (1923; p. 311), but exactly the same proof is valid when Y is any Banach space. This result is often incorrectly attributed to Banach or Banach and Steinhaus, although certain special cases were known earlier (see Dunford and Schwartz (1958; pp. 80–81)). Banach and Steinhaus (1927) showed that the uniform boundedness theorem is still valid when X and Y are Banach spaces and (8.14.1) holds only for x in a second-category subset of X. For an elegant proof of the uniform boundedness theorem, based only on completeness and avoiding the notion of category, see Hausdorff (1932; p. 304).

Banach (1929a) and (1929b) first proved the bounded inverse theorem (Theorem 8.19) for the general case where X and Y are Banach spaces. For another proof of Theorem 8.19, see Schauder (1930). Banach (1932; Theorem 7, p. 41) also proved the closed graph theorem (Theorem 8.23) in the general case where both X and Y are Banach spaces.

The notion of the *adjoint* of a bounded linear operator originated with matrix theory and with the theories of differential and integral equations. In the L_p spaces ($p > 1$) and ℓ_2, Riesz (1910; p. 478), (1913; p. 85) made use of this notion. Banach (1929a; p. 235) introduced the adjoint operator in a general Banach space, and proved Theorem 8.31 and Lemma 8.33(2) in this context. The closed range theorem (Theorem 8.18) is due to Hestenes (1961; Theorem 3.3).

What we have called a "generalized solution" is also called a "least squares solution" or "extremal solution" by some authors. The notion of a generalized inverse goes back at least to Fredholm (1903), where a particular generalized inverse (called a "pseudoinverse") of an integral operator was considered. (A nice outline of the history behind generalized inverses can be found in the Introduction of the book by Ben-Israel and Greville (1974).) Generalized inverses of differential operators were implicit in the study of generalized Green's functions by Hilbert (1904). For a good historical survey of this subject up to 1968, see Reid's book (1968). We should mention that some authors use other notation for generalized inverses, for example, $A^\dagger$ and A^+.

In a short abstract, Moore (1920) (see also Moore (1935) for more complete details) defined a unique generalized inverse (called a "general reciprocal") for *every* rectangular matrix. Moore's definition, in the more general context of bounded linear operators with closed range, is the following: If $A \in \mathcal{B}(X, Y)$ has closed range, then A^- is the unique operator in $\mathcal{B}(Y, X)$ that satisfies $AA^- = P_{\mathcal{R}(A)}$ and $A^- A = P_{\mathcal{R}(A^-)}$. That is, owing to Theorem 8.43, Moore's definition of generalized inverse agrees with the "variational" definition (Definition 8.36) given here. We preferred Definition 8.36 since it seemed to us more natural and geometrically intuitive, while Moore's "definition" (as well as others mentioned below) actually required first *proving* that the object even existed and was unique! Of course, with Definition 8.36, it may not have been a priori obvious that the generalized inverse was linear. We note that Definition 8.36 of generalized inverse was also preferred in the book of Groetsch (1977).

A number of authors, evidently unaware of Moore's work, also considered generalized inverses in different contexts, for example, Siegel (1937) for matrices, and Tseng (1933) (see also (1949a), (1949b), (1949c), and (1956)), Murray and von Neumann (1936), and Atkinson (1951), (1953) for operators. Bjerhammar (1951) defined the generalized inverse of a rectangular matrix, and Penrose (1955) refined Bjerhammar's notion, which, in the more general context of operators, may be stated as follows. If $A \in \mathcal{B}(X, Y)$ has closed range, then A^- is the unique operator in $\mathcal{B}(Y, X)$ that satisfies the following four properties: (1) $AA^- = (AA^-)^*$, (2) $A^-A = (A^-A)^*$, (3) $AA^-A = A$, and (4) $A^-AA^- = A^-$. Rado (1956) pointed out the equivalence of the definitions of Moore and Penrose, and this is one of the reasons why this generalized inverse is often called the "Moore–Penrose" inverse. That is, Rado (1956) proved the equivalence of statements (2) and (3) in Theorem 8.43. For an operator $A \in \mathcal{B}(X, Y)$ that has closed range, Desoer and Whalen (1963) defined the generalized inverse of A to be the unique operator $A^- \in \mathcal{B}(Y, X)$ that satisfies (1) $A^-Ax = x$ for all $x \in \mathcal{N}(A)^\perp$ and (2) $A^-y = 0$ for all $y \in \mathcal{R}(A)^\perp$. By Theorem 8.43, the definitions of Moore, Penrose, and Desoer and Whalen are all equivalent to Definition 8.36.

Anderson and Duffin (1969; Theorem 8) have proved the following interesting result: If M and N are closed subspaces of a Hilbert space, then $P_{M \cap N} = 2P_M(P_M + P_N)^- P_N$.

While the Moore–Penrose inverse is generally agreed to be *the* generalized inverse par excellence, not all definitions of generalized inverses that have been studied are equivalent. For example, Rao (1962) defined a "g-inverse" of an $m \times n$ matrix A to be an $n \times m$ matrix G that satisfies only the third property of Penrose's definition, namely, $AGA = A$. (See also the book of Rao and Mitra (1971), where this is more fully developed.)

The main reason for the "closed range" hypothesis (8.39.4) for several of the fundamental results of this chapter is to ensure the *existence* of generalized solutions, and hence the existence of generalized inverses. To see this, first note that the range of a matrix operator, being finite-dimensional, is always closed, so the closed range hypothesis is automatically satisfied. However, for more general operators between Hilbert spaces, this is not always the case. If $\mathcal{R}(A)$ were not closed in the Hilbert space Y, then $\mathcal{R}(A)$ would not be Chebyshev, and hence there would exist some $y \in Y$ that failed to have a best approximation in $\mathcal{R}(A)$. It follows by Theorem 8.35(3) that the set of generalized solutions $G(y)$ to the equation $Ax = y$ is empty! In particular, the generalized inverse could *not* be defined at y. In the case where the range of the operator is not closed, the generalized inverse turns out to be an *unbounded* linear operator (see Groetsch (1977; Chapter III)).

There have been many books and countless papers devoted to the theory and computation of generalized inverses. For example, a small sample of books includes Boullion and Odell (1971), Rao and Mitra (1971), Ben-Israel and Greville (1974), Nashed (1976), and Groetsch (1977). In particular, Groetsch (1977; Chapter II) has described various methods for actually computing the generalized inverse of operators. Finally, we should note that a very general abstract theory of generalized inverses for linear operators between two linear topological spaces has been studied by Nashed and Votruba ([1974I], [1974II]).

CHAPTER 9

THE METHOD OF ALTERNATING PROJECTIONS

The Case of Two Subspaces

In this chapter we will describe a theoretically powerful method for computing best approximations from a closed convex set K that is the intersection of a finite number of closed convex sets, $K = \cap_1^r K_i$. This method is an iterative algorithm that reduces the problem to finding best approximations from the *individual* sets K_i. The efficacy of the method thus depends on whether the given set K can be represented as the intersection of a finite number of sets K_i from which it is "easy" to compute best approximations. This will be the case, for example, when the K_i are either half-spaces, hyperplanes, finite-dimensional subspaces, or certain cones. Several applications will be made to a variety of problems including solving linear equations, solving linear inequalities, computing the best isotone and best convex regression functions, and solving the general shape-preserving interpolation problem.

The *rate of convergence* of the method of alternating projections will also be studied, at least when all the K_i's are subspaces or, more generally, affine sets. This rate will be described in terms of the *angles* between the various subspaces involved.

In its simplest abstract formulation (namely, the intersection of two subspaces), the method of alternating projections originated with von Neumann in 1933. Because the theory is complete, simpler, and more elegant in the two-subspace case, we shall study this case first in some detail to motivate the more general results to come later.

We begin with two simple observations. The first characterizes the bounded linear operators that are orthogonal projections, and the second characterizes exactly when the product of two orthogonal projections is again an orthogonal projection.

9.1 Lemma. *Let X be an inner product space and $A \in B(X, X)$. Then A is idempotent and self-adjoint (i.e., $A^2 = A$ and $A^* = A$) if and only if $A = P_M$ for some Chebyshev subspace M. (In fact, $M = \mathcal{R}(A)$, the range of A.)*

Proof. If $A = P_M$, then by Theorem 5.13, $A = A^*$ and $A^2 = A$.

Conversely, suppose $A = A^*$ and $A^2 = A$. Let $M = \mathcal{R}(A)$. If $y \in M$, $y = Ax$ for some $x \in X$, which implies $y = Ax = A^2x = A(Ax) = Ay$. That is, $Ay = y$ for all $y \in M$. For any $x \in X$ and $y \in M$,

$$\langle x - Ax, y \rangle = \langle x, y \rangle - \langle Ax, y \rangle = \langle x, y \rangle - \langle x, Ay \rangle \quad (\text{since } A^* = A)$$
$$= \langle x, y \rangle - \langle x, y \rangle = 0.$$

Thus $x - Ax \in M^\perp$. By Theorem 4.9, $Ax = P_M x$. Since x was arbitrary, $A = P_M$. ∎

Next we characterize when orthogonal projections commute.

9.2 Lemma. *Let M and N be Chebyshev subspaces of the inner product space X. Then the following statements are equivalent:*

(1) P_M *and* P_N *"commute"; i.e.,* $P_M P_N = P_N P_M$;
(2) $P_N(M) \subset M$;
(3) $P_M(N) \subset N$;
(4) $P_M P_N = P_{M \cap N}$;
(5) $P_M P_N$ *is the orthogonal projection onto a Chebyshev subspace.*

In particular, if $M \subset N$ or $N \subset M$, then P_M and P_N commute.

Proof. $(1) \Longrightarrow (2)$. Suppose P_M and P_N commute. Then for any $x \in M$, we have that

$$P_N(x) = P_N P_M(x) = P_M P_N(x) \in M.$$

Thus (2) holds.

$(2) \Longrightarrow (3)$. Suppose (2) holds, $x \in N$, and $y \in N^\perp$. Then

$$\begin{aligned}
\langle P_M(x), y \rangle &= \langle P_M P_N(x), y \rangle = \langle P_N(x), P_M(y) \rangle \\
&= \langle x, P_N P_M(y) \rangle = \langle x, P_M P_N P_M(y) \rangle \ (\text{since } P_N P_M(y) \in M \text{ by } (2)) \\
&= \langle P_M(x), P_N P_M(y) \rangle = \langle P_N P_M(x), P_M(y) \rangle \\
&= \langle P_M P_N P_M(x), y \rangle = \langle P_N P_M(x), y \rangle \text{ by } (2) \\
&= 0 \ (\text{since } y \in N^\perp).
\end{aligned}$$

This proves that $P_M(x) \in N^{\perp\perp}$. But $N^{\perp\perp} = N$ by Theorem 4.5(8). Hence $P_M(x) \in N$ for every $x \in N$ and (3) holds.

$(3) \Longrightarrow (4)$. Suppose $P_M(N) \subset N$. Let $x \in X$ and $x_0 := P_M P_N(x)$. To show $x_0 = P_{M \cap N}(x)$, it suffices by Theorem 4.9 to show that $x_0 \in M \cap N$ and $x - x_0 \in (M \cap N)^\perp$. For every $y \in M \cap N$,

$$\begin{aligned}
\langle x - x_0, y \rangle &= \langle x, y \rangle - \langle P_M P_N(x), y \rangle = \langle x, y \rangle - \langle P_N(x), P_M(y) \rangle \\
&= \langle x, y \rangle - \langle P_N(x), y \rangle = \langle x, y \rangle - \langle x, P_N(y) \rangle = \langle x, y \rangle - \langle x, y \rangle = 0.
\end{aligned}$$

Thus $x - x_0 \in (M \cap N)^\perp$. Clearly, $x_0 = P_M P_N(x) \in M$. Also, for every $z \in N^\perp$,

$$\begin{aligned}
\langle x_0, z \rangle &= \langle P_M P_N(x), z \rangle = \langle P_N P_M P_N(x), z \rangle \quad (\text{since } P_M P_N(x) \in N \text{ by } (3)) \\
&= \langle P_M P_N(x), P_N(z) \rangle = \langle P_M P_N(x), 0 \rangle \quad (\text{by Theorem 5.8(4)}) \\
&= 0.
\end{aligned}$$

Thus $x_0 \in N^{\perp\perp} = N$ using Theorem 4.5(8). Hence $x_0 \in M \cap N$, and this proves $x_0 = P_{M \cap N}(x)$.

The implication $(4) \Longrightarrow (5)$ is obvious.

$(5) \Longrightarrow (1)$. Let $P = P_M P_N$ be the orthogonal projection onto some Chebyshev subspace. By Lemma 9.1, $(P_M P_N)^* = P_M P_N$ and $(P_M P_N)^2 = P_M P_N$. Using Lemma 8.30, we obtain $P_M P_N = (P_M P_N)^* = P_N^* P_M^* = P_N P_M$ so (1) holds.

The last statement of the theorem follows from the equivalence of statements (1) through (3). ∎

This theorem says that if the projections P_M and P_N commute, the product $P_M P_N$ is the projection $P_{M \cap N}$ onto $M \cap N$. What can be said if P_M and P_N do *not* commute? We will see that if $P_M P_N$ is *iterated*, then the iterates of $P_M P_N$ converge pointwise to $P_{M \cap N}$.

More precisely, let M_1 and M_2 be closed subspaces of the Hilbert space X, let $x \in X$ be arbitrary, and define the "alternating" sequence $\{x_n\}$ as follows:

$$x_0 = x,$$

(9.2.1) $$x_{2n-1} = P_{M_1}(x_{2n-2}),$$

$$x_{2n} = P_{M_2}(x_{2n-1}) \quad (= (P_{M_2} P_{M_1})^n(x)) \quad (n = 1, 2, \dots).$$

(see Figure 9.2.3 below). Then the next theorem shows that

(9.2.2) $$\lim x_n = P_{M_1 \cap M_2}(x).$$

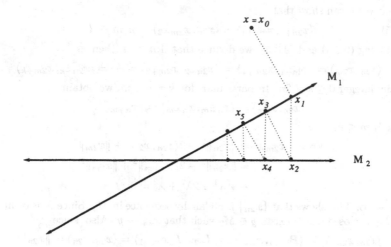

Figure 9.2.3. Alternating projections

9.3 Von Neumann Theorem. *Let M_1 and M_2 be closed subspaces in the Hilbert space X. Then for each $x \in X$,*

(9.3.1) $$\lim_{n \to \infty} (P_{M_2} P_{M_1})^n(x) = P_{M_1 \cap M_2}(x).$$

Proof. For simplicity of notation, let $P_i = P_{M_i}$ $(i = 1, 2)$. Fix any $x \in X$ and let

$$x_0 = x,$$

(9.3.2) $$x_{2n-1} = P_1(x_{2n-2}) \quad (= P_1(P_2 P_1)^{n-1}x),$$

$$x_{2n} = P_2(x_{2n-1}) \quad (= (P_2 P_1)^n x),$$

for $n = 1, 2, \dots$. To complete the proof, it suffices to show that $x_{2n} \to P_{M_1 \cap M_2}(x)$. In fact, we will prove that the *whole sequence* $\{x_n\}$ converges to $P_{M_1 \cap M_2}(x)$.

Note that since $\|P_i\| \leq 1$,

$$\|x_{2n}\| = \|P_2 x_{2n-1}\| \leq \|x_{2n-1}\| = \|P_1 x_{2n-2}\| \leq \|x_{2n-2}\|.$$

This proves that the sequence of norms $\{\|x_n\|\}$ is decreasing, and hence

(9.3.3) $$\lambda := \lim_{n \to \infty} \|x_n\|$$

exists.

Using the facts that the P_i are self-adjoint (Theorem 5.13), $x_{2n} \in M_2$, and $x_{2n-1} \in M_1$, we deduce that

$$\langle x_{2n}, x_{2m} \rangle = \langle P_2 x_{2n-1}, x_{2m} \rangle = \langle x_{2n-1}, P_2 x_{2m} \rangle = \langle x_{2n-1}, x_{2m} \rangle$$
$$= \langle P_1 x_{2n-1}, x_{2m} \rangle = \langle x_{2n-1}, P_1 x_{2m} \rangle = \langle x_{2n-1}, x_{2m+1} \rangle.$$

That is,

(9.3.4) $$\langle x_{2n}, x_{2m} \rangle = \langle x_{2n-1}, x_{2m+1} \rangle, \quad n, m \in \mathbb{N}.$$

Similarly, we can show that

(9.3.5) $$\langle x_{2n+1}, x_{2m+1} \rangle = \langle x_{2n}, x_{2m+2} \rangle, \quad n, m \in \mathbb{N}.$$

Combining (9.3.4) and (9.3.5), we deduce that if $n \geq m$, then

$$\langle x_{2n}, x_{2m} \rangle = \langle x_{2n-1}, x_{2m+1} \rangle = \langle x_{2n-2}, x_{2m+2} \rangle = \cdots = \langle x_{2n-k}, x_{2m+k} \rangle$$

for any integer $0 \leq k \leq 2n$. In particular, for $k = n - m$, we obtain

$$\langle x_{2n}, x_{2m} \rangle = \langle x_{n+m}, x_{n+m} \rangle = \|x_{n+m}\|^2.$$

Thus, if $m \leq n$,

$$\|x_{2n} - x_{2m}\|^2 = \|x_{2n}\|^2 - 2\langle x_{2n}, x_{2m} \rangle + \|x_{2m}\|^2$$
$$= \|x_{2n}\|^2 - 2\|x_{n+m}\|^2 + \|x_{2m}\|^2$$
$$\longrightarrow \lambda^2 - 2\lambda^2 + \lambda^2 = 0$$

as $m \to \infty$. This shows that $\{x_{2n}\}$ is a Cauchy sequence in M_2. Since X is complete and M_2 is closed, there exists $y \in M_2$ such that $x_{2n} \to y$. Also, since

$$\langle x_{2n}, x_{2n-1} \rangle = \langle P_2 x_{2n}, x_{2n-1} \rangle = \langle x_{2n}, P_2 x_{2n-1} \rangle = \langle x_{2n}, x_{2n} \rangle = \|x_{2n}\|^2$$

implies that

$$\|x_{2n} - x_{2n-1}\|^2 = \|x_{2n}\|^2 - 2\langle x_{2n}, x_{2n-1} \rangle + \|x_{2n-1}\|^2$$
$$= -\|x_{2n}\|^2 + \|x_{2n-1}\|^2 \longrightarrow -\lambda^2 + \lambda^2 = 0$$

as $n \to \infty$, it follows that

$$\|x_{2n-1} - y\| \leq \|x_{2n-1} - x_{2n}\| + \|x_{2n} - y\| \longrightarrow 0.$$

Since $x_{2n-1} \in M_1$ for all n and M_1 is closed, $y \in M_1$. Thus, we have shown that

(9.3.6) $$\lim x_n = y \in M_1 \cap M_2.$$

To verify that $y = P_{M_1 \cap M_2}(x)$, it suffices, by Theorem 4.9, to show that $x - y \perp M_1 \cap M_2$. To this end, let $z \in M_1 \cap M_2$. Then, using Theorem 4.9,

$$\langle x_{2n}, z \rangle = \langle x_{2n} - x_{2n-1} + x_{2n-1}, z \rangle = \langle P_1 x_{2n-1} - x_{2n-1} + x_{2n-1}, z \rangle = \langle x_{2n-1}, z \rangle.$$

Similarly,

$$\langle x_{2n-1}, z \rangle = \langle x_{2n-2}, z \rangle.$$

By induction, it follows that $\langle x_{2n}, z \rangle = \langle x_0, z \rangle = \langle x, z \rangle$ for all n. Thus

$$\langle x - y, z \rangle = \langle x, z \rangle - \langle y, z \rangle = \langle x, z \rangle - \lim \langle x_{2n}, z \rangle = \langle x, z \rangle - \langle x, z \rangle = 0. \quad \blacksquare$$

Remarks. (1) An inspection of the proof shows that the theorem holds, more generally, whenever M_1 and M_2 are any *complete* (hence Chebyshev) subspaces in any inner product space X that is complete or not.

(2) Geometrically, it appears that the rate of convergence of the alternating sequence $\{x_n\}$ defined in the theorem (see Figure 9.2.3) depends on the "angle" between the subspaces M_1 and M_2. We shall see below that this is indeed the case.

Angle Between Two Subspaces

9.4 Definition. *The* **angle** *between two subspaces M and N is the angle $\alpha(M,N)$ between 0 and $\pi/2$ whose cosine, $c(M,N) = \cos\alpha(M,N)$, is defined by the expression*

(9.4.1)
$$c(M,N) := \sup\{|\langle x,y\rangle| \mid x \in M \cap (M \cap N)^{\perp}, \|x\| \leq 1,$$
$$y \in N \cap (M \cap N)^{\perp}, \|y\| \leq 1\}.$$

Some basic facts about the angle between two subspaces are recorded next.

9.5 Lemma. *Let M and N be closed subspaces in the Hilbert space X. Then:*

(1) $0 \leq c(M,N) \leq 1$.

(2) $c(M,N) = c(N,M)$.

(3) $c(M,N) = c\left(M \cap (M \cap N)^{\perp}, N \cap (M \cap N)^{\perp}\right)$.

(4) $c(M,N) = \sup\{|\langle x,y\rangle| \mid x \in M, \|x\| \leq 1, y \in N \cap (M \cap N)^{\perp}, \|y\| \leq 1\}$.

(5) $c(M,N) = \sup\{|\langle x,y\rangle| \mid x \in M \cap (M \cap N)^{\perp}, \|x\| \leq 1, y \in N, \|y\| \leq 1\}$.

(6) $|\langle x,y\rangle| \leq c(M,N) \|x\| \|y\|$ whenever $x \in M$, $y \in N$, and at least one of x or y is in $(M \cap N)^{\perp}$.

(7) $c(M,N) = \|P_N P_M - P_{M\cap N}\| = \|P_N P_M P_{(M\cap N)^{\perp}}\|$
$$= \|P_{M\cap(M\cap N)^{\perp}} P_{N\cap(M\cap N)^{\perp}}\|.$$

(8) $c(M,N) = 0$ if and only if P_M and P_N commute. (In this case, $P_N P_M = P_{M\cap N}$.) In particular, $c(M,N) = 0$ if $M \subset N$ or $N \subset M$.

Proof. Statements (1) and (2) are obvious.

(3) Let $S = M \cap (M \cap N)^{\perp}$ and $T = N \cap (M \cap N)^{\perp}$. Then $S \cap T = \{0\}$, so $(S \cap T)^{\perp} = X$ and

$$c(M,N) = \sup\{|\langle x,y\rangle| \mid x \in S, \|x\| \leq 1, y \in T, \|y\| \leq 1\}$$
$$= \sup\{|\langle x,y\rangle| \mid x \in S \cap (S \cap T)^{\perp}, \|x\| \leq 1, y \in T \cap (S \cap T)^{\perp}, \|y\| \leq 1\}$$
$$= c(S,T).$$

(4) Using the readily verified facts that

$$M = M \cap N + M \cap (M \cap N)^{\perp}$$

and

$$P_M = P_{M\cap N} + P_{M\cap(M\cap N)^{\perp}}$$

(see Exercise 18 at the end of the chapter), we deduce that

$$\sup\{|\langle x, y\rangle| \mid x \in M, \|x\| \le 1, \ y \in N \cap (M \cap N)^{\perp}, \|y\| \le 1\}$$
$$= \sup\{|\langle P_M x, P_{N \cap (M \cap N)^{\perp}} y\rangle| \mid \|x\| \le 1, \|y\| \le 1\}$$
$$= \sup\{|\langle P_{M \cap N} x + P_{M \cap (M \cap N)^{\perp}} x, \ P_{N \cap (M \cap N)^{\perp}} y\rangle| \mid \|x\| \le 1, \|y\| \le 1\}$$
$$= \sup\{|\langle P_{M \cap (M \cap N)^{\perp}} x, \ P_{N \cap (M \cap N)^{\perp}} y\rangle| \mid \|x\| \le 1, \|y\| \le 1\}$$
$$= \sup\{|\langle x, y\rangle| \mid x \in M \cap (M \cap N)^{\perp}, \|x\| \le 1, \ y \in N \cap (M \cap N)^{\perp}, \|y\| \le 1\}$$
$$= c(M, N).$$

(5) Interchange the roles played by M and N in (4) and use (2).

(6) This follows easily from (4) and (5).

(7) Clearly,

$$c(M, N)$$
$$= \sup\{|\langle x, y\rangle| \mid x \in M \cap (M \cap N)^{\perp}, \|x\| \le 1, \ y \in N \cap (M \cap N)^{\perp}, \|y\| \le 1\}$$
$$= \sup\{|\langle P_{M \cap (M \cap N)^{\perp}} x, P_{N \cap (M \cap N)^{\perp}} y\rangle| \mid \|x\| \le 1, \|y\| \le 1\}$$
$$= \sup\{|\langle P_{N \cap (M \cap N)^{\perp}} P_{M \cap (M \cap N)^{\perp}} x, y\rangle| \mid \|x\| \le 1, \|y\| \le 1\}$$
$$= \|P_{N \cap (M \cap N)^{\perp}} P_{M \cap (M \cap N)^{\perp}}\|,$$

where the last equality follows from (8.0.5). Using (2), this verifies the third equality of (7).

Since $M \cap N \subset M$, $P_{M \cap N} = P_M P_{M \cap N} = P_{M \cap N} P_M$ by Theorem 5.14. Hence

$$P_M P_{(M \cap N)^{\perp}} = P_M (I - P_{M \cap N}) = P_M - P_M P_{M \cap N} = P_M - P_{M \cap N} P_M$$
$$= (I - P_{M \cap N}) P_M = P_{(M \cap N)^{\perp}} P_M.$$

That is, P_M commutes with $P_{(M \cap N)^{\perp}}$. Similarly, P_N commutes with $P_{(M \cap N)^{\perp}}$. From Lemma 9.2, $P_{M \cap (M \cap N)^{\perp}} = P_M P_{(M \cap N)^{\perp}}$ and $P_{N \cap (M \cap N)^{\perp}} = P_N P_{(M \cap N)^{\perp}}$. Thus

$$c(M, N) = \|P_{N \cap (M \cap N)^{\perp}} P_{M \cap (M \cap N)^{\perp}}\| = \|P_N P_{(M \cap N)^{\perp}} P_M P_{(M \cap N)^{\perp}}\|$$
$$= \|P_N P_M P_{(M \cap N)^{\perp}}\|,$$

which proves the second equality of (7). But

$$P_N P_M P_{(M \cap N)^{\perp}} = P_N P_M (I - P_{M \cap N}) = P_N P_M - P_N P_M P_{M \cap N} = P_N P_M - P_{M \cap N},$$

since $M \cap N \subset M$ and $M \cap N \subset N$. This proves the first equality of (7).

(8) From (7) we see that $c(M, N) = 0$ if and only if $P_N P_M = P_{M \cap N}$. By Lemma 9.2, this means that P_N and P_M commute. The last statement also follows from Lemma 9.2. ∎

9.6 Examples. (1) Consider the subspaces (i.e., "lines") in $\ell_2(2)$ given by

$$M = \{x \in \ell_2(2) \mid x(1) = x(2)\} \qquad \text{and} \qquad N = \{x \in \ell_2(2) \mid x(2) = 0\}$$

(see Figure 9.6.1). Then $c(M, N) = 1/\sqrt{2}$, and hence $\alpha(M, N) = \pi/4$.

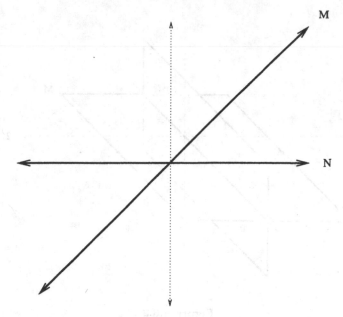

Figure 9.6.1

To see this, note that $M \cap N = \{0\}$, and if $x \in M \cap (M \cap N)^{\perp} = M$, $\|x\| \le 1$, and $y \in N \cap (M \cap N)^{\perp} = N$, $\|y\| \le 1$, then $|x(1)| = |x(2)| \le 1/\sqrt{2}$, $|y(1)| \le 1$, and

$$\langle x, y \rangle = x(1)y(1) \le 1/\sqrt{2},$$

with equality holding when $x(1) = x(2) = 1/\sqrt{2}$ and $y(1) = 1$.

(2) Consider the subspaces (i.e., "planes") in $\ell_2(3)$ given by $M = \{x \in \ell_2(3) \mid x(3) = 0\}$ and $N = \{x \in \ell_2(3) \mid x(2) = 0\}$ (see Figure 9.6.2). Then $c(M, N) = 0$ and hence $\alpha(M, N) = \pi/2$. We leave the verification of this as an exercise.

Remark. Note that when $c(M, N) < 1$, Theorem 9.5(6) represents a strengthening of the Schwarz inequality for certain pairs of vectors x and y. Moreover, the rate of convergence of the method of alternating projections given in Theorem 9.3 is governed by the factor $c(M, N)^{2n-1}$ (see Theorem 9.8 below). For these reasons, it is important to know when $c(M, N) < 1$. It will be verified later in Theorem 9.35 that $c(M, N) < 1$ if and only if $M + N$ is closed. To prove this, we need first to discuss some results concerning "weak convergence." However, at this point we can establish another characterization of when $M + N$ is closed.

9.7 Theorem. *Let M and N be closed subspaces of the Hilbert space X such that $M + N$ is closed. Then*

(9.7.1) $$(M \cap N)^{\perp} = M^{\perp} + N^{\perp}.$$

In particular, $M + N$ is closed if and only if $M^{\perp} + N^{\perp}$ is closed.

Proof. First observe that by Theorem 4.6(4),

(9.7.2) $$(M \cap N)^{\perp} = \overline{M^{\perp} + N^{\perp}}.$$

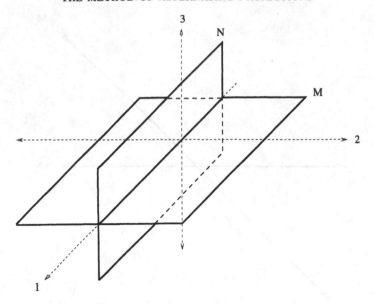

Figure 9.6.2

Thus to verify (9.7.1), it suffices to show that

$$(9.7.3) \qquad\qquad (M \cap N)^{\perp} \subset M^{\perp} + N^{\perp}.$$

To this end, let $x \in (M \cap N)^{\perp}$. Then $Z := M + N$ is a closed subspace of X so is a Hilbert space in its own right. If some $z \in Z$ has two representations $z = u_1 + v_1 = u_2 + v_2$, where $u_i \in M$, $v_i \in N$, then $u_1 - u_2 = v_2 - v_1 \in M \cap N$, so $\langle u_1 - u_2, x \rangle = 0$ or $\langle u_1, x \rangle = \langle u_2, x \rangle$. It follows that the functional f, defined on Z by

$$(9.7.4) \qquad f(u + v) := \langle u, x \rangle \quad \text{for each } u \in M \text{ and } v \in N,$$

is well-defined, linear, and satisfies

$$(9.7.5) \qquad\qquad f(v) = 0 \quad \text{for every } v \in N.$$

We next show that f is *bounded* on Z, i.e., $f \in Z^*$. First we observe that $A : M \times N \to M + N$, the "addition" mapping on $M \times N$ defined by $A(u,v) = u + v$, is bounded, linear, and onto. The linearity and the surjectivity of A are clear. To see that A is bounded, we note that for each $(u, v) \in M \times N \setminus \{(0,0)\}$,

$$\frac{\|A(u,v)\|}{\|(u,v)\|} = \frac{\|u + v\|}{(\|u\|^2 + \|v\|^2)^{1/2}} \leq \frac{\|u\| + \|v\|}{(\|u\|^2 + \|v\|^2)^{1/2}} \leq \frac{\sqrt{2}(\|u\|^2 + \|v\|^2)^{1/2}}{(\|u\|^2 + \|v\|^2)^{1/2}} = \sqrt{2}.$$

This proves that A is bounded (and, in fact, $\|A\| \leq \sqrt{2}$). Since $M + N$ is closed, hence complete, in X, it follows by the open mapping theorem (Theorem 8.17) that there exists $\delta > 0$ such that $A(B(M \times N)) \supset \delta B(M + N)$, where $B(E)$ denotes the unit ball in E. In particular, for each $u \in M$, $v \in N$ with $\|u + v\| \leq 1$, there exists $(\bar{x}, \bar{y}) \in B(M \times N)$ such that $\bar{x} + \bar{y} = A(\bar{x}, \bar{y}) = \delta(u + v)$. Then

$$|f(u + v)| = \frac{1}{\delta}|f(\bar{x} + \bar{y})| = \frac{1}{\delta}|\langle \bar{x}, x \rangle| \leq \frac{1}{\delta}\|\bar{x}\| \, \|x\| \leq \frac{1}{\delta}\|x\|.$$

Thus $\|f\| \leq \frac{1}{\delta}\|x\|$ and f is bounded on Z.

By the Fréchet-Riesz representation theorem (Theorem 6.10), there exists $\phi \in Z$ such that

(9.7.6) $$f(z) = \langle z, \phi \rangle \qquad \text{for each } z \in Z.$$

In particular,

(9.7.7) $$f(u + v) = \langle u + v, \phi \rangle \qquad \text{for each } u \in M \text{ and } v \in N.$$

By (9.7.5) it follows that $\phi \in N^\perp$. That is,

(9.7.8) $$f(u + v) = \langle u, \phi \rangle \qquad \text{for each } u \in M \text{ and } v \in N.$$

Combining (9.7.4) and (9.7.8), we deduce $\langle u, x \rangle = \langle u, \phi \rangle$ for every $u \in M$, and so $x - \phi \in M^\perp$. Hence

$$x = (x - \phi) + \phi \in M^\perp + N^\perp,$$

and this completes the proof of (9.7.3). This proves (9.7.1).

In particular, since $(M \cap N)^\perp$ is closed, $M^\perp + N^\perp$ must be closed whenever $M + N$ is closed. Conversely, if $M^\perp + N^\perp$ is closed, then the above proof shows that $M^{\perp\perp} + N^{\perp\perp}$ is closed. However, by Theorem 4.5(8), we have $M^{\perp\perp} = M$ and $N^{\perp\perp} = N$. This proves that $M + N$ is closed when $M^{\perp\perp} + N^{\perp\perp}$ is. ∎

Rate of Convergence for Alternating Projections (Two Subspaces)

The next theorem shows that the rate of convergence of the method of alternating projections is fast (respectively slow) if the angle between the subspaces is large (respectively small).

9.8 Rate of Convergence. *Let M_1 and M_2 be closed subspaces in the Hilbert space X and $c = c(M_1, M_2)$. Then, for each $x \in X$,*

(9.8.1) $$\|(P_{M_2}P_{M_1})^n(x) - P_{M_1 \cap M_2}(x)\| \leq c^{2n-1}\|x - P_{M_1 \cap M_2}(x)\| \leq c^{2n-1}\|x\|$$

for $n = 1, 2, \dots$. Moreover, the constant c^{2n-1} is the smallest possible independent of x.

Proof. Let $M = M_1 \cap M_2$, $P_i = P_{M_i}$ $(i = 1, 2)$, and $P = P_2 P_1$. We first will prove that

(9.8.2) $$P_i(M^\perp) \subset M^\perp \qquad (i = 1, 2)$$

and hence

(9.8.3) $$P(M^\perp) \subset M^\perp.$$

To see (9.8.2), let $x \in M^\perp$ and $y \in M$. Then, since P_i is self-adjoint and idempotent,

$$\langle P_i(x), y \rangle = \langle x, P_i(y) \rangle = \langle x, y \rangle = 0.$$

Thus $P_i(x) \in M^\perp$, which proves (9.8.2). Relation (9.8.3) follows, since

$$P(M^\perp) = P_2 P_1(M^\perp) \subset P_2(M^\perp) \subset M^\perp.$$

Next we verify that

(9.8.4) $\|P(x)\| \leq c\|P_1(x)\| \leq c\|x\|$ for all $x \in M^\perp$.

To see this, let $x \in M^\perp$. Then

$$
\begin{aligned}
\|P(x)\|^2 &= \langle P(x), P(x) \rangle = \langle P(x) - P_1(x) + P_1(x), P(x) \rangle \\
&= \langle P_1(x), P(x) \rangle \text{ since } P(x) - P_1(x) \in M_2^\perp \, , \, P(x) \in M_2 \\
&\leq c\|P_1(x)\| \, \|P(x)\| \text{ by Lemma 9.5, (9.8.2), and (9.8.3)} \\
&\leq c\|x\| \, \|P(x)\| \text{ since } \|P_i\| \leq 1.
\end{aligned}
$$

Dividing both sides of the inequality by $\|P(x)\|$, we obtain (9.8.4).

Now suppose $x \in P(M^\perp)$. Then $x \in M_2 \cap M^\perp$ by (9.8.3). Since $P_1(x) \in M_1 \cap M^\perp$ by (9.8.2), we obtain, using Lemma 9.5 (6),

$$\|P_1(x)\|^2 = \langle P_1(x), P_1(x) \rangle = \langle x, P_1(x) \rangle \leq c\|x\| \, \|P_1(x)\|.$$

Thus $\|P_1(x)\| \leq c\|x\|$. Substituting this into (9.8.4) yields

(9.8.5) $\|P(x)\| \leq c^2\|x\|$ for all $x \in P(M^\perp)$.

From (9.8.3), (9.8.4), and an induction on (9.8.5), we get that

$$
\begin{aligned}
\|(P_2 P_1)^n(x) - P_M(x)\| &= \|P^n(x) - P^n P_M(x)\| = \|P^n[x - P_M(x)]\| \\
&= \|P^{n-1}[P(x - P_M(x))]\| \leq c^{2(n-1)}\|P(x - P_M(x))\| \\
&\leq c^{2n-1}\|x - P_M(x)\| \leq c^{2n-1}\|x\|.
\end{aligned}
$$

This verifies (9.8.1).

The fact that the constant c^{2n-1} is smallest possible, independent of x, will be established in Theorem 9.31 below. ∎

Remarks. If $c(M_1, M_2) = 0$, then Lemma 9.5 implies that $P_{M_2} P_{M_1} = P_{M_1 \cap M_2}$, and hence the iteration converges in one step ($n = 1$).

(2) We will see later (Theorem 9.34) that $c(M_1, M_2) < 1$ if and only if $M_1 + M_2$ is closed. This will be the case, for example, if either M_1 or M_2 is finite-dimensional.

(3) We shall also see later (Corollary 9.28) that Theorem 9.3 holds for more than two subspaces. However, if the subspaces are replaced by arbitrary closed convex sets, the analogous result fails.

To see this, consider the following example, in the plane $\ell_2(2)$ (see Figure 9.8.6 below). Let

$$C_1 = \{x \in \ell_2(2) \mid x(2) \geq x(1) \geq 0\} \quad \text{and} \quad C_2 = \{x \in \ell_2(2) \mid x(2) = 1\}.$$

Thus C_1 is a convex cone (the upper half of the first quadrant), and C_2 is a horizontal line (one unit above the horizontal axis). Setting $C = C_1 \cap C_2$, we see that C is the line segment joining the points $(0, 1)$ and $(1, 1)$.

Let $x_0 = (1, -1)$. Then $x_1 = P_{C_1}(x_0) = (0, 0)$ and $x_2 = P_{C_2}(x_1) = (0, 1)$. Since $x_2 \in C$, it follows that all the terms of the sequence

$$x_{2n-1} := P_{C_1}(x_{2n-2}) \text{ and } x_{2n} := P_{C_2}(x_{2n-1}) \qquad (n = 2, 3, \dots)$$

are equal to x_2. In particular,

$$\lim_{n \to \infty} (P_{C_2} P_{C_1})^n (x_0) = x_2.$$

However, it is easy to verify that

$$P_{C_1 \cap C_2}(x_0) = P_C(x_0) = (1, 1) \neq x_2.$$

This shows that Theorem 9.3 fails if the two subspaces are replaced by more general convex sets.

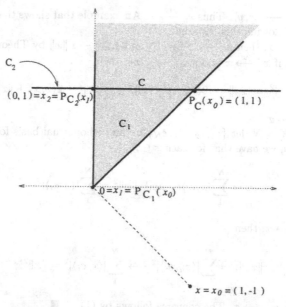

Figure 9.8.6

In spite of this example, there is a variant of Theorem 9.3 that *is* valid for general closed convex sets, not necessarily subspaces. Before turning to this, we need to discuss some aspects of *weak convergence* in a Hilbert space.

Weak Convergence

9.9 Definition. *A sequence $\{x_n\}$ in an inner product space X is said to* **converge weakly** *to x, written $x_n \xrightarrow{w} x$ or $x_n \longrightarrow x$ weakly, if*

$$\langle x_n, y \rangle \longrightarrow \langle x, y \rangle \qquad \text{for every } y \in X.$$

Equivalently, $x_n \longrightarrow x$ weakly if and only if $x^*(x_n) \longrightarrow x^*(x)$ for each $x^* \in X^*$. This follows obviously from Theorem 6.10 if X is complete, and slightly less obviously from Theorem 6.8 if X is not complete.

It is an easy consequence of the definition that weak limits are unique, and if $x_n \xrightarrow{w} x$ and $y_n \xrightarrow{w} y$, then $\alpha x_n + \beta y_n \xrightarrow{w} \alpha x + \beta y$. The relationship between ordinary (or "strong") convergence and weak convergence can be summarized as follows.

9.10 Theorem (Weak vs Strong Convergence).
(1) If $x_n \longrightarrow x$, then $x_n \xrightarrow{w} x$ (but not conversely in general).
(2) $x_n \longrightarrow x$ if and only if $x_n \xrightarrow{w} x$ and $\|x_n\| \longrightarrow \|x\|$.
(3) If X is finite-dimensional, then $x_n \xrightarrow{w} x$ if and only if $x_n \longrightarrow x$.
Proof. (1) Suppose $x_n \longrightarrow x$. Then for each $y \in X$,

$$|\langle x_n, y \rangle - \langle x, y \rangle| = |\langle x_n - x, y \rangle| \le \|x_n - x\| \, \|y\| \longrightarrow 0,$$

so that $\langle x_n, y \rangle \longrightarrow \langle x, y \rangle$. Thus $x_n \xrightarrow{w} x$. An example that shows that the converse fails is given following this theorem.

(2) If $x_n \longrightarrow x$, then $x_n \xrightarrow{w} x$ by (1), and $\|x_n\| \longrightarrow \|x\|$ by Theorem 1.12. Conversely, if $x_n \xrightarrow{w} x$ and $\|x_n\| \longrightarrow \|x\|$, then

$$\|x_n - x\|^2 = \|x_n\|^2 - 2\langle x_n, x \rangle + \|x\|^2 \longrightarrow \|x\|^2 - 2\langle x, x \rangle + \|x\|^2 = 0.$$

That is, $x_n \longrightarrow x$.

(3) If $\dim X = N$, let $\{e_1, e_2, \ldots, e_N\}$ be an orthonormal basis for X. Then by Corollary 4.10, we have that for each $x \in X$,

$$x = \sum_1^N \langle x, e_i \rangle e_i \quad \text{and} \quad \|x\|^2 = \sum_1^N |\langle x, e_i \rangle|^2.$$

Hence if $x_n \xrightarrow{w} x$, then

$$\|x_n\|^2 = \sum_1^N |\langle x_n, e_i \rangle|^2 \longrightarrow \sum_1^N |\langle x, e_i \rangle|^2 = \|x\|^2.$$

Hence, by (2), $x_n \longrightarrow x$. The converse follows by (1). ∎

9.11 Example (Weak convergence does not imply norm convergence).
Consider the space ℓ_2 and the orthonormal sequence $\{x_n\}$ defined by $x_n(i) = \delta_{ni}$ for all $n, i \in \mathbb{N}$. Then for each $y \in \ell_2$,

$$\langle x_n, y \rangle = y(n) \longrightarrow 0 = \langle 0, y \rangle,$$

since $\sum_1^\infty |y(n)|^2 = \|y\|^2 < \infty$. Thus $x_n \xrightarrow{w} 0$. But $\|x_n\| = 1$ for all n, so that $\{x_n\}$ does not converge to 0.

The most important property of weak convergence, for our purposes, is the following result, which states that bounded sets satisfy a certain "weak compactness" criterion.

9.12 Weak Compactness Theorem. *Every bounded sequence in a Hilbert space has a weakly convergent subsequence.*

Proof. Let $\{x_n\}$ be a sequence in the Hilbert space X, with $\|x_n\| \le c$ for all n. Consider the sequence of real numbers $\{\langle x_1, x_n \rangle\}$. This sequence is bounded (since $|\langle x_1, x_n \rangle| \le \|x_1\| \, \|x_n\| \le c^2$), so that there exists a subsequence that converges: $\langle x_1, x_{1(n)} \rangle \to \xi_1 \in \mathbb{R}$ as $n \to \infty$. Similarly, $\{\langle x_2, x_{1(n)} \rangle\}$ is bounded, so it also has a convergent subsequence: $\langle x_2, x_{2(n)} \rangle \to \xi_2$ as $n \to \infty$. Continuing in this fashion, we see that for each $k \in \mathbb{N}$, there is a subsequence $\{x_{k(n)}\}$ of $\{x_{(k-1)(n)}\}$ such that $\langle x_k, x_{k(n)} \rangle \to \xi_k$ as $n \to \infty$. Then the "diagonal" sequence $\{x_{n(n)}\}$ has the property that $\{x_{n(n)}\}$ is a subsequence of $\{x_n\}$ such that for each $k \in \mathbb{N}$,

$$(9.12.1) \qquad \langle x_k, x_{n(n)} \rangle \longrightarrow \xi_k \ \text{ as } n \longrightarrow \infty.$$

Let

$$M = \left\{ x \in X \mid \lim_n \langle x, x_{n(n)} \rangle \text{ exists} \right\}$$

and define f on M by

$$(9.12.2) \qquad f(x) := \lim_n \langle x, x_{n(n)} \rangle, \quad x \in M.$$

By (9.12.1), $x_k \in M$ $(k = 1, 2, \dots)$. Also, if $x, y \in M$ and $\alpha, \beta \in \mathbb{R}$, then

$$\lim_n \langle \alpha x + \beta y, x_{n(n)} \rangle = \lim_n [\alpha \langle x, x_{n(n)} \rangle + \beta \langle y, x_{n(n)} \rangle] = \alpha f(x) + \beta f(y).$$

Thus $\alpha x + \beta y \in M$. This proves that M is a linear subspace and that f is linear on M. Also, for any $x \in M$,

$$|f(x)| = \lim_n |\langle x, x_{n(n)} \rangle| \le \limsup_n \|x\| \, \|x_{n(n)}\| \le c\|x\|.$$

Thus f is bounded on M and $\|f\| \le c$.

Next we show that M is closed. Fix any $y \in \overline{M}$. Then there exists $y_m \in M$ such that $y_m \to y$. Since

$$|f(y_n) - f(y_m)| \le \|f\| \, \|y_n - y_m\| \le c\|y_n - y_m\|$$

and $\{y_n\}$ is Cauchy, it follows that $\{f(y_n)\}$ is Cauchy in $\mathbb{R}$. Thus $\lim_n f(y_n)$ exists. If also $y_n' \in M$ and $y_n' \to y$, then $y_n - y_n' \to 0$ and

$$|f(y_n) - f(y_n')| \le \|f\| \, \|y_n - y_n'\| \longrightarrow 0.$$

This proves that $F(y) := \lim_n f(y_n)$ exists and is independent of the particular sequence $\{y_n\}$ in M that converges to y. Given $\epsilon > 0$, choose an integer N_1 such that

$$(9.12.3) \qquad c\|y - y_{N_1}\| < \epsilon/3 \ \text{ and } \ |f(y_{N_1}) - F(y)| < \epsilon/3.$$

Then choose an integer N such that for all $n \ge N$,

$$(9.12.4) \qquad |\langle y_{N_1}, x_{n(n)} \rangle - f(y_{N_1})| < \epsilon/3.$$

Thus if $n \geq N$, then (9.12.3) and (9.12.4) imply that

$$|\langle y, x_{n(n)} \rangle - F(y)| \leq |\langle y, x_{n(n)} \rangle - \langle y_{N_1}, x_{n(n)} \rangle| + |\langle y_{N_1}, x_{n(n)} \rangle - f(y_{N_1})|$$
$$+ |f(y_{N_1}) - F(y)|$$
$$< \|y - y_{N_1}\| \, \|x_{n(n)}\| + 2\epsilon/3 \leq c\|y - y_{N_1}\| + 2\epsilon/3 < \epsilon.$$

This shows that $F(y) = \lim_n \langle y, x_{n(n)} \rangle$ exists and thus $y \in M$. Hence M is closed.

Since M is a closed subspace of the complete space X, M is a Hilbert space in its own right, and M is a Chebyshev subspace of X. Theorem 6.10 implies that f has a representer $x_0 \in M$; that is,

$$(9.12.5) \qquad\qquad f(x) = \langle x, x_0 \rangle, \quad x \in M.$$

Moreover, Theorem 5.8 implies that each $x \in X$ may be written as $x = P_M(x) + P_{M^\perp}(x)$. Since x_0 and $x_{n(n)}$ are in M for all n, we obtain from (9.12.2) and (9.12.5) that

$$\langle x, x_0 \rangle = \langle P_M(x), x_0 \rangle = f(P_M(x)) = \lim_n \langle P_M(x), x_{n(n)} \rangle = \lim_n \langle x, x_{x(n)} \rangle.$$

This shows that $x_{n(n)} \xrightarrow{w} x_0$ and completes the proof. ∎

Next we show that weakly convergent sequences are bounded and the norm is "weakly lower semicontinuous."

9.13 Lemma. *Let the sequence $\{x_n\}$ converge weakly to x. Then $\{x_n\}$ is bounded and $\|x\| \leq \liminf_n \|x_n\|$.*

Proof. For any $y \in X$ define $\widetilde{y}$ on X^* by

$$\widetilde{y}(x^*) := x^*(y), \quad x^* \in X^*.$$

It is clear that $\widetilde{y} : X^* \to \mathbb{R}$ is linear and

$$|\widetilde{y}(x^*)| = |x^*(y)| \leq \|x^*\| \, \|y\|, \quad x^* \in X^*.$$

Thus $\widetilde{y}$ is bounded and $\|\widetilde{y}\| \leq \|y\|$. By Remark (2) after Theorem 6.23, choose $x^* \in X^*$ with $\|x^*\| = 1$ and $x^*(y) = \|y\|$. Then $\widetilde{y}(x^*) = x^*(y) = \|y\|$ implies that $\|\widetilde{y}\| \geq \|y\|$ and hence $\|\widetilde{y}\| = \|y\|$. We have shown that $\widetilde{y} \in X^{**} = \mathcal{B}(X^*, \mathbb{R})$ and $\|\widetilde{y}\| = \|y\|$. Since X^* is a Hilbert space (by Exercise 7 or 8 of Chapter 6), and since

$$\widetilde{x}_n(x^*) = x^*(x_n) \to x^*(x) = \widetilde{x}(x^*)$$

for each $x^* \in X^*$, it follows by Corollary 8.15 that

$$\sup_n \|x_n\| = \sup_n \|\widetilde{x}_n\| < \infty.$$

Thus $\{x_n\}$ is bounded.

Again using Remark (2) after Theorem 6.23, choose $x^* \in X^*$ with $\|x^*\| = 1$ and $x^*(x) = \|x\|$. Thus

$$\|x\| = x^*(x) = \lim x^*(x_n) \leq \liminf \|x^*\| \, \|x_n\| = \liminf \|x_n\|. \quad ∎$$

The following lemma shows that each continuous linear operator is also "weakly continuous."

9.14 Lemma. *Let X and Y be inner product spaces with X complete, and let $A \in B(X,Y)$. Then $Ax_n \to Ax$ weakly whenever $x_n \to x$ weakly.*

Proof. Let $x_n \to x$ weakly. Then for each $y \in Y$,

$$\langle Ax_n, y \rangle = \langle x_n, A^*y \rangle \longrightarrow \langle x, A^*y \rangle = \langle Ax, y \rangle.$$

That is, $Ax_n \to Ax$ weakly. ∎

9.15 Definition. *A subset K of an inner product space X is called* **weakly closed** *if $x_n \in K$ and $x_n \xrightarrow{w} x$ implies $x \in K$.*

The last result we need that concerns the weak topology is that a convex set is closed if and only if it is weakly closed.

9.16 Theorem. (1) *Every weakly closed set is closed.*

(2) *A convex set is closed if and only if it is weakly closed.*

Proof. (1) Let K be weakly closed, $x_n \in K$, and $x_n \to x$. To show that K is closed, it suffices to show that $x \in K$. By Theorem 9.10 (1), $x_n \xrightarrow{w} x$. Since K is weakly closed, $x \in K$.

(2) Owing to (1), if suffices to show that if K is closed and convex, then K is weakly closed. Let $x_n \in K$ and $x_n \xrightarrow{w} x$. If $x \notin K$, then by Theorem 6.23, there exists $x^* \in X^*$ with $\|x^*\| = 1$ and $x^*(x) > \sup x^*(K)$. It follows that

$$x^*(x) > \sup_n x^*(x_n) \geq \lim_n x^*(x_n) = x^*(x),$$

which is absurd. Thus $x \in K$, and hence K is weakly closed. ∎

Dykstra's Algorithm

Now we can describe, and prove the convergence of, an iterative algorithm called Dykstra's algorithm that allows one to compute best approximations from an intersection $\bigcap_1^r K_i$ of a finite number of closed convex sets K_i (not necessarily subspaces), by reducing the computation to a problem of finding best approximations from the individual sets K_i.

Let K_i be a closed convex subset of the Hilbert space X $(i = 1, 2, \ldots, r)$, and let $K := \bigcap_1^r K_i$. We assume $K \neq \emptyset$. Given any $x \in X$, we would like to compute $P_K(x)$.

For each $n \in \mathbb{N}$, let $[n]$ denote "$n \bmod r$"; that is,

$$[n] := \{1, 2, \ldots, r\} \cap \{n - kr \mid k = 0, 1, 2, \ldots\}.$$

Thus $[1] = 1$, $[2] = 2, \ldots, [r] = r$, $[r+1] = 1$, $[r+2] = 2, \ldots, [2r] = r, \ldots$. For a fixed $x \in X$, set

$$(9.16.1) \quad \begin{aligned} & x_0 := x, \quad e_{-(r-1)} = \cdots = e_{-1} = e_0 = 0, \\ & x_n := P_{K_{[n]}}(x_{n-1} + e_{n-r}), \\ & e_n := x_{n-1} + e_{n-r} - x_n \\ & \quad = x_{n-1} + e_{n-r} - P_{K_{[n]}}(x_{n-1} + e_{n-r}) \quad (n = 1, 2, \ldots) \end{aligned}$$

(see Figure 9.16.2). We will show that

$$(9.16.3) \quad \lim_n \|x_n - P_K(x)\| = 0.$$

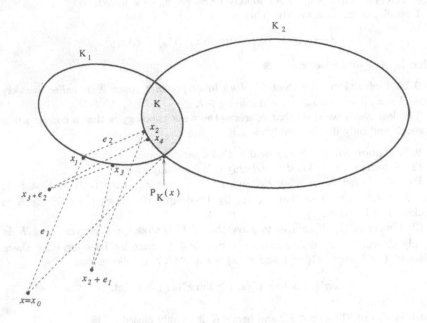

Figure 9.16.2. Dykstra's algorithm ($r = 2$)

It is convenient to decompose the proof of (9.16.3) into several smaller pieces, which we designate as lemmas. For notational simplicity, we set $P_i := P_{K_i}$ ($i = 1, 2, \ldots, r$).

9.17 Lemma. *For each n,*

$$(9.17.1) \qquad \langle x_n - y, e_n \rangle \geq 0 \quad \text{for all } y \in K_{[n]}.$$

Proof. Using Theorem 4.1 with $x = x_{n-1} + e_{n-r}$, $y_0 = P_{[n]}(x_{n-1} + e_{n-r})$, and $K = K_{[n]}$, we obtain, for all $y \in K_{[n]}$,

$$\langle x_n - y, e_n \rangle = \langle P_{[n]}(x_{n-1} + e_{n-r}) - y, \; x_{n-1} + e_{n-r} - P_{[n]}(x_{n-1} + e_{n-r}) \rangle \geq 0. \quad \blacksquare$$

9.18 Lemma. *For each $n \geq 0$,*

$$(9.18.1) \qquad x - x_n = e_{n-(r-1)} + e_{n-(r-2)} + \cdots + e_{n-1} + e_n.$$

Proof. By induction on n. For $n = 0$, $x - x_0 = x - x = 0$ and $e_{-(r-1)} + e_{-(r-2)} + \cdots + e_{-1} + e_0 = 0$ by (9.16.1). Now assume that the result is valid for some $n \geq 0$. Then

$$\begin{aligned}
x - x_{n+1} &= (x - x_n) + (x_n - x_{n+1}) \\
&= (e_{n-(r-1)} + e_{n-(r-2)} + \cdots + e_{n-1} + e_n) + (e_{n+1} - e_{n+1-r}) \\
&= e_{n-(r-2)} + e_{n-(r-3)} + \cdots + e_n + e_{n+1} \\
&= e_{n+1-(r-1)} + e_{n+1-(r-2)} + \cdots + e_n + e_{n+1},
\end{aligned}$$

which shows that (9.18.1) is valid when n is replaced by $n + 1$. $\blacksquare$

9.19 Lemma. *For each $n \in \mathbb{N}$, $0 \leq m \leq n$, and $y \in K$,*

$$\|x_m - y\|^2 = \|x_n - y\|^2 + \sum_{k=m+1}^{n} \|x_k - x_{k-1}\|^2 + 2 \sum_{k=m+1}^{n} \langle e_{k-r}, x_{k-r} - x_k \rangle$$

(9.19.1)

$$+ 2 \sum_{k=n-(r-1)}^{n} \langle e_k, x_k - y \rangle - 2 \sum_{k=m-(r-1)}^{m} \langle e_k, x_k - y \rangle.$$

Proof. For any set of vectors $\{y_m, y_{m+1}, \ldots, y_{n+1}\}$ in X, we have the identity

$$\|y_m - y_{n+1}\|^2 = \|(y_m - y_{m+1}) + (y_{m+1} - y_{m+2}) + \cdots + (y_n - y_{n+1})\|^2$$

$$= \sum_{k=m+1}^{n+1} \|y_{k-1} - y_k\|^2 + 2 \sum_{m+1 \leq i < j \leq n+1} \langle y_{i-1} - y_i, y_{j-1} - y_j \rangle$$

(9.19.2)

$$= \|y_n - y_{n+1}\|^2 + \sum_{k=m+1}^{n} \|y_{k-1} - y_k\|^2$$

$$+ 2 \sum_{i=m+1}^{n} \left(\sum_{j=i+1}^{n+1} \langle y_{i-1} - y_i, y_{j-1} - y_j \rangle \right).$$

But

$$\sum_{i=m+1}^{n} \left(\sum_{j=i+1}^{n+1} \langle y_{i-1} - y_i, y_{j-1} - y_j \rangle \right) = \sum_{i=m+1}^{n} \left\langle y_{i-1} - y_i, \sum_{j=i+1}^{n+1} (y_{j-1} - y_j) \right\rangle$$

(9.19.3)

$$= \sum_{i=m+1}^{n} \langle y_{i-1} - y_i, y_i - y_{n+1} \rangle.$$

Substituting $y_i = x_i$ for all $i \leq n$ and $y_{n+1} = y$ into (9.19.3), we get

$$\sum_{i=m+1}^{n} \langle y_{i-1} - y_i, y_i - y_{n+1} \rangle = \sum_{i=m+1}^{n} \langle x_{i-1} - x_i, x_i - y \rangle$$

$$= \sum_{i=m+1}^{n} \langle e_i - e_{i-r}, x_i - y \rangle$$

$$= \sum_{i=m+1}^{n} \langle e_i, x_i - y \rangle - \sum_{i=m+1}^{n} \langle e_{i-r}, x_i - y \rangle$$

$$= \sum_{i=m+1}^{n} \langle e_i, x_i - y \rangle - \sum_{i=m+1}^{n} [\langle e_{i-r}, x_i - x_{i-r} \rangle + \langle e_{i-r}, x_{i-r} - y \rangle]$$

$$= \sum_{i=m+1}^{n} \langle e_i, x_i - y \rangle - \sum_{i=m+1}^{n} \langle e_{i-r}, x_{i-r} - y \rangle + \sum_{i=m+1}^{n} \langle e_{i-r}, x_{i-r} - x_i \rangle$$

$$= \sum_{i=m+1}^{n} \langle e_i, x_i - y \rangle - \sum_{i=m+1-r}^{n-r} \langle e_i, x_i - y \rangle + \sum_{i=m+1}^{n} \langle e_{i-r}, x_{i-r} - x_i \rangle$$

$$= \sum_{i=n-r+1}^{n} \langle e_i, x_i - y \rangle - \sum_{i=m-r+1}^{m} \langle e_i, x_i - y \rangle + \sum_{i=m+1}^{n} \langle e_{i-r}, x_{i-r} - x_i \rangle.$$

Substituting these same values for y_i into (9.19.2) and using the above relation, we get

$$\|x_m - y\|^2 = \|x_n - y\|^2 + \sum_{k=m+1}^{n} \|x_{k-1} - x_k\|^2 + 2 \sum_{i=n-r+1}^{n} \langle e_i, x_i - y \rangle$$
$$+ 2 \sum_{i=m+1}^{n} \langle e_{i-r}, x_{i-r} - x_i \rangle - 2 \sum_{i=m-r+1}^{m} \langle e_i, x_i - y \rangle.$$

This verifies (9.19.1). ∎

9.20 Lemma. $\{x_n\}$ *is a bounded sequence and*

(9.20.1) $$\sum_{1}^{\infty} \|x_{k-1} - x_k\|^2 < \infty.$$

In particular,

(9.20.2) $$\|x_{n-1} - x_n\| \longrightarrow 0 \ \text{as } n \longrightarrow \infty.$$

Proof. Setting $m = 0$ in (9.19.1), we note that the third and fourth terms on the right of (9.19.1) are nonnegative by Lemma 9.17, while the last term is zero using (9.16.1). Thus we obtain

$$\|x_0 - y\|^2 \geq \|x_n - y\|^2 + \sum_{k=1}^{n} \|x_k - x_{k-1}\|^2$$

for all n. It follows that $\{\|x_n - y\|\}$ is bounded and (9.20.1) holds. Hence $\{x_n\}$ is bounded as well. Finally, (9.20.1) implies (9.20.2). ∎

9.21 Lemma. *For any* $n \in \mathbb{N}$,

(9.21.1) $$\|e_n\| \leq \sum_{k=1}^{n} \|x_{k-1} - x_k\|.$$

Proof. We induct on n. For $n = 1$,

$$\|e_1\| = \|x_0 - x_1 - e_{1-r}\| = \|x_0 - x_1\|,$$

since $e_i = 0$ for all $i \leq 0$ by (9.16.1). Now assume that (9.21.1) holds for all $m \leq n$. Then

$$\|e_{n+1}\| = \|x_n - x_{n+1} + e_{n+1-r}\| \leq \|x_n - x_{n+1}\| + \|e_{n+1-r}\|$$
$$\leq \|x_n - x_{n+1}\| + \sum_{k=1}^{n+1-r} \|x_{k-1} - x_k\| \leq \sum_{k=1}^{n+1} \|x_{k-1} - x_k\|$$

implies that (9.21.1) holds when n is replaced by $n + 1$. ∎

9.22 Lemma.

$$(9.22.1) \qquad \liminf_n \sum_{k=n-(r-1)}^{n} |\langle x_k - x_n, e_k \rangle| = 0.$$

Proof. Using Schwarz's inequality and Lemma 9.21, we get

$$\sum_{k=n-(r-1)}^{n} |\langle x_k - x_n, e_k \rangle| \le \sum_{k=n-(r-1)}^{n} \|e_k\| \, \|x_k - x_n\|$$

$$\le \sum_{k=n-(r-1)}^{n} \left(\sum_{j=1}^{k} \|x_{j-1} - x_j\| \right) \left(\sum_{i=k+1}^{n} \|x_{i-1} - x_i\| \right)$$

$$\le r \sum_{j=1}^{n} \|x_{j-1} - x_j\| \left(\sum_{i=n-(r-2)}^{n} \|x_{i-1} - x_i\| \right).$$

Letting $a_i = \|x_{i-1} - x_i\|$, we see that it suffices to show that

$$(9.22.2) \qquad \liminf_n \sum_{j=1}^{n} a_j \left(\sum_{i=n-(r-2)}^{n} a_i \right) = 0.$$

But

$$A := \sum_{1}^{\infty} a_i^2 = \sum_{1}^{\infty} \|x_{i-1} - x_i\|^2 < \infty$$

by Lemma 9.20. Using the Schwarz inequality, we get

$$\sum_{1}^{n} a_j \le \sqrt{n} \left(\sum_{1}^{n} a_j^2 \right)^{1/2} \le \sqrt{n} A^{1/2}$$

for each n. Hence to verify (9.22.2), it is enough to verify

$$(9.22.3) \qquad \liminf_n \sqrt{n} \sum_{i=n-(r-2)}^{n} a_i = 0.$$

To establish (9.22.3), let

$$\rho := \liminf_n \sqrt{n} \sum_{i=n-(r-2)}^{n} a_i.$$

If (9.22.3) were false, then we would have $\rho > 0$. We may assume $\rho < \infty$. (The proof for $\rho = \infty$ is similar.) Then

$$\sqrt{n} \sum_{i=n-(r-2)}^{n} a_i > \frac{1}{2} \rho \qquad \text{``eventually''}$$

(i.e., for n sufficiently large). Thus $\sum_{i=n-(r-2)}^{n} a_i > \rho/(2\sqrt{n})$ eventually. Using Schwarz's inequality, we deduce that

$$\frac{\rho^2}{4n} < \left(\sum_{i=n-(r-2)}^{n} a_i \right)^2 \leq (r-1) \sum_{i=n-(r-2)}^{n} a_i^2 \quad \text{eventually.}$$

Thus for some integer N,

$$\frac{\rho^2}{4n} < (r-1) \sum_{i=n-(r-2)}^{n} a_i^2 \quad \text{for all } n \geq N.$$

Hence

$$\infty = \sum_{n=N}^{\infty} \frac{\rho^2}{4n} \leq (r-1) \sum_{n=N}^{\infty} \sum_{i=n-(r-2)}^{n} a_i^2$$

$$\leq (r-1) \sum_{n=N}^{\infty} [a_{n-(r-2)}^2 + a_{n-(r-3)}^2 + \cdots + a_n^2] \leq (r-1)^2 \sum_{1}^{\infty} a_i^2 < \infty,$$

which is absurd. This contradiction proves that (9.22.3) must hold. ∎

9.23 Lemma. *There exists a subsequence $\{x_{n_j}\}$ of $\{x_n\}$ such that*

(9.23.1) $$\limsup_{j} \langle y - x_{n_j}, x - x_{n_j} \rangle \leq 0 \quad \text{for each } y \in K, \text{ and}$$

(9.23.2) $$\lim_{j} \sum_{k=n_j-(r-1)}^{n_j} |\langle x_k - x_{n_j}, e_k \rangle| = 0.$$

Proof. Using Lemma 9.18, we have for all $y \in K$, $n \geq 0$, that

$$\langle y - x_n, x - x_n \rangle = \langle y - x_n, e_{n-(r-1)} + e_{n-(r-2)} + \cdots + e_n \rangle$$

(9.23.3) $$= \sum_{k=n-(r-1)}^{n} \langle y - x_n, e_k \rangle$$

$$= \sum_{k=n-(r-1)}^{n} \langle y - x_k, e_k \rangle + \sum_{k=n-(r-1)}^{n} \langle x_k - x_n, e_k \rangle.$$

By Lemma 9.17, the first sum is no more than 0. Hence

(9.23.4) $$\langle y - x_n, x - x_n \rangle \leq \sum_{k=n-(r-1)}^{n} \langle x_k - x_n, e_k \rangle.$$

By Lemma 9.22, we deduce that there is a subsequence $\{n_j\}$ of $\mathbb{N}$ such that

(9.23.5) $$\lim_{j} \sum_{k=n_j-(r-1)}^{n_j} |\langle x_k - x_{n_j}, e_k \rangle| = 0.$$

Note that the right side of (9.23.4) does not depend on y. In view of (9.23.4), it follows that (9.23.5) implies that (9.23.1) and (9.23.2) hold. ∎

Now we are ready to state and prove the main theorem. Recall our setup: X is a Hilbert space, $K_1, K_2, \ldots, K_r$ are closed convex subsets, and $K = \cap_1^r K_i \neq \emptyset$. For each $x \in X$, define

(9.23.6)
$$x_0 = x, \qquad e_{-(r-1)} = \cdots = e_{-1} = e_0 = 0,$$
$$x_n = P_{K_{[n]}}(x_{n-1} + e_{n-r}),$$
$$e_n = x_{n-1} + e_{n-r} - x_n \qquad (n = 1, 2, \ldots),$$

where
$$[n] = \{1, 2, \ldots, r\} \cap \{n - kr \mid k = 0, 1, 2, \ldots\}.$$

9.24 Boyle–Dykstra Theorem. *Let* $K_1, K_2, \ldots, K_r$ *be closed convex subsets of the Hilbert space* X *such that* $K := \cap_1^r K_i \neq \emptyset$. *For each* $x \in X$, *define the sequence* $\{x_n\}$ *as in (9.23.6). Then*

(9.24.1)
$$\lim_n \|x_n - P_K(x)\| = 0.$$

Proof. By Lemma 9.23, there exists a subsequence $\{x_{n_j}\}$ such that

(9.24.2)
$$\limsup_j \langle y - x_{n_j}, x - x_{n_j} \rangle \leq 0 \qquad \text{for each} \quad y \in K.$$

Since $\{x_n\}$ is bounded by Lemma 9.20, it follows by Theorem 9.12 (by passing to a further subsequence if necessary), that there is $y_0 \in X$ such that

(9.24.3)
$$x_{n_j} \xrightarrow{w} y_0,$$

and

(9.24.4)
$$\lim_j \|x_{n_j}\| \text{ exists.}$$

By Theorem 9.13,

(9.24.5)
$$\|y_0\| \leq \liminf_j \|x_{n_j}\| = \lim_j \|x_{n_j}\|.$$

Since there are only a finite number of sets K_i, an infinite number of the x_{n_j}'s must lie in a single set K_{i_0}. Since K_{i_0} is closed and convex, it is weakly closed by Theorem 9.16, and hence $y_0 \in K_{i_0}$. By (9.20.2), $x_n - x_{n-1} \to 0$. By a repeated application of this fact, we see that all the sequences $\{x_{n_j+1}\}, \{x_{n_j+2}\}, \{x_{n_j+3}\}, \ldots$ converge weakly to y_0, and hence $y_0 \in K_i$ for every i. That is,

(9.24.6)
$$y_0 \in K.$$

For any $y \in K$, (9.24.5) and (9.24.2) imply that

$$\langle y - y_0, x - y_0 \rangle = \langle y, x \rangle - \langle y, y_0 \rangle - \langle y_0, x \rangle + \|y_0\|^2$$
$$\leq \lim_j [\langle y, x \rangle - \langle y, x_{n_j} \rangle - \langle x_{n_j}, x \rangle + \|x_{n_j}\|^2]$$
$$= \lim_j \langle y - x_{n_j}, x - x_{n_j} \rangle \leq 0.$$

By Theorem 4.1,

(9.24.7)
$$y_0 = P_K(x).$$

Moreover, putting $y = y_0$ in the above inequalities, we get equality in the chain of inequalities and hence

(9.24.8)
$$\lim_j \|x_{n_j}\|^2 = \|y_0\|^2$$

and

(9.24.9)
$$\lim_j \langle y_0 - x_{n_j}, x - x_{n_j} \rangle = 0.$$

By (9.24.3) and (9.24.8), it follows from Theorem 9.10 (2) that

(9.24.10)
$$\|x_{n_j} - y_0\| \longrightarrow 0.$$

Hence

(9.24.11)
$$\|x_{n_j} - P_K(x)\| = \|x_{n_j} - y_0\| \longrightarrow 0.$$

To complete the proof, we must show that the whole sequence $\{x_n\}$ converges to y_0. From equation (9.23.3) with $y = y_0$ and $n = n_j$, we get

(9.24.12)
$$\langle y_0 - x_{n_j}, x - x_{n_j} \rangle = \sum_{k=n_j-(r-1)}^{n_j} \langle y_0 - x_k, e_k \rangle + \sum_{k=n_j-(r-1)}^{n_j} \langle x_k - x_{n_j}, e_k \rangle.$$

The left side of (9.24.12) tends to zero as $j \to \infty$ by (9.24.9), while the second sum on the right tends to zero by (9.23.4). Hence

(9.24.13)
$$\lim_j \sum_{k=n_j-(r-1)}^{n_j} \langle y_0 - x_k, e_k \rangle = 0.$$

Using Lemmas 9.19 and 9.17 with $m = n_j$ and $y = y_0$, we see that for all $n \geq n_j$,

$$\|x_{n_j} - y_0\|^2 \geq \|x_n - y_0\|^2 - 2 \sum_{k=n_j-(r-1)}^{n_j} \langle e_k, x_k - y_0 \rangle,$$

or

(9.24.14)
$$\|x_n - y_0\|^2 \leq \|x_{n_j} - y_0\|^2 + 2 \sum_{k=n_j-(r-1)}^{n_j} \langle e_k, x_k - y_0 \rangle.$$

But both terms on the right of inequality (9.24.14) tend to zero as $j \to \infty$ by (9.24.10) and (9.24.13). It follows that $\lim_n \|x_n - y_0\| = 0$, and this completes the proof. ∎

Remark. A close inspection of the proof of the Boyle–Dykstra theorem reveals the following more general result, which does *not* require completeness of X, but only that of the K_i.

9.25 Theorem. Let $K_1, K_2, \ldots, K_r$ be complete convex sets in an inner product space X such that $K := \cap_1^r K_i \neq \emptyset$. Then

(1) K is a convex Chebyshev set, and
(2) For each $x \in X$, the sequence $\{x_n\}$ defined by (9.23.5) satisfies

$$\lim_n \|x_n - P_K(x)\| = 0.$$

The Case of Affine Sets

The Dykstra algorithm described in (9.16.1)–(9.16.3) substantially simplifies in the special case where all the sets K_i are subspaces or, more generally, where all the K_i are affine sets (i.e., translates of subspaces). Specifically, the computation of the residuals e_n *may be omitted* entirely from the algorithm!

To see this, we first record the following result that characterizes best approximations from affine sets (see also Exercise 2 at the end of Chapter 4).

9.26 Characterization of Best Approximations from Affine Sets. *Let V be an affine set in the inner product space X. Thus $V = M + v$, where M is a subspace and v is any given element of V. Let $x \in X$ and $y_0 \in V$. Then the following statements are equivalent:*

(1) $y_0 = P_V(x)$;

(2) $x - y_0 \in M^\perp$;

(3) $\langle x - y_0, y - v \rangle = 0$ for all $y \in V$.

Moreover,

(9.26.1) $P_V(x + e) = P_V(x)$ *for all $x \in X$, $e \in M^\perp$.*

Proof. Using Theorem 2.7 (1) (ii), we see that $y_0 = P_V(x) \Longleftrightarrow y_0 = P_{M+v}(x) = P_M(x - v) + v \Longleftrightarrow y_0 - v = P_M(x - v)$. Using Theorem 4.9, we deduce that $y_0 - v = P_M(x-v) \Longleftrightarrow x - y_0 = (x-v) - (y_0-v) \in M^\perp$. This proves the equivalence of (1) and (2). The equivalence of (2) and (3) is obvious, since $M = V - v$.

To prove (9.26.1), let $x \in X$ be arbitrary and $e \in M^\perp$. Using Theorem 2.7 (1) (ii) again, along with the facts (Theorems 5.7 and 5.12) that P_M is linear and $P_M(e) = 0$, we see that

$$P_V(x + e) = P_{M+v}(x + e) = P_M(x + e - v) + v$$
$$= P_M(x - v) + v = P_{M+v}(x) = P_V(x). \ \blacksquare$$

The geometric interpretation of Theorem 9.26 is that the error vector $x - P_V(x)$ must be orthogonal to the translate of V that passes through the origin (see Figure 9.26.2). Note that in the special case where $V = M$ is a subspace, we recover Theorem 4.9.

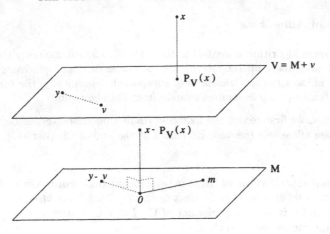

Figure 9.26.2

Let us return now to Dykstra's algorithm in the particular case where one of the closed convex sets K_i is an affine set. For definiteness, suppose K_1 is a closed affine set. Then

$$K_1 = M_1 + v_1,$$

where M_1 is a closed subspace and $v_1 \in K_1$.

If n is any integer with $[n] = 1$, Theorem 9.26 implies that

$$e_n = x_{n-1} + e_{n-r} - P_{K_{[n]}}(x_{n-1} + e_{n-r})$$
$$= x_{n-1} + e_{n-r} - P_{K_1}(x_{n-1} + e_{n-r}) \in M_1^\perp.$$

Since $[n-r] = 1$ also, it follows that $e_{n-r} \in M_1^\perp$, and (9.26.1) implies that

$$x_n = P_{K_{[n]}}(x_{n-1} + e_{n-r}) = P_{K_1}(x_{n-1} + e_{n-r}) = P_{K_1}(x_{n-1}) = P_{K_{[n]}}(x_{n-1}).$$

That is, to determine x_n, we need only project x_{n-1} onto $K_{[n]}$; it is not necessary first to add e_{n-r} to x_{n-1} and then project the sum $x_{n-1} + e_{n-r}$ onto $K_{[n]}$.

In particular, if *all* the sets K_i are closed affine sets, then it is not necessary to compute *any* of the e_n's; each term x_n in Dykstra's algorithm may be determined by projecting x_{n-1} onto $K_{[n]}$:

(9.26.3) $$x_n = P_{K_{[n]}}(x_{n-1}) \qquad (n = 1, 2, \dots).$$

Summarizing all this, we can state the following generalization of Theorem 9.3 to the case of more than two subspaces, or even when the subspaces are replaced by affine sets.

9.27 Theorem. *Let* $V_1, V_2, \dots, V_r$ *be* r *closed affine sets in the Hilbert space* X *such that* $\cap_1^r V_i \neq \emptyset$. *Then for each* $x \in X$,

(9.27.1) $$\lim_{n \to \infty} (P_{V_r} P_{V_{r-1}} \cdots P_{V_1})^n(x) = P_{\cap_1^r V_i}(x)$$

Proof. Fix $x \in X$. By Dykstra's algorithm, the sequence $\{x_n\}$ determined by (9.16.1)–(9.16.3) converges to $P_{\cap_1^r V_i}(x)$. But by (9.26.3), we have that $x_n = P_{V_{[n]}}(x_{n-1})$ for each $n \geq 1$. In particular, replacing n by nr and inducting on this relation, we get

$$
\begin{aligned}
x_{nr} &= P_{V_{[nr]}}(x_{nr-1}) = P_{V_r}[P_{V_{[nr-1]}}(x_{nr-2})] \\
&= P_{V_r}P_{V_{r-1}}[P_{V_{[nr-2]}}(x_{nr-3})] = \cdots \\
&= P_{V_r}P_{V_{r-1}} \cdots P_{V_1}(x_{nr-r}) = P_{V_r}P_{V_{r-1}} \cdots P_{V_1}(x_{(n-1)r}) \\
&= (P_{V_r}P_{V_{r-1}} \cdots P_{V_1})^2(x_{(n-2)r}) = \cdots \\
&= (P_{V_r}P_{V_{r-1}} \cdots P_{V_1})^n(x).
\end{aligned}
$$

Since $\{x_n\}$ converges to $P_{\cap_1^r V_i}(x)$, so does the subsequence $\{x_{nr}\}$. Thus (9.27.1) follows. ∎

The subspace case of this theorem is important enough to be stated as a separate corollary.

9.28 Corollary (Halperin). *Let $M_1, M_2, \ldots, M_r$ be closed subspaces in the Hilbert space X and $M = \cap_1^r M_i$. Then*

$$
(9.28.1) \qquad \lim_{n \to \infty} (P_{M_r}P_{M_{r-1}} \cdots P_{M_2}P_{M_1})^n(x) = P_M(x)
$$

for each $x \in X$.

Note that this corollary generalizes Theorem 9.3 to any finite number of subspaces, not just two.

Rate of Convergence for Alternating Projections

Now we turn to the question of the *rate of convergence* of the method of alternating projections. If $M_1, M_2, \ldots, M_r$ are r closed subspaces in the Hilbert space X and $M = \cap_1^r M_i$, then Corollary 9.28 states that

$$
\lim_n \left\| (P_{M_r}P_{M_{r-1}} \cdots P_{M_1})^n x - P_M x \right\| = 0
$$

for every $x \in X$. Since

$$
(9.28.2) \qquad \left\| (P_{M_r}P_{M_{r-1}} \cdots P_{M_1})^n x - P_M x \right\| \leq B_r(n)\|x\|
$$

for all x, where

$$
(9.28.3) \qquad B_r(n) := \left\| (P_{M_r}P_{M_{r-1}} \cdots P_{M_1})^n - P_M \right\|,
$$

it follows that $B_r(n)$ is the sharpest (i.e., smallest) bound available in (9.28.2) that is independent of x.

We will compute $B_2(n)$ exactly and give upper bounds on $B_r(n)$ when $r > 2$. To do this, we first note that there are alternative ways of describing $B_r(n)$.

9.29 Lemma. *Let $M_1, M_2, \ldots, M_r$ be closed subspaces of the Hilbert space X and $M = \cap_1^r M_i$. Then:*

(1) $P_{M^\perp}$ *is idempotent and self-adjoint.*

(2) $P_{M^\perp}$ *commutes with each P_{M_i} and with $P_{M_r} P_{M_{r-1}} \cdots P_{M_1}$.*

(3) *For $i = 1, 2, \ldots, r$, $P_{M_i} x = x$ for each $x \in M$ and $P_{M_i}(M^\perp) \subset M^\perp$. Hence*

$$P_{M_r} P_{M_{r-1}} \cdots P_{M_1} x = x \quad \text{for each } x \in M$$

and

$$P_{M_r} P_{M_{r-1}} \cdots P_{M_1}(M^\perp) \subset M^\perp.$$

Proof. Let $P_i = P_{M_i}$.

(1) This follows from Theorem 5.13.

(2) Since $M \subset M_i$ for each i, it follows from Theorem 9.2 that P_M commutes with P_i. Moreover, using Theorem 5.8, we obtain

$$P_{M^\perp} P_i = (I - P_M) P_i = P_i - P_M P_i = P_i - P_i P_M = P_i (I - P_M) = P_i P_{M^\perp},$$

and thus $P_{M^\perp}$ commutes with each P_i. It follows that $P_{M^\perp}$ commutes with $P_r P_{r-1} \cdots P_1$.

(3) If $x \in M$, then $x \in M_i$ for all i, so that $P_i x = x$. The rest of the proof follows from (2) and Theorem 9.2. ∎

9.30 Lemma. *Let $M_1, M_2, \ldots, M_r$ be closed subspaces in the Hilbert space X and $M = \cap_1^r M_i$. Then for each $n \in \mathbb{N}$,*

$$
\begin{aligned}
(P_{M_r} P_{M_{r-1}} \cdots P_{M_1})^n - P_M &= (P_{M_r} P_{M_{r-1}} \cdots P_{M_1})^n P_{M^\perp} \\
&= (P_{M_r} P_{M_{r-1}} \cdots P_{M_1} P_{M^\perp})^n = (Q_r Q_{r-1} \cdots Q_1)^n,
\end{aligned}
$$
(9.30.1)

where $Q_j = P_{M_j \cap M^\perp}$ $(j = 1, 2, \ldots, r)$.

In particular,

$$
\begin{aligned}
\|(P_{M_r} P_{M_{r-1}} \cdots P_{M_1})^n - P_{\cap_1^r M_i}\| &= \|(P_{M_r} P_{M_{r-1}} \cdots P_{M_1})^n P_{M^\perp}\| \\
&= \|(P_{M_r} P_{M_{r-1}} \cdots P_{M_1} P_{M^\perp})^n\| \\
&= \|(Q_r Q_{r-1} \cdots Q_1)^n\| = \|(Q_1 Q_2 \cdots Q_r)^n\|.
\end{aligned}
$$
(9.30.2)

Proof. Let $P_i = P_{M_i}$ and $P = P_r P_{r-1} \cdots P_1$. Since the adjoint of $(Q_r \cdots Q_1)^n$ is $(Q_1 \cdots Q_r)^n$, the last equality in (9.30.2) follows from Theorem 8.25 (3). From (8.25.2), it suffices to prove (9.30.1). Using Lemma 9.29 and Theorem 5.14, we obtain $P_M = P^n P_M$ and

$$
\begin{aligned}
P^n - P_M &= P^n - P^n P_M = P^n (I - P_M) = P^n P_{M^\perp} \\
&= P^n P_{M^\perp}^n \quad \text{(since $P_{M^\perp}$ is idempotent)} \\
&= (P P_{M^\perp})^n \quad \text{(since P commutes with $P_{M^\perp}$)} \\
&= [(P_r P_{M^\perp})(P_{r-1} P_{M^\perp}) \cdots (P_1 P_{M^\perp})]^n,
\end{aligned}
$$

since P_i commutes with $P_{M^\perp}$ and $P_{M^\perp}$ is idempotent. Finally, Lemma 9.2 implies that $P_i P_{M^\perp} = P_{M_i \cap M^\perp} = Q_i$ $(i = 1, 2, \ldots, r)$ and this completes the proof. ∎

Now we consider the special case where $r = 2$.

9.31 Theorem. *For each $n \in \mathbb{N}$,*

$$(9.31.1) \qquad \|(P_{M_2} P_{M_1})^n - P_{M_1 \cap M_2}\| = c(M_1, M_2)^{2n-1}.$$

Proof. From Lemma 9.30, we have

$$(9.31.2) \qquad \|(P_{M_2} P_{M_1})^n - P_{M_1 \cap M_2}\| = \|(Q_1 Q_2)^n\|,$$

where $Q_i = P_{M_i \cap (M_1 \cap M_2)^\perp}$. Since the adjoint of $(Q_1 Q_2)^n$ is $(Q_2 Q_1)^n$, it follows from Theorem 8.25 that

$$(9.31.3) \qquad \|(Q_1 Q_2)^n\|^2 = \|(Q_1 Q_2)^n (Q_2 Q_1)^n\|.$$

Further, by a simple inductive argument (using the fact that Q_i is idempotent), we obtain

$$(9.31.4) \qquad (Q_1 Q_2)^n (Q_2 Q_1)^n = (Q_1 Q_2 Q_1)^{2n-1}.$$

Since $Q_1 Q_2 Q_1$ is self-adjoint, it follows from (9.31.4) and Lemma 8.32 that

$$(9.31.5) \qquad \|(Q_1 Q_2)^n\|^2 = \|(Q_1 Q_2 Q_1)^{2n-1}\| = \|Q_1 Q_2 Q_1\|^{2n-1}.$$

Using Theorem 8.25 and Lemma 9.5(7), we deduce

$$\begin{aligned}
\|Q_1 Q_2 Q_1\| &= \|(Q_1 Q_2)(Q_2 Q_1)\| = \|(Q_1 Q_2)(Q_1 Q_2)^*\| \\
&= \|Q_1 Q_2\|^2 = \|P_{M_1 \cap (M_1 \cap M_2)^\perp} P_{M_2 \cap (M_1 \cap M_2)^\perp}\|^2 \\
&= c(M_1, M_2)^2.
\end{aligned}$$

Thus (9.31.5) implies that $\|(Q_1 Q_2)^n\|^2 = c(M_1, M_2)^{2(2n-1)}$, and substituting this into (9.31.2) yields the result. ∎

9.32 Remarks. (1) Proposition 9.31 shows that the bound in Theorem 9.8 is sharp.

(2) A crude bound on $B_r(n)$ can be immediately obtained from (9.30.2):

$$(9.32.1) \qquad B_r(n) = \|(P_{M_r} P_{M_{r-1}} \cdots P_{M_1})^n - P_M\| \le \|P_{M_r} P_{M_{r-1}} \cdots P_{M_1} P_{M^\perp}\|^n$$

for all n, where $M := \cap_1^r M_i$.

(3) In particular, for $r = 2$, we obtain from (9.30.2) the inequality

$$B_2(n) = \|(Q_1 Q_2)^n\| \le \|Q_1 Q_2\|^n = c(M_1, M_2)^n.$$

Thus the extra care taken in the proof of Proposition 9.31 resulted in an improvement of this cruder bound by the multiplicative factor $c(M_1, M_2)^{n-1}$.

For $r > 2$, we do not compute $B_r(n)$ exactly, but we are able to get an upper bound for it.

9.33 Theorem. *Let $M_1, M_2, \ldots, M_r$ be closed subspaces in the Hilbert space X, let $M = \cap_1^r M_i$, and let*

$$(9.33.1) \qquad c_i = c\left(M_i, \bigcap_{j=i+1}^{r} M_j\right) \qquad (i = 1, 2, \ldots, r-1).$$

Then for each $x \in X$,

$$(9.33.2) \qquad \|(P_{M_r} P_{M_{r-1}} \cdots P_{M_1})^n x - P_M x\| \leq c^n \|x\|,$$

where

$$(9.33.3) \qquad c := \left[1 - \prod_{i=1}^{r-1} (1 - c_i^2)\right]^{1/2}.$$

Proof. Let $P_i = P_{M_i}$ and $P = P_r P_{r-1} \cdots P_1$, and write $X = M \oplus M^\perp$. We first observe that from Lemma 9.29(3),

$$(9.33.4) \qquad Px = x \quad \text{for all } x \in M,$$

and

$$(9.33.5) \qquad P(M^\perp) \subset M^\perp.$$

We will next show that for every $x \in M^\perp$,

$$(9.33.6) \qquad \|Px\|^2 \leq c^2 \|x\|^2.$$

We proceed by induction on r. For $r = 1$, $M = M_1$, $P = P_1 = P_M$, and hence the left side of (9.33.6) is 0.

Now assume that the result holds for $r - 1$ subspaces, where $r \geq 2$. Let

$$M' = M_2 \cap M_3 \cap \cdots \cap M_r \quad \text{and} \quad P' = P_r P_{r-1} \cdots P_2.$$

Then by Lemma 9.29(3), we have

$$(9.33.7) \qquad P'(x') = x' \quad \text{for every } x' \in M',$$

and

$$(9.33.8) \qquad P'(M'^\perp) \subset M'^\perp.$$

For any $x \in M^\perp$, write $x = x_1 + x'$, where $x_1 \in M_1$ and $x' \in M_1^\perp \subset M^\perp$. Since $x' \in M^\perp$, we see that

$$x_1 = x - x' \in M^\perp \cap M_1.$$

Next write $x_1 = x_1' + x_1''$, where $x_1' \in M'$ and $x_1'' \in M'^\perp$. Since $x_1'' \in M^\perp$, we have

$$x_1' = x_1 - x_1'' \in M^\perp \cap M'.$$

Using (9.33.7), we get

$$P'(x_1) = P'(x_1' + x_1'') = x_1' + P'(x_1'').$$

Since $P'(x_1'') \in M'^\perp$ by (9.33.8), we obtain

(9.33.9) $$\|P'(x_1)\|^2 = \|x_1'\|^2 + \|P'(x_1'')\|^2.$$

The induction hypothesis yields

$$\|P'(x_1'')\|^2 \leq \left[1 - \prod_{i=2}^{r-1}(1 - c_i^2)\right]\|x_1''\|^2.$$

Substituting this and the equation $\|x_1''\|^2 = \|x_1\|^2 - \|x_1'\|^2$ into (9.33.9), we get

$$\|P'(x_1)\|^2 \leq \|x_1'\|^2 + \left[1 - \prod_{i=2}^{r-1}(1 - c_i^2)\right](\|x_1\|^2 - \|x_1'\|^2)$$

$$= \left[\prod_{i=2}^{r-1}(1 - c_i^2)\right]\|x_1'\|^2 + \left[1 - \prod_{i=2}^{r-1}(1 - c_i^2)\right]\|x_1\|^2.$$

Now,

$$\|x_1'\|^2 = \langle x_1', x_1'\rangle = \langle x_1 - x_1'', x_1'\rangle = \langle x_1, x_1'\rangle$$
$$\leq c(M_1, M')\|x_1\|\,\|x_1'\| = c_1\|x_1\|\,\|x_1'\|$$

implies $\|x_1'\| \leq c_1\|x_1\|$. Hence, from above,

$$\|P'(x_1)\|^2 \leq \left[\prod_{i=2}^{r-1}(1 - c_i^2)\right]c_1^2\|x_1\|^2 + \left[1 - \prod_{i=2}^{r-1}(1 - c_i^2)\right]\|x_1\|^2$$

$$= \left[1 - (1 - c_1^2)\prod_{i=2}^{r-1}(1 - c_i^2)\right]\|x_1\|^2$$

$$= \left[1 - \prod_{i=1}^{r-1}(1 - c_i^2)\right]\|x_1\|^2 = c^2\|x_1\|^2$$

$$= c^2(\|x\|^2 - \|x_1'\|^2) \leq c^2\|x\|^2.$$

Since $x = x_1 + x_1'$ with $x_1 \in M_1$ and $x_1' \in M_1^\perp$, we obtain $Px_1' = P_r \cdots P_2 P_1 x_1' = 0$ and hence

$$P(x) = Px_1 = P_r \cdots P_2 P_1 x_1 = P_r \cdots P_2 x_1 = P'(x_1).$$

This proves that

$$\|P(x)\|^2 = \|P'(x_1)\|^2 \leq c^2\|x\|^2,$$

which completes the induction and proves (9.33.6). In particular, this proves

(9.33.10) $$\|P\| \leq c.$$

Further, we have

$$P^n - P_M = P^n - P^n P_M = P^n(I - P_M) = P^n P_{M^\perp}.$$

Hence, for any $x \in X$, we have that $P_{M^\perp}x \in M^\perp$ and $P^{n-1}(P_{M^\perp}x) \in M^\perp$ by (9.33.5). Thus, by (9.33.6),

$$
\begin{aligned}
\|P^n x - P_M x\| = \|P^n P_{M^\perp}x\| &= \|P\, P^{n-1}(P_{M^\perp}x)\| \\
&\le c\|P^{n-1}(P_{M^\perp}x)\| \le c^2\|P^{n-2}(P_{M^\perp}x)\| \\
&\le \cdots \le c^n\|P_{M^\perp}x\| \le c^n\|x\|.
\end{aligned}
$$

This completes the proof of (9.33.2). ∎

A more general result than Theorem 9.33 can be deduced from it. Let $V_1, \ldots, V_r$ denote r closed affine sets in X such that $\cap_1^r V_i \ne \emptyset$. Then (see, e.g., Exercise 13 at the end of Chapter 2) there exist unique closed subspaces $M, M_1, \ldots, M_r$ such that

$$
V_i = M_i + v_0 \qquad (i = 1, 2, \ldots, r),
$$

and

$$
\bigcap_1^r V_i = M + v_0
$$

for any $v_0 \in \cap_1^r V_i$.

9.34 Corollary. *For each x in the Hilbert space X,*

(9.34.1) $$\|(P_{V_r} \cdots P_{V_1})^n x - P_{\cap_1^r V_i} x\| \le c^n \|x - v_0\|$$

$(n = 1, 2, \ldots)$, *where*

(9.34.2) $$c = \left[1 - \prod_{i=1}^{r-1}(1 - c_i^2)\right]^{1/2}$$

and

(9.34.3) $$c_i = c\left(M_i, \bigcap_{j=i+1}^{r} M_j\right) \qquad (i = 1, 2, \ldots, r-1).$$

The proof follows from the fact that $P_{V_i}x = P_{M_i}(x - v_0) + v_0$ (see Theorem 2.7 (ii) and Theorem 9.33). We leave the details as an exercise.

Before we turn to some applications of these theorems, we need a deeper property of the angle between subspaces. It characterizes when the angle between two subspaces is positive. We shall prove it now.

9.35 Theorem. *Let M and N be closed subspaces of the Hilbert space X. Then the following statements are equivalent.*
   (1) $c(M, N) < 1$;
   (2) $M \cap (M \cap N)^\perp + N \cap (M \cap N)^\perp$ *is closed;*
   (3) $M + N$ *is closed.*
   (4) $M^\perp + N^\perp$ *is closed.*

Proof. For notational simplicity, let $Y := M \cap (M \cap N)^\perp$ and $Z := N \cap (M \cap N)^\perp$. We first verify the following two subspace identities.
   Claim.   (a) $Y + Z = (M + N) \cap (M \cap N)^\perp$.

(b) $M + N = Y + Z + M \cap N$.

To verify (a), first note that $Y + Z \subset (M + N) \cap (M \cap N)^{\perp}$ is obvious. Next fix any $x \in (M + N) \cap (M \cap N)^{\perp}$. Then $x = y + z$, where $y \in M$, $z \in N$. Since $x \in (M \cap N)^{\perp}$, it follows that $P_{M \cap N} x = 0$. Thus

$$x = x - P_{M \cap N} x = (y - P_{M \cap N} y) + (z - P_{M \cap N} z)$$
$$\in M \cap (M \cap N)^{\perp} + N \cap (M \cap N)^{\perp} = Y + Z$$

using Theorem 4.9. This proves (a).

To prove (b), first note that $Y + Z + M \cap N \subset M + N$ is obvious. Next fix any $x \in M + N$. Then $x = y + z$ for some $y \in M$ and $z \in N$. By Theorem 5.8(2),

$$x = P_{M \cap N}(x) + P_{(M \cap N)^{\perp}}(x) = P_{M \cap N}(y) + P_{M \cap N}(z) + P_{(M \cap N)^{\perp}}(y) + P_{(M \cap N)^{\perp}}(z).$$

But since $M \cap N \subset M$, $P_{M \cap N}$ commutes with P_M by Lemma 9.2. It follows that $P_{(M \cap N)^{\perp}} = I - P_{M \cap N}$ also commutes with P_M. By Lemma 9.2 again, $P_{(M \cap N)^{\perp}}(y) \in M \cap (M \cap N)^{\perp} = Y$. Similarly, $P_{(M \cap N)^{\perp}}(z) \in N \cap (M \cap N)^{\perp} = Z$. This proves that $x \in M \cap N + Y + Z$ and verifies (b).

(2) $\Longleftrightarrow$ (3). If $M + N$ is closed, then since $(M \cap N)^{\perp}$ is always closed, $Y + Z$ is closed by Claim (a).

Conversely, suppose $Y + Z$ is closed. Since $Y + Z \subset (M \cap N)^{\perp}$, the sum $(Y + Z) + M \cap N$ is closed because it is a sum of orthogonal closed subspaces (see Exercise 14 at the end of the chapter). Using Claim (b), it follows that $M + N$ is closed.

(1) $\Longrightarrow$ (2). Let $c = c(M, N) < 1$. We must show that $Y + Z$ is closed. By Lemma 9.5(6),

$$(9.35.1) \qquad |\langle y, z \rangle| \leq c \|y\| \, \|z\|$$

for all $y \in Y$ and $z \in Z$. Now let $x_n \in Y + Z$ and $x_n \to x$. It suffices to show that $x \in Y + Z$. Write $x_n = y_n + z_n$, where $y_n \in Y$, $z_n \in Z$. Then $\{x_n\}$ is bounded, so there exists a constant $\rho > 0$ such that

$$\begin{aligned} \rho \geq \|x_n\|^2 &= \|y_n + z_n\|^2 = \|y_n\|^2 + 2\langle y_n, z_n \rangle + \|z_n\|^2 \\ &\geq \|y_n\|^2 - 2|\langle y_n, z_n \rangle| + \|z_n\|^2 \\ &\geq \|y_n\|^2 - 2c\|y_n\| \, \|z_n\| + \|z_n\|^2 \qquad \text{(using (9.35.1))} \\ &= (\|y_n\| - \|z_n\|)^2 + 2(1 - c)\|y_n\| \, \|z_n\|. \end{aligned}$$

Since $0 \leq c < 1$, it follows that both sequences $\{\|y_n\| - \|z_n\|\}$ and $\{\|y_n\| \, \|z_n\|\}$ are bounded. From this it follows that both sequences $\{\|y_n\|\}$ and $\{\|z_n\|\}$ are bounded. By passing to a subsequence, if necessary, Theorem 9.12 implies that $\{y_n\}$ converges weakly to some y, and $\{z_n\}$ converges weakly to some z. Since Y and Z are closed subspaces, they are weakly closed by Theorem 9.16(2). Thus $y \in Y$, $z \in Z$ and $x_n = y_n + z_n \to y + z$ weakly. But $x_n \to x$ implies $x_n \to x$ weakly, and hence $x = y + z \in Y + Z$. This proves (2).

(2) $\Longrightarrow$ (1). Suppose that $Y + Z$ is closed. Then $H := Y + Z$ is a Hilbert space and $Y \cap Z = \{0\}$. By Proposition 8.24, the linear mapping $Q : H \to Y$ defined by

$$Q(y + z) = y \quad , \quad y \in Y, \quad z \in Z,$$

is continuous.

If (1) failed, then $c := c(M, N) = 1$, and there would exist $y_n \in Y$, $z_n \in Z$ with $\|y_n\| = 1 = \|z_n\|$ and $1 = c = \lim_{n \to \infty} \langle y_n, z_n \rangle$. Then

$$\|y_n - z_n\|^2 = \|y_n\|^2 - 2\langle y_n, z_n \rangle + \|z_n\|^2 = 2 - 2\langle y_n, z_n \rangle \longrightarrow 0$$

implies that

$$y_n = Q(y_n - z_n) \longrightarrow Q(0) = 0,$$

which is absurd. Thus $c < 1$ and this proves (1).

We have shown the equivalence of the first three statements. But Theorem 9.7 already established the equivalence of (3) and (4). ∎

Remark. It is also true that $c(M^\perp, N^\perp) = c(M, N)$. However, since an elementary proof of this fact seems fairly lengthy, and since we use this fact in only one example below (Example 9.40), we have omitted it.

9.36 Lemma. *Let M be a closed subspace and N a finite-dimensional subspace in the inner product space X. Then $M + N$ is a closed subspace.*

Proof. The proof is by induction on $k = \dim N$. For $k = 1$, $N = \operatorname{span}\{b_1\}$ for some $b_1 \in X \setminus \{0\}$. If $b_1 \in M$, then $M + N = M$ is closed. Thus we may assume $b_1 \notin M$. Let $x_n \in M + N$ and $x_n \to x$. We must show that $x \in M + N$. Then $x_n = y_n + \alpha_n b_1$ for some $y_n \in M$ and $\alpha_n \in \mathbb{R}$. We have

$$d(x_n, M) = d(x_n - y_n, M) = d(\alpha_n b_1, M) = |\alpha_n| d(b_1, M),$$

so that

$$|\alpha_n| = \frac{d(x_n, M)}{d(b_1, M)} \longrightarrow \frac{d(x, M)}{d(b_1, M)},$$

which implies that $\{\alpha_n\}$ is bounded. Thus there is a subsequence $\{\alpha_{n_j}\}$ and $\alpha \in \mathbb{R}$ such that $\alpha_{n_j} \to \alpha$. Hence

$$y_{n_j} = x_{n_j} - \alpha_{n_j} b_1 \longrightarrow x - \alpha b_1 =: y \in M,$$

since M is closed, and $x = y + \alpha b_1 \in M + N$.

Now assume that the result is true for all subspaces of dimension at most k. If N has dimension $k + 1$, then we can write $N = N_k + N_1$, where $\dim N_k = k$ and $\dim N_1 = 1$. By the induction hypothesis, $M_k := M + N_k$ is closed. By the proof when $k = 1$, $M_k + N_1$ is closed. Since $M + N = M + N_k + N_1 = M_k + N_1$, it follows that $M + N$ is closed. ∎

9.37 Corollary. *Let M and N be closed subspaces in the Hilbert space X such that M or N has either finite dimension or finite codimension. Then $c(M, N) < 1$.*

Proof. If $\dim N < \infty$, then $M + N$ is closed by Lemma 9.36. If $\operatorname{codim} N < \infty$, then $\dim N^\perp < \infty$ by Theorem 7.12, and $M^\perp + N^\perp$ is closed by Lemma 9.36. Thus either $M + N$ or $M^\perp + N^\perp$ is closed. By Theorem 9.35, $c(M, N) < 1$. ∎

Before turning to some possible applications of the von Neumann and Dykstra algorithms, we will state two lemmas that will prove useful in these applications.

9.38 Lemma. *Let M be a closed subspace of the Hilbert space X and $x \in X \setminus \{0\}$. Then*

$$(9.38.1) \qquad c\left(\operatorname{span}\{x\}, M\right) = \begin{cases} \|P_M(x)\|/\|x\| & \text{if } x \notin M, \\ 0 & \text{if } x \in M. \end{cases}$$

In particular, $c\left(\operatorname{span}\{x\}, M\right) < 1$.

Proof. If $x \notin M$, $\operatorname{span}\{x\} \cap M = \{0\}$, so that

$$c\left(M, \operatorname{span}\{x\}\right) = \sup\{|\langle y, \alpha x\rangle| \mid y \in M, \|y\| \leq 1, \alpha \in \mathbb{R}, |\alpha| \leq \|x\|^{-1}\}$$

$$= \frac{1}{\|x\|} \sup\{|\langle y, x\rangle| \mid y \in M, \|y\| \leq 1\}$$

$$= \frac{1}{\|x\|} \sup\{|\langle P_M z, x\rangle| \mid z \in X, \|z\| \leq 1\}$$

$$= \frac{1}{\|x\|} \sup\{|\langle z, P_M x\rangle| \mid z \in X, \|z\| \leq 1\}$$

$$= \frac{1}{\|x\|} \|P_M x\|.$$

If $x \in M$, then $\operatorname{span}\{x\} \subset M$ and Lemma 9.5 (8) implies that $c\left(\operatorname{span}\{x\}, M\right) = 0$. This proves (9.38.1).

If $x \notin M$, then since $x = P_M(x) + P_{M^\perp}(x)$, we have

$$\|x\|^2 = \|P_M(x)\|^2 + \|P_{M^\perp}(x)\|^2 > \|P_M(x)\|^2,$$

so that $\|P_M(x)\|/\|x\| < 1$. This proves the last statement. $\blacksquare$

It is convenient to restate here a special consequence of Theorem 6.31 concerning best approximations from half-spaces.

9.39 Theorem (Best approximation from half-spaces). *Let X be a Hilbert space, $z \in X \setminus \{0\}$, $c \in \mathbb{R}$, and*

$$(9.39.1) \qquad H = \{y \in X \mid \langle y, z\rangle \leq c\}.$$

Then:

(1) *H is a Chebyshev half-space.*

(2) *For any $x \in X$,*

$$(9.39.2) \qquad P_H(x) = x - \frac{[\langle x, z\rangle - c]^+ z}{\|z\|^2}$$

and

$$(9.39.3) \qquad d(x, H) = \frac{[\langle x, z\rangle - c]^+}{\|z\|},$$

where

$$\alpha^+ = \begin{cases} \alpha & \text{if } \alpha \geq 0, \\ 0 & \text{if } \alpha < 0. \end{cases}$$

What formula (9.39.2) implies, in particular, is that it is easy to compute best approximations from the closed half-spaces (9.39.1), and hence Dykstra's algorithm is effective for the intersection of a finite number of closed half-spaces (i.e., for a polyhedron) in a Hilbert space.

Examples

We now list several examples, which can all be effectively handled by Dykstra's algorithm or von Neumann's alternating projections algorithm.

9.40 Example (Linear equations).

Let X be a Hilbert space, $\{y_1, y_2, \ldots, y_r\}$ in $X \setminus \{0\}$, and $\{c_1, c_2, \ldots, c_r\} \subset \mathbb{R}$. We want to find an $x \in X$ that satisfies the equations

$$(9.40.1) \qquad \langle x, y_i \rangle = c_i \qquad (i = 1, 2, \ldots, r).$$

Setting

$$H_i = \{y \in X \mid \langle y, y_i \rangle = c_i\} \qquad (i = 1, 2, \ldots, r)$$

and

$$(9.40.2) \qquad K = \bigcap_1^r H_i,$$

we see that x satisfies (9.40.1) if and only if $x \in K$. Assuming $K \neq \emptyset$, we see that since the H_i are closed hyperplanes and, by Theorem 6.17,

$$P_{H_i}(z) = z - \frac{1}{\|y_i\|^2} [\langle z, y_i \rangle - c_i] y_i$$

for any $z \in X$. It follows from Corollary 9.34 that $(P_{H_r} P_{H_{r-1}} \cdots P_{H_1})^n (x)$ converges to $P_K(x)$, for any $x \in X$, at the rate

$$(9.40.3) \qquad \|(P_{H_r} P_{H_{r-1}} \cdots P_{H_1})^n (x) - P_K(x)\| \leq c^n \|x - v_0\|,$$

where $v_0 \in K$ is arbitrary,

$$(9.40.4) \qquad c = \left[1 - \prod_{i=1}^{r-1} (1 - c_i^2) \right]^{1/2},$$

$$(9.40.5) \qquad c_i = c\left(M_i, \bigcap_{j=i+1}^r M_j \right) \qquad (i = 1, 2, \ldots, r-1),$$

and

$$(9.40.6) \qquad M_i = H_i - v_0 = \{y \in X \mid \langle y, y_i \rangle = 0\}.$$

But using Theorem 4.6(5) and the remark preceding Lemma 9.36, we obtain

$$c_i = c\left(M_i^\perp, \left(\bigcap_{i+1}^r M_j \right)^\perp \right) = c\left(M_i^\perp, \overline{M_{i+1}^\perp + \cdots + M_r^\perp} \right)$$

$$= c\left(M_i^\perp, M_{i+1}^\perp + \cdots + M_r^\perp \right),$$

since each $M_j^\perp$ is finite-dimensional so that the sum $M_{i+1}^\perp + \cdots + M_r^\perp$ is closed (Lemma 9.36). Since $M_j = y_j^\perp = [\text{span}\{y_j\}]^\perp$, it follows by Theorem 4.5(8) that

$$(9.40.7) \qquad M_j^\perp = [\text{span}\{y_j\}]^{\perp\perp} = \text{span}\{y_j\}.$$

Thus by Lemma 9.38,

$$
\begin{aligned}
c_i &= c\,(\text{span}\{y_i\},\ \text{span}\{y_{i+1}, y_{i+2}, \ldots, y_r\}) \\
&= \begin{cases} 0 & \text{if } y_i \in \text{span}\{y_{i+1}, \ldots, y_r\}, \\ \|P_{\text{span}\{y_{i+1},\ldots,y_r\}}(y_i)\|/\|y_i\| & \text{otherwise}, \end{cases}
\end{aligned}
$$

and $c_i < 1$ for all i. This shows that $c < 1$.

In particular, if $X = \ell_2(N)$ and $y_i = (a_{i1}, a_{i2}, \ldots, a_{iN})$ $(i = 1, 2, \ldots, r)$, equations (9.40.1) can be rewritten as

$$(9.40.8) \qquad \sum_{j=1}^{N} a_{ij} x(j) = c_i \qquad (i = 1, 2, \ldots, r).$$

Thus we can use von Neumann's alternating projections algorithm to compute a particular solution to any consistent system of linear equations (9.40.8). In particular, starting with $x_0 = 0$, the von Neumann algorithm produces a sequence of points $\{x_n\}$ in $\ell_2(N)$ that converges geometrically to the unique *minimal norm* solution of (9.40.8).

9.41 Example (Linear inequalities).

Let X be a Hilbert space, $\{y_1, y_2, \ldots, y_r\} \subset X \setminus \{0\}$, and $\{c_1, c_2, \ldots, c_r\} \subset \mathbb{R}$. We want to find an $x \in X$ that satisfies the inequalities

$$(9.41.1) \qquad \langle x, y_i \rangle \leq c_i \qquad (i = 1, 2, \ldots, r).$$

Setting $K = \cap_1^r H_i$, where

$$(9.41.2) \qquad H_i = \{y \in X \mid \langle y, y_i \rangle \leq c_i\},$$

we see that x solves (9.41.1) if and only if $x \in K$. We assume $K \neq \emptyset$.

To obtain a point in K, we start with any $x_0 \in X$ and generate the sequence $\{x_n\}$ according to Dykstra's algorithm that converges to $P_K(x_0)$. This algorithm is effective, since it is easy to obtain best approximations from the half-spaces H_i by Lemma 9.39.

In particular, taking $X = \ell_2(N)$ and $y_i = (a_{i1}, a_{i2}, \ldots, a_{iN})$, the inequalities (9.41.1) become

$$(9.41.3) \qquad a_{i1} x(1) + a_{i2} x(2) + \cdots + a_{iN} x(N) \leq c_i \qquad (i = 1, 2, \ldots, r).$$

Thus Dykstra's algorithm can be used effectively to solve a system of linear inequalities.

9.42 Example (Isotone regression). An important problem in statistical inference is to find the isotone regression for a given function defined on a finite ordered set of points. In the language of best approximation, a typical such problem may be rephrased as follows. Given any $x \in \ell_2(N)$, find the best approximation to x from the set of all increasing functions in $\ell_2(N)$:

$$(9.42.1) \qquad C = \{ y \in \ell_2(N) \mid y(1) \leq y(2) \leq \cdots \leq y(N) \}.$$

Note that we can write C as

$$(9.42.2) \qquad C = \bigcap_{i=1}^{N-1} H_i \,,$$

where

$$H_i = \{ y \in \ell_2(N) \mid y(i) \leq y(i+1) \} = \{ y \in \ell_2(N) \mid \langle y, y_i \rangle \leq 0 \}$$

is a closed half-space, and

$$y_i(j) = \begin{cases} 1 & \text{if } j = i, \\ -1 & \text{if } j = i+1, \\ 0 & \text{otherwise.} \end{cases}$$

Thus we are seeking $P_C(x)$. Since C is the intersection of a finite number of half-spaces, this reduces to an application of Dykstra's algorithm just as in Example 9.41 beginning with $x_0 = x$.

9.43 Example (Convex regression). Another problem in statistical inference is to find the convex regression for a given function defined on a finite subset of the real line. In the language of best approximation, the standard such problem can be rephrased as follows. Given any $x \in \ell_2(N)$, find the best approximation to x from the set of all functions in $\ell_2(N)$ that are convex, that is, from the set

$$(9.43.1) \quad C = \bigcap_{i=1}^{N-2} \{ y \in \ell_2(N) \mid y(i+1) - y(i) \leq y(i+2) - y(i+1) \} = \bigcap_{i=1}^{N-2} H_i,$$

where

$$H_i = \{ y \in \ell_2(N) \mid y(i+1) - y(i) \leq y(i+2) - y(i+1) \}.$$

But

$$H_i = \{ y \in \ell_2(N) \mid -y(i) + 2y(i+1) - y(i+2) \leq 0 \} = \{ y \in \ell_2(N) \mid \langle y, y_i \rangle \leq 0 \}$$

is a closed half-space, where

$$y_i(j) = \begin{cases} -1 & \text{if } j \in \{i, i+2\}, \\ 2 & \text{if } j = i+1, \\ 0 & \text{otherwise.} \end{cases}$$

Since we are seeking $P_C(x)$, and C is the intersection of a finite number of closed half-spaces, we again can apply Dykstra's algorithm (just as in Example 9.41) starting with $x_0 = x$.

9.44 Example (Shape-Preserving Interpolation). In the next chapter we will see that a large class of problems, which fall under the general heading of "shape-preserving interpolation," can be put into the following form.

Let $X = L_2[0,1]$, $\{y_1, y_2, \ldots, y_r\} \subset X \setminus \{0\}$, and $\{c_1, c_2, \ldots, c_r\} \subset \mathbb{R}$. Assume that the set

$$(9.44.1) \qquad K = \{y \in X \mid y \geq 0, \ \langle y, y_i \rangle = c_i \quad (i = (1, 2, \ldots, r)\}$$

is nonempty. We want to find the element of K having minimal norm; that is, we seek $P_K(0)$.

Setting

$$K_i = \{y \in X \mid \langle y, y_i \rangle = c_i\} \qquad (i = 1, 2, \ldots, r)$$

and

$$K_{r+1} = \{y \in X \mid y \geq 0\},$$

we see that K_{r+1} is a closed convex cone, $K_1, \ldots, K_r$ are closed hyperplanes, and

$$(9.44.2) \qquad K = \bigcap_1^{r+1} K_i.$$

Since best approximations from K_{r+1} are given by the formula (Theorem 4.8)

$$P_{K_{r+1}}(x) = x^+ := \max\{x, 0\},$$

and best approximations from the K_i $(i = 1, 2, \ldots, r)$ are given by (Theorem 6.17)

$$P_{K_i}(x) = x - \frac{1}{\|y_i\|^2}[\langle x, y_i \rangle - c_i]y_i,$$

we see that this problem can be effectively handled by Dykstra's algorithm starting with $x_0 = 0$. In fact, since each of the K_i for $i = 1, 2, \ldots, r$ is an affine set, we can ignore all the residuals when projecting onto $K_1, K_2, \ldots, K_r$ (see (9.26.3)). We need only keep track of the residuals obtained when projecting onto K_{r+1}. This is a substantial simplification of Dykstra's algorithm for this case.

More precisely, set

$$[n] = \{1, 2, \ldots, r+1\} \cap \{n - k(r+1) \mid k = 0, 1, 2, 3 \ldots\},$$
$$x_0 = 0, \qquad e_{-r} = e_{-(r-1)} = \cdots = e_{-1} = e_0 = 0,$$
$$x_n = P_{K_{[n]}}(x_{n-1}) \quad \text{if} \quad [n] \neq r+1,$$

and

$$x_n = P_{K_{r+1}}(x_{n-1} + e_{n-(r+1)}) \quad \text{and} \quad e_n = x_{n-1} + e_{n-(r+1)} - x_n$$

if $[n] = r+1$.

Then the sequence $\{x_n\}$ converges to $P_K(0)$. More generally, start with any $x \in X$ and put $x_0 = x$. Then the sequence $\{x_n\}$ converges to $P_K(x)$.

Exercises

1. Let $\{x_n\}$ be the sequence defined in the von Neumann theorem (Theorem 9.3). Verify the following statements:
   (a) $\|x_{2n}\| \leq \|x_{2n-1}\|$ for all n and $\|x_{2n}\| = \|x_{2n-1}\|$ if and only if $x_{2n-1} \in M_2$.
   (b) $\|x_{2n+1}\| \leq \|x_{2n}\|$ for all n and $\|x_{2n+1}\| = \|x_{2n}\|$ if and only if $x_{2n} \in M_1$.
   (c) If $x_{2n} \in M_1$ for some n, then $x_k = x_{2n}$ for all $k \geq 2n$ and $x_{2n} = P_{M_1 \cap M_2}(x)$.
   (d) If $x_{2n-1} \in M_2$ for some n, then $x_k = x_{2n-1}$ for all $k \geq 2n-1$ and $x_{2n-1} = P_{M_1 \cap M_2}(x)$.
   (e) $\langle x_{2n+1}, x_{2m+1} \rangle = \langle x_{2n}, x_{2m+2} \rangle$ for all $n, m \in \mathbb{N}$.
2. Verify Example 9.6 (2). That is, show that the angle between the subspaces $M = \{x \in \ell_2(3) \mid x(3) = 0\}$ and $N = \{x \in \ell_2(3) \mid x(1) = 0\}$ is $\pi/2$ (radians).
3. What is the angle between the following pairs of subspaces?
   (a) M and $M^\perp$ (M an arbitrary subspace),
   (b) $M = \{x \in \ell_2 \mid x(2n) = 0 \quad (n = 1, 2, \ldots)\}$
   and
   $N = \{x \in \ell_2 \mid x(2n-1) = 0 \quad (n = 1, 2, \ldots)\}$.
4. Some authors have defined the "minimal angle" between subspaces M and N as the angle between 0 and $\pi/2$ whose "cosine" is

$$c_0(M, N) := \sup\{|\langle x, y \rangle| \mid x \in M, \|x\| \leq 1, y \in N, \|y\| \leq 1\}.$$

   Verify the following statements.
   (a) $0 \leq c_0(M, N) \leq 1$.
   (b) $c_0(M, N) = c_0(N, M)$.
   (c) $c(M, N) \leq c_0(M, N)$ and $c(M, N) = c_0(M, N)$ if $M \cap N = \{0\}$.
   (d) $c_0(M, N) < 1$ if and only if $M \cap N = \{0\}$ and $M + N$ is closed. [Hint: $c_0(M, N) = c(M, N)$ in the cases of relevance.]
   (e) If $M + N$ is closed and $M \cap N \neq \{0\}$, then $c(M, N) < c_0(M, N) = 1$.
5. Verify the following statements.
   (a) Weak limits are unique; that is, if $x_n \xrightarrow{w} x$ and $x_n \xrightarrow{w} y$, then $x = y$.
   (b) If $x_n \xrightarrow{w} x$ and $y_n \xrightarrow{w} y$, then $x_n + y_n \xrightarrow{w} x + y$.
   (c) If $x_n \xrightarrow{w} x$ and $\alpha_n \longrightarrow \alpha$ (in $\mathbb{R}$), then $\alpha_n x_n \xrightarrow{w} \alpha x$.
   (d) $x_n \xrightarrow{w} x$ weakly if and only if $x^*(x_n) \longrightarrow x^*(x)$ for all $x^* \in X^*$. [Hint: Theorem 6.8.]
6. In a Hilbert space, show that a sequence $\{x_n\}$ is bounded if and only if each subsequence of $\{x_n\}$ has a subsequence that converges weakly. [Hint: Theorem 9.12 and Lemma 9.13.]
7. Let X be a Hilbert space and $\{x_n\}$ a sequence in X. Show that the following statements are equivalent.
   (a) $\{x_n\}$ converges weakly to x.
   (b) Every subsequence of $\{x_n\}$ has a subsequence that converges weakly to x. [Hint: Lemma 9.13 and Exercise 6.]
8. Call a set K **weakly compact** if each sequence in K has a subsequence that converges weakly to a point in K. Prove the following statements.
   (a) Every weakly compact set is weakly closed.

(b) In a Hilbert space, a set is weakly compact if and only if it is weakly closed and bounded.

[Hint: Exercise 6.]

9. Let M be a Chebyshev subspace in X, $v \in X$, and $V = M + v$. Prove that V is an affine Chebyshev set, and for all $x, y \in X$ and $\alpha \in \mathbb{R}$,

$$P_V(x + y) = P_V(x) + P_M(y)$$

and

$$P_V(\alpha x) = \alpha P_M(x) + P_V(0).$$

10. In the text, Corollary 9.28 was deduced from Theorem 9.27. Assuming Corollary 9.28, deduce Theorem 9.27.

[Hint: Translate $\cap_1^r V_i$ by an element in this set.]

11. In the proof of Theorem 9.31, we used the fact that if Q_1 and Q_2 are idempotent, then

$$(Q_1 Q_2)^n (Q_2 Q_1)^n = (Q_1 Q_2 Q_1)^{2n-1} \qquad (n = 1, 2, \dots).$$

Prove this by induction on n.

12. Let $A, B \in \mathcal{B}(X, X)$. Show that if A and B commute, then

$$A(\mathcal{R}(B)) \subset \mathcal{R}(B) \text{ and } A(\mathcal{N}(B)) \subset \mathcal{N}(B).$$

In words, *the range and null space of any one of two commuting operators are invariant under the other operator.*

13. Prove Corollary 9.34.

[Hint: Use Theorem 2.7 (ii), Exercise 13 at the end of Chapter 2, and induction to deduce that

$$(P_{V_r} \cdots P_{V_1})^n(x) = (P_{M_r} \cdots P_{M_1})^n(x - v_0) + v_0$$

and

$$P_{\cap_1^r V_i}(x) = P_{\cap_1^r M_i}(x)(x - v_0) + v_0.$$

Then apply Theorem 9.33.]

14. (**Sum of orthogonal subspaces is closed**) Suppose that M and N are closed orthogonal subspaces of the Hilbert space X, i.e., $M \subset N^\perp$. Show that $M + N$ is closed. More generally, $M + N$ is closed if M and N are orthogonal Chebyshev subspaces in the inner product space X.

15. Suppose X and Y are Hilbert spaces and $A \in \mathcal{B}(X, Y)$. Show that A is bounded below if and only if A is one-to-one and has closed range. [Hint: Lemma 8.18 for the "if" part, and Theorem 9.12 and Lemma 9.14 for the "only if" part.]

16. An operator $A \in \mathcal{B}(X, X)$ is called **nonnegative**, denoted by $A \geq 0$, if $\langle Ax, x \rangle \geq 0$ for all $x \in X$. If A, B are in $\mathcal{B}(X, X)$, we write $A \geq B$ or $B \leq A$ if $A - B \geq 0$.

If M and N are Chebyshev subspaces in X, show that the following three statements are equivalent:

(a) $P_M \leq P_N$;

(b) $M \subset N$;

(c) $\|P_M x\| \leq \|P_N x\|$ for all $x \in X$.

17. Let M and N be closed subspaces of the Hilbert space X and $A \in \mathcal{B}(X, X)$. Show that

$$\|P_N P_M A\|^2 \leq c^2 \|P_M A\|^2 + (1 - c^2) \|P_{M \cap N} A\|^2,$$

where $c = c(M, N)$.

18. Let M and N be closed subspaces of the Hilbert space X. Show that

(a) $M = M \cap N + M \cap (M \cap N)^{\perp}$,

(b) $P_M = P_{M \cap N} + P_{M \cap (M \cap N)^{\perp}}$.

[Hint: For (a), let $x \in M$, $y = P_{M \cap N} x$, and $z = x - y$. Note that $z \in M \cap (M \cap N)^{\perp}$ and $x \in M \cap N + M \cap (M \cap N)^{\perp}$. For (b), use (a) and show that $x - (P_{M \cap N} x + P_{M \cap (M \cap N)^{\perp}} x) \in M^{\perp}$ for any x.]

19. For closed subspaces M and N in a Hilbert space X, prove that the following five statements are equivalent.

(a) P_M and P_N commute;

(b) P_M and $P_{N^{\perp}}$ commute;

(c) $P_{M^{\perp}}$ and P_N commute;

(d) $P_{M^{\perp}}$ and $P_{N^{\perp}}$ commute;

(e) $M = M \cap N + M \cap N^{\perp}$.

20. Let M_1, M_2, and M_3 be closed subspaces in the Hilbert space X with $M_2 \cup M_3 \subset M_1$ and $M_2 \perp M_3$. Suppose $P_M := P_{M_1} - P_{M_2} - P_{M_3}$ is an orthogonal projection. Show that $M = \{0\}$ if and only if $M_1 = M_2 + M_3$.

21. Let M and N denote the following subspaces of $X = \ell_2(4)$:

$$M = \text{span}\{e_1, e_2 + e_3\} \quad \text{and} \quad N = \text{span}\{e_1 + e_2, e_3 + e_4\},$$

where e_i are the canonical unit basis vectors for X: $e_i(j) = \delta_{ij}$. It was shown in Chapter 4 (Exercise 26 at the end of Chapter 4) that M and N are 2-dimensional Chebyshev subspaces in X with $M \cap N = \{0\}$, and for every $x = \sum_1^4 x(i) e_i$,

$$P_M(x) = x(1) e_1 + \frac{1}{2} [x(2) + x(3)](e_2 + e_3)$$

and

$$P_N(x) = \frac{1}{2} [x(1) + x(2)](e_1 + e_2) + \frac{1}{2} [x(3) + x(4)](e_3 + e_4).$$

(a) Compute $c(M, N)$.

(b) Compute $P_N P_M(x)$ for any $x \in X$.

(c) Compute $(P_N P_M)^n(x)$ for any $x \in X$ and every $n \in \mathbb{N}$.

(e) Give a *direct* proof that

$$(P_N P_M)^n(x) \to 0 \text{ as } n \to \infty$$

for every $x \in X$. That is, do *not* appeal to Theorem 9.3.

22. **(Opial's condition)** Let X be an inner product space and let $\{x_n\}$ be a sequence in X that converges weakly to x. Show that for each $y \in X \setminus \{x\}$,

$$\liminf_n \|x_n - y\| > \liminf_n \|x_n - x\|.$$

[Hint: Expand $\|x_n - y\|^2 = \|(x_n - x) + (x - y)\|^2$.]

Historical Notes

From Lemma 9.2, we see that $P_{M_2}P_{M_1} = P_{M_1 \cap M_2}$ if P_{M_1} and P_{M_2} commute, that is, the product of two commuting projections is a projection. Von Neumann (1933) was interested in what could be said in the case where P_{M_1} and P_{M_2} do *not* commute. He subsequently proved Theorem 9.3, which states that *the sequence of higher iterates of the product of the two projections* $\{(P_{M_2}P_{M_1})^n \mid n = 1, 2, \dots\}$ *converges pointwise to* $P_{M_1 \cap M_2}$. This theorem was rediscovered by several authors including Aronszajn (1950), Nakano (1953), Wiener (1955), Powell (1970), Gordon, Bender, and Herman (1970), and Hounsfield (1973)—the Nobel Prize winning inventor of the EMI scanner.

There are at least ten different areas of mathematics in which the method of alternating projections has found applications. They include solving linear equations, the Dirichlet problem, probability and statistics, computing Bergman kernels, approximating multivariate functions by sums of univariate ones, least change secant updates, multigrid methods, conformal mapping, image restoration, and computed tomography (see the survey by Deutsch (1992) for a more detailed description of these areas of applications and a list of references).

There are several posssible ways of extending the von Neumann theorem (Theorem 9.3). In a strictly convex reflexive Banach space X, every closed convex subset C is a Chebyshev set (see Riesz (1934), who proved it in case X is a Hilbert space, and Day (1941; p. 316) for the general case) so P_C is well-defined. Stiles (1965a) proved the following result.

Theorem (Stiles). *If X is a strictly convex reflexive Banach space having dimension at least 3, and if for each pair of closed subspaces M and N in X and each $x \in X$,*

$$\lim_{n \to \infty} (P_M P_N)^n(x) = P_{M \cap N}(x),$$

then X must be a Hilbert space.

Thus trying to extend Theorem 9.3 to more general Banach spaces seems fruitless. However, by replacing the subspaces by their orthogonal complements in Theorem 9.3, and using the relationship between their orthogonal projections (see Theorem 5.8(2)), we obtain the following *equivalent* version of Theorem 9.3.

Theorem. *Let X be a Hilbert space and let M and N be closed subspaces. Then, for each $x \in X$,*

$$(9.44.3) \qquad \lim_{n \to \infty} [(I - P_N)(I - P_M)]^n(x) = (I - P_{\overline{M+N}})(x).$$

Interestingly enough, in contrast to the equivalent Theorem 9.3, this theorem *does* extend to more general Banach spaces. Indeed, Klee (1963) exhibited a 2-dimensional non-Hilbert space X in which (9.44.3) holds for all closed subspaces M and N. (Klee's result answered negatively a question of Hirschfeld (1958), who asked, If X is strictly convex and reflexive and (9.44.3) holds for all closed subspaces M and N, must X be a Hilbert space?) Extending this even further, Stiles (1965c) showed that (9.44.3) *holds if X is any finite-dimensional smooth and strictly convex space*. For a variant of the Stiles (1965c) theorem when strict convexity is dropped, see Pantelidis (1968). Franchetti (1973), Atlestam and Sullivan (1976), and Deutsch (1979) have each generalized the Stiles (1965c) theorem. The most inclusive of these is in the latter paper, and it may be stated as follows.

Theorem (Deutsch). *Let X be a uniformly smooth and uniformly convex Banach space (i.e., both X and X^* are uniformly convex). Let M and N be closed subspaces of X such that $M + N$ is closed. Then*

$$(9.44.4) \qquad \lim_{n \to \infty} [(I - P_N)(I - P_M)]^n (x) = (I - P_{\overline{M+N}})(x) \quad \text{for each } x \in X.$$

Boznay (1986) stated that this theorem holds *without* the restriction that $M + N$ be closed, but his proof is not convincing. Franchetti and Light (1984) proved that (9.44.4) holds for *all pairs* of closed subspaces M and N if, in addition, the modulus of convexity δ_X of X (respectively δ_{X^*} of X^*) satisfies $\inf_{\epsilon > 0} \epsilon^{-2} \delta_X(\epsilon) > 0$ (respectively $\inf_{\epsilon > 0} \epsilon^{-2} \delta_{X^*}(\epsilon) > 0$). They also conjecture that there exists a uniformly convex and uniformly smooth Banach space X such that (9.44.4) *fails* for some pair of closed subspaces M and N (whose sum is not closed). Finally, they proved that in the Deutsch theorem, the condition that X be uniformly convex could be weakened to X is an "E-space." (An E-space is a strictly convex Banach space such that every weakly closed set is approximatively compact.)

If A and B are closed convex subsets of a Hilbert space X, Bauschke and Borwein (1993) made a thorough study of the convergence of the iterates $\{(P_B P_A)^n(x)\}$, even when A and B have empty intersection.

The angle between two subpaces given in Definition 9.4 is due to Friedrichs (1937). Dixmier (1949) gave a different definition (obtained by deleting the factors $(M \cap N)^\perp$ in the Friedrichs definition). These definitions agree when $M \cap N = \{0\}$, but not in general.

Lemma 9.5(7)–(8) is from Deutsch (1984). Theorem 9.7, even in more general Banach spaces, can be found in Kato (1984; Theorem 4.8). The rate of convergence in Theorem 9.8 is essentially due to Aronszajn (1950), while the fact that the constant c^{2n-1} is best possible is due to Kayalar and Weinert (1988).

The weak compactness theorem (Theorem 9.12) is due to Hilbert (1906) in the space ℓ_2 (who called it a "principle of choice"), and to Banach (1932) in any reflexive space.

Dykstra's algorithm was first described, and convergence proved, by Dykstra (1983) in the case where X is Euclidean n-space and where all the closed convex sets K_i were actually convex cones. Later, Boyle and Dykstra (1985) proved their general convergence theorem (Theorem 9.24) by essentially the same proof as described in this chapter. Dykstra's algorithm is applicable for problems involving isotone regression, convex regression, linear inequalities, quadratic programming, and linear programming (see the exposition by Deutsch (1995) for a more detailed description of these problems).

Corollary 9.28 was first proved by Halperin (1962) by a more direct approach. A short proof of Halperin's theorem was given by Stiles (1965b). Halperin's proof was adapted by Smarzewski (1996) (see also Bauschke, Deutsch, Hundal, and Park (1999)) to yield the following generalization.

Theorem (Smarzewski). *Let $T_1, T_2, \ldots, T_k$ be self-adjoint, nonnegative, and nonexpansive bounded linear operators on the Hilbert space X. Let $T = T_1 T_2 \cdots T_k$ and $M = \text{Fix}\, T$. Then*

$$\lim_{n \to \infty} T^n(x) = P_M(x) \quad \text{for each } x \in X.$$

Aronszajn (1950) proved the inequality

$$(9.46.3) \qquad \|(P_{M_2} P_{M_1})^n - P_{M_1 \cap M_2}\| \le c (M_1, M_2)^{2n-1} ,$$

and Kayalar and Weinert (1988) sharpened this by showing that *equality* actually holds in (9.46.3) (i.e., they proved Theorem 9.31). The rate of convergence theorem (Theorem 9.33) is due to Smith, Solmon, and Wagner (1977). Sharper rates of convergence theorems were later established by Kayalar and Weinert (1988) and Deutsch and Hundal (1997). Also, in contrast to Theorem 9.31 for two subspaces, Deutsch and Hundal (1997) showed that *none* of the bounds given by Smith, Solmon, and Wagner (1977), by Kayalar and Weinert (1988), and by Deutsch and Hundal (1997) is sharp for more than two subspaces!

The equivalence of the first two statements in Theorem 9.35 is due to Deutsch (1984), while the equivalence of the second two is due to Simonič, who privately communicated his results to Bauschke and Borwein (1993; Lemma 4.10). Related to this and to the error bound (9.32.1), we have the following result: *Let $M_1, M_2, \ldots, M_r$ be closed subspaces of a Hilbert space and $M = \cap_1^r M_i$. Then $\|P_{M_r} P_{M_{r-1}} \cdots P_{M_1} P_{M^\perp}\| < 1$ if and only if $M_1^\perp + M_2^\perp + \cdots + M_r^\perp$ is closed.* This follows from a combination of results of Bauschke and Borwein (1996; Lemma 5.18 and Theorem 5.19) and Bauschke, Borwein, and Lewis (1996; Proposition 3.7.3 and Theorem 3.7.4). (See also Bauschke (1996; Theorem 5.5.1).)

Hundal and Deutsch (1997) extended Dykstra's algorithm in two directions. First they allowed the number of sets K_i to be *infinite*; second, they allowed a *random*, rather than cyclic, ordering of the sets K_i.

Finally, we should mention that the main practical drawback of the method of alternating projections (MAP), at least for some applications, is that it is slowly convergent. Both Gubin, Polyak, and Raik (1967) and Gearhart and Koshy (1989) have considered a geometrically appealing method of *accelerating* the MAP that consists in adding a line search at each step (but no proofs were given of convergence of the acceleration scheme in either paper). Bauschke, Deutsch, Hundal, and Park (1999) considered a general class of iterative methods, which includes the MAP as a special case, and they studied the same kind of acceleration scheme. They proved that the acceleration method for the MAP is actually faster in the case of two subspaces, but not faster in general for more than two subspaces. They also showed that the acceleration method for a *symmetric* version of the MAP is always faster (for any number of subspaces). (See also the brief survey by Deutsch (2001) of line-search methods for accelerating the MAP.) Whether a similar acceleration scheme works for the Dykstra algorithm is an interesting open question.

CHAPTER 10

CONSTRAINED INTERPOLATION FROM A CONVEX SET

Shape-Preserving Interpolation

In many problems that arise in applications, one is given certain function values or "data" along with some reliable evidence that the unknown function that generated the data has a certain shape. For example, the function may be nonnegative or nondecreasing or convex. The problem is to recover the unknown function from this information.

A natural way to approximate the unknown function is to choose a specific function from the class of functions that have the same shape and that also interpolate the data. However, in general there may be more than one such function that interpolates the data and has the same shape. Thus an additional restriction is usually imposed to guarantee uniqueness. This additional condition can often be justified from physical considerations.

To motivate the general theory, let us consider a specific example.

10.1 Example. Fix any integers $k \geq 0$ and $m \geq 1$, and let $L_2^{(k)}[0,1]$ denote the space of all real-valued functions f on $[0,1]$ with the property that the kth derivative $f^{(k)}$ of f exists and is in $L_2[0,1]$. Consider the prescribed $m+k$ points

$$0 \leq t_1 < t_2 < \cdots < t_{m+k} \leq 1.$$

Suppose some $f_0 \in L_2^{(k)}[0,1]$ satisfies $f_0^{(k)} \geq 0$ on $[0,1]$. Thus if $k = 0, 1$, or 2, f_0 would be nonnegative, nondecreasing, or convex, respectively.

Now let

$$G = G_k(f_0) := \left\{ g \in L_2^{(k)}[0,1] \mid g(t_i) = f_0(t_i) \ (i = 1, 2, \ldots, m+k) \text{ and } g^{(k)} \geq 0 \right\}.$$

That is, G is the set of all functions in $L_2^{(k)}[0,1]$ that interpolate f_0 at the points t_i and that have the "same shape" as f_0. Suppose all that we know about the function f_0 are its values at the $m+k$ points t_i and that its kth derivative is nonnegative. To approximate f_0 on the basis of this limited information, it is natural to seek an element of G. But *which* element of G?

Certainly, G is nonempty (since $f_0 \in G$), and if there is more than one element of G, then there are infinitely many, since G is a convex set. For definiteness, we shall seek the element $g_0 \in G$ whose kth derivative has minimal L_2-norm:

$$(10.1.1) \qquad \qquad \|g_0^{(k)}\| = \inf \left\{ \|g^{(k)}\| \mid g \in G \right\}.$$

It can be shown, by using integration by parts, that this problem is equivalent to one of the following type. Determine the element h_0 of minimal norm from the set

$$(10.1.2) \qquad H := \{ h \in L_2[0,1] \mid h \geq 0, \quad \langle h, x_i \rangle = b_i \ (i = 1, 2, \ldots, m) \}.$$

That is, choose $h_0 \in H$ such that

$$(10.1.3) \qquad\qquad \|h_0\| = \inf\{\|h\| \mid h \in H\}.$$

Here the scalars b_i and the functions $x_i \in L_2[0,1]$ $(i = 1, 2, \ldots, m)$ are completely determined by the points t_i and the function values $f_0(t_i)$ $(i = 1, 2, \ldots, m + k)$. In fact, if h_0 is a solution to (10.1.3), then the element $g_0 \in L_2^{(k)}[0,1]$ with the property that $g_0^{(k)} = h_0$ solves the problem (10.1.1).

Let us rephrase the problem (10.1.3). Let $\{x_1, x_2, \ldots, x_m\}$ be a set in $L_2 = L_2[0,1]$, $b = (b_1, b_2, \ldots, b_m) \in \ell_2(m)$, $C = \{x \in L_2 \mid x \geqslant 0\}$, and

$$(10.1.4) \qquad K = \{x \in C \mid \langle x, x_i \rangle = b_i, \quad i = 1, 2, \ldots, m\}.$$

Clearly, K is a closed convex subset of the Hilbert space L_2, so that if K is nonempty, best approximations in K to any $x \in L_2$ always exist and are unique. Briefly, K is Chebyshev (Theorem 3.5). The problem (10.1.3) can then be stated as "determine the minimum norm element $P_K(0)$ in K."

If we define a linear mapping $A : L_2 \to \ell_2(m)$ by

$$Ax = (\langle x, x_1 \rangle, \langle x, x_2 \rangle, \ldots, \langle x, x_m \rangle), \quad x \in L_2,$$

then we see that K may be rewritten in the compact form

$$(10.1.5) \qquad K = \{x \in C \mid Ax = b\} = C \cap A^{-1}(b).$$

In this chapter we shall consider the following generalization of this problem.

Let X be a Hilbert space, C a closed convex subset of X, $b \in \ell_2(m)$, A a bounded linear operator from X to $\ell_2(m)$, and

$$K = K(b) := \{x \in C \mid Ax = b\} = C \cap A^{-1}(b).$$

Our problem can be stated as follows. Given any $x \in X$, determine its best approximation $P_{K(b)}(x)$ in $K(b)$. We shall be mainly interested in establishing a useful *characterization* of best approximations from $K(b)$. What we shall see is that $P_{K(b)}(x)$ is equal to the best approximation to a perturbation $x + A^*y$ of x from C (or from a convex extremal subset C_b of C). The merit of this characterization is threefold. First, it is generally much *easier* to compute best approximations from C (or C_b) than from $K(b)$. Second, when X is infinite-dimensional, the problem of computing $P_{K(b)}(x)$ is intrinsically an *infinite-dimensional* problem, but the computation of the $y \in \ell_2(m)$ for which $P_C(x + A^*y) \in A^{-1}(b)$ depends only on a *finite* number m of parameters. Third, in many cases there are standard methods for solving the latter problem.

Stong Conical Hull Intersection Property (strong CHIP)

Before proceeding with establishing the characterization of best approximations from $K(b) = C \cap A^{-1}(b)$, it is convenient first to make a definition of a property (the "strong CHIP") that will prove fundamental to our analysis. Loosely speaking, when the strong CHIP is present (respectively is not present), then $P_{K(b)}(x) = P_C(x + A^*y)$ (respectively $P_{K(b)}(x) = P_{C_b}(x + A^*y)$) for some $y \in \ell_2(m)$. Then we give some examples, and establish some general results, concerning this property.

10.2 Definition. *A collection of closed convex sets* $\{C_1, C_2, \ldots, C_m\}$ *in* X, *which has a nonempty intersection, is said to have the* **strong conical hull intersection property,** *or the* **strong CHIP,** *if*

$$(10.2.1) \qquad \left(\bigcap_1^m C_i - x\right)^\circ = \sum_1^m (C_i - x)^\circ \quad \text{for each} \quad x \in \bigcap_1^m C_i.$$

There are useful alternative ways of describing the strong CHIP. We state one next, and include others in the exercises.

10.3 Lemma. *Let* $\{C_1, C_2, \ldots, C_m\}$ *be a collection of closed convex sets in a Hilbert space* X. *Then the following statements are equivalent:*

(1) $\{C_1, C_2, \ldots, C_m\}$ *has the strong CHIP;*
(2) *For each* $x \in \cap_1^m C_i$,

$$(10.3.1) \qquad \overline{\text{con}} \left(\bigcap_1^m C_i - x\right) = \bigcap_1^m \overline{\text{con}} \, (C_i - x)$$

and

$$(10.3.2) \qquad \sum_1^m (C_i - x)^\circ \quad \text{is closed}$$

(3) *For each* $x \in \cap_1^m C_i$,

$$(10.3.3) \qquad \overline{\text{con}} \left(\bigcap_1^m C_i - x\right) \supset \bigcap_1^m \overline{\text{con}} \, (C_i - x)$$

and (10.3.2) *holds.*

Proof. (1) $\Longrightarrow$ (2). Suppose (1) holds. Then for each $x \in \cap_1^m C_i$, (10.2.1) holds. Taking dual cones of both sides of (10.2.1), we obtain

$$\left(\bigcap_1^m C_i - x\right)^{\circ\circ} = \left[\sum_1^m (C_i - x)^\circ\right]^\circ.$$

Using 4.5(9) and 4.6(3), we deduce

$$\overline{\text{con}} \left(\bigcap_1^m C_i - x\right) = \bigcap_1^m (C_i - x)^{\circ\circ} = \bigcap_1^m \overline{\text{con}} \, (C_i - x),$$

which proves (10.3.1). Also, (10.3.2) is a consequence of (10.2.1) and 4.5(1).
(2) $\Longrightarrow$ (3) is obvious.

(3) $\implies$ (1). If (3) holds, then since the inclusion $\overline{\mathrm{con}}\left(\cap_1^m C_i - x\right) \subset \cap_1^m \overline{\mathrm{con}}\left(C_i - x\right)$ always holds, it follows that (2) holds. Taking dual cones of both sides of (10.3.1), and using 4.5(3) and 4.6(4), we obtain

$$
\left(\bigcap_1^m C_i - x\right)^\circ = \left[\overline{\mathrm{con}}\left(\bigcap_1^m C_i - x\right)\right]^\circ = \left[\bigcap_1^m \overline{\mathrm{con}}\left(C_i - x\right)\right]^\circ
$$

$$
= \overline{\sum_1^m \left[\overline{\mathrm{con}}\left(C_i - x\right)\right]^\circ} = \overline{\sum_1^m (C_i - x)^\circ}
$$

$$
= \sum_1^m (C_i - x)^\circ,
$$

since the latter set is closed. Hence $\{C_1, C_2, \ldots, C_m\}$ has the strong CHIP. ∎

When one is approximating from an intersection of sets having the strong CHIP, a strengthened characterization of best approximations can be given.

10.4 Theorem. *Let $\{C_1, C_2, \ldots, C_m\}$ be a collection of closed convex sets with the strong CHIP in X, $x \in X$, and $x_0 \in C := \cap_1^m C_i$. Then $x_0 = P_C(x)$ if and only if*

$$(10.4.1) \qquad x - x_0 \in \sum_1^m (C_i - x_0)^\circ.$$

Proof. By Theorem 4.3, $x_0 = P_C(x)$ if and only if $x - x_0 \in (C - x_0)^\circ$. By the strong CHIP, the result follows. ∎

The main usefulness of this result is in the case where the sets C_i have the additional property that $(C_i - x_0)^\circ$ does *not* depend on the particular $x_0 \in \cap_1^m C_i$. The next few examples we give are of this type.

10.5 Example of the Strong CHIP (Subspaces). *Let $\{M_1, M_2, \ldots, M_m\}$ be a collection of closed subspaces in the Hilbert space X. If each of the M_i, except possibly one, has finite codimension, then*

$$(10.5.1) \qquad \left(\bigcap_1^m M_i - x\right)^\circ = \sum_1^m M_i^\perp = \sum_1^m (M_i - x)^\perp = \sum_1^m (M_i - x)^\circ$$

for each $x \in \cap_1^m M_i$ and so $\{M_1, M_2, \ldots, M_m\}$ has the strong CHIP.

In particular, if X is finite-dimensional, then every finite collection of subspaces in X has the strong CHIP.

Proof. Let $x \in \cap_1^m M_i$. Since $M_i - x = M_i$ and $\cap_1^m M_i - x = \cap_1^m M_i$, we obtain

$$
\left(\bigcap_1^m M_i - x\right)^\circ = \left(\bigcap_1^m M_i\right)^\circ = \overline{\sum_1^m M_i^\circ} = \overline{\sum_1^m M_i^\perp} = \sum_1^m M_i^\perp
$$

$$
= \sum_1^m (M_i - x)^\perp = \sum_1^m (M_i - x)^\circ,
$$

where the second equality follows from Theorem 4.6(4), the third from Theorem 4.5(6), the fourth from Theorem 9.36, the fifth from $M_i = M_i - x$, and the seventh from Theorem 4.5(6). ∎

Before giving our next example of the strong CHIP, it is convenient to restate here Lemma 6.38, which is a result about when the conical hull operation commutes with the operation of intersection.

10.6 Lemma. *Let* $\{C_1, C_2, \ldots, C_m\}$ *be a collection of convex sets with* $0 \in \cap_1^m C_i$. *Then*

$$(10.6.1) \qquad \mathrm{con}\left(\bigcap_1^m C_i\right) = \bigcap_1^m \mathrm{con}\, C_i.$$

Remark. We note that if the conical hull "con" is replaced by the closed conical hull "$\overline{\mathrm{con}}$" in Lemma 10.6, the result is no longer valid! (See Exercise 25 at the end of Chapter 4.)

10.7 Example of the Strong CHIP (Half-Spaces). *Let* X *be a Hilbert space,* $h_i \in X\backslash\{0\}$, $\alpha_i \in \mathbb{R}$, *and*

$$H_i := \{x \in X \mid \langle x, h_i \rangle \le \alpha_i\} \qquad (i = 1, 2, \ldots, m).$$

Then, for each $x \in \cap_1^m H_i$,

$$(10.7.1) \qquad \left(\bigcap_1^m H_i - x\right)^{\circ} = \mathrm{con}\,\{h_i \mid i \in I(x)\} = \sum_1^m (H_i - x)^{\circ},$$

and so $\{H_i, H_2, \ldots, H_m\}$ *has the strong CHIP. Here* $I(x)$ *denotes the set of "active" indices for* x; *that is,* $I(x) := \{i \mid \langle x, h_i \rangle = \alpha_i\}$.

Proof. This result is just a restatement of Theorem 6.40. ∎

Other examples of the strong CHIP can be constructed from these examples by using the next lemma.

10.8 Lemma. *Let* $\{C_1, C_2, \ldots, C_m\}$ *have the strong CHIP. Let* $J_1, \ldots, J_n$ *be nonempty disjoint sets of indices whose union is* $\{1, 2, \ldots, m\}$. *Then*

$$\left\{\bigcap_{i \in J_1} C_i, \bigcap_{i \in J_2} C_i, \ldots, \bigcap_{i \in J_n} C_i\right\}$$

has the strong CHIP.

Proof. Let $A_j := \cap_{i \in J_j} C_i$ $(j = 1, 2, \ldots, n)$. To show that $\{A_1, A_2, \ldots, A_n\}$ has the strong CHIP. Let $x \in \cap_1^n A_j$. Then $x \in \cap_1^m C_i$ and

$$\left(\bigcap_1^n A_j - x\right)^\circ = \left(\bigcap_1^m C_i - x\right)^\circ = \sum_1^m (C_i - x)^\circ$$

$$= \sum_{i \in J_1} (C_i - x)^\circ + \cdots + \sum_{i \in J_n} (C_i - x)^\circ$$

$$\subset \left[\bigcap_{J_1} (C_i - x)\right]^\circ + \cdots + \left[\bigcap_{J_n} (C_i - x)\right]^\circ \qquad \text{(by Theorem 4.6(2))}$$

$$= (A_1 - x)^\circ + \cdots + (A_n - x)^\circ = \sum_1^n (A_j - x)^\circ$$

$$\subset \left(\bigcap_1^n A_j - x\right)^\circ \qquad \text{(by Theorem 4.6(2))}.$$

It follows that the two subset symbols "$\subset$" may be replaced by equality symbols "$=$", and hence

$$\left(\bigcap_1^n A_j - x\right)^\circ = \sum_1^n (A_j - x)^\circ.$$

This proves that $\{A_1, A_2, \ldots, A_n\}$ has the strong CHIP. ∎

10.9 Example of the Strong CHIP (Hyperplanes). *Let X be a Hilbert space, $h_i \in X \backslash \{0\}$, $\alpha_i \in \mathbb{R}$, and*

$$H_i := \{x \in X \mid \langle x, h_i \rangle = \alpha_i\} \qquad (i = 1, 2, \ldots, m).$$

Then, for each $x \in \cap_1^m H_i$,

$$(10.9.1) \qquad \left(\bigcap_1^m H_i - x\right)^\circ = \operatorname{span}\{h_1, \ldots, h_m\} = \sum_1^m (H_i - x)^\circ,$$

and so $\{H_1, H_2, \ldots, H_m\}$ has the strong CHIP.

Proof. Each H_i is the intersection of two half-spaces, $H_i = \tilde{H}_{2i-1} \cap \tilde{H}_{2i}$, where

$$\tilde{H}_{2i-1} = \{x \in \mid \langle x, h_i \rangle \le \alpha_i\} \quad \text{and} \quad \tilde{H}_{2i} = \{x \in \mid \langle x, -h_i \rangle \le -\alpha_i\}.$$

Since the collection $\{\tilde{H}_1, \tilde{H}_2, \ldots, \tilde{H}_{2m-1}, \tilde{H}_{2m}\}$ has the strong CHIP by Theorem 10.7, it follows by Theorem 10.8 that $\{H_1, \ldots, H_m\} = \{\tilde{H}_1 \cap \tilde{H}_2, \ldots, \tilde{H}_{2m-1} \cap \tilde{H}_{2m}\}$ has the strong CHIP, and hence

$$\left(\bigcap_1^m H_i - x\right)^\circ = \sum_1^m (H_i - x)^\circ.$$

Moreover, Theorem 10.7 implies that for each $x \in \cap_1^m H_i$, we have $x \in \cap_1^{2m} \tilde{H}_i$, so the active index set is $I(x) = \{1, 2, \ldots, 2m\}$. Thus

$$\left(\bigcap_1^m H_i - x\right)^\circ = \operatorname{con}\{\pm h_i \mid i = 1, 2, \ldots, m\} = \operatorname{span}\{h_i \mid i = 1, 2, \ldots, m\}. \quad \blacksquare$$

Recall that a *polyhedral* set is the intersection of a finite number of closed half-spaces. In particular, every half-space and every hyperplane is a polyhedral set.

10.10 Example of the Strong CHIP (Polyhedral Sets). *Any finite collection of polyhedral sets in a Hilbert space has the strong CHIP.*

Proof. Any collection of half-spaces has the strong CHIP by Example 10.7. Using Lemma 10.8, the result follows. ∎

10.11 Lemma. *Suppose that X is a Hilbert space and $\{C_0, C_1, \ldots, C_m\}$ is a collection of closed convex subsets such that $\{C_1, \ldots, C_m\}$ has the strong CHIP. Then the following statements are equivalent:*

(1) *$\{C_0, C_1, \ldots, C_m\}$ has the strong CHIP;*
(2) *$\{C_0, \cap_1^m C_i\}$ has the strong CHIP.*

Proof. (1) $\Longrightarrow$ (2). This is a consequence of Lemma 10.8.

(2) $\Longrightarrow$ (1). Suppose that $\{C_0, \cap_1^m C_i\}$ has the strong CHIP. Then for each $x \in \overset{m}{\underset{0}{\cap}} C_i$, we have that

$$\left(\overset{m}{\underset{0}{\cap}} C_i - x\right)^\circ = \left[C_0 \cap \left(\overset{m}{\underset{1}{\cap}} C_i\right) - x\right]^\circ$$

$$= (C_0 - x)^\circ + \left(\overset{m}{\underset{1}{\cap}} C_i - x\right)^\circ \qquad \text{by (2)}$$

$$= (C_0 - x)^\circ + \sum_1^m (C_i - x)^\circ \qquad \text{by hypothesis}$$

$$= \sum_0^m (C_i - x)^\circ.$$

Thus $\{C_0, C_1, \ldots, C_m\}$ has the strong CHIP. ∎

Let us now return to the main problem of this chapter. Unless explicitly stated otherwise, the standing *assumptions* for the remainder of the chapter are that X is a Hilbert space, C is a closed convex subset of X, A is a bounded linear operator from X into $\ell_2(m)$, $b = (b_1, b_2, \ldots, b_m) \in \ell_2(m)$, and

$$(10.11.1) \qquad K = K(b) := C \bigcap A^{-1}(b) \neq \emptyset.$$

Our main goal is to *characterize* $P_{K(b)}(x)$, and to describe methods to *compute* $P_{K(b)}(x)$, for any $x \in X$.

By Theorem 8.28, A has a representation of the form

$$(10.11.2) \qquad Ax = (\langle x, x_1\rangle, \langle x, x_2\rangle, \ldots, \langle x, x_m\rangle), \quad x \in X,$$

for some $x_i \in X$, and $A^* : \ell_2(m) \to X$ has the representation

$$(10.11.3) \qquad A^* y = \sum_1^m y(i) x_i, \quad y \in \ell_2(m).$$

If $x_i = 0$ for some i, then we must have $b_i = 0$, since $A^{-1}(b) \neq \emptyset$ by (10.11.1). Thus, without loss of generality, we may assume that $x_i \neq 0$ for every i. Letting $H_i := \{x \in X \mid \langle x, x_i\rangle = b_i\}$ $(i = 1, 2, \ldots, m)$, we can rewrite $K(b)$ in the form

$$(10.11.4) \qquad K(b) = C \cap A^{-1}(b) = C \cap \left(\overset{m}{\underset{1}{\cap}} H_i\right).$$

Since the H_i are hyperplanes, it follows by Example 10.9 and Lemma 10.11 that $\{C, A^{-1}(b)\}$ has the strong CHIP if and only if $\{C, H_1, \ldots, H_m\}$ has the strong CHIP.

By specializing Lemma 10.3 to the situation at hand, we can deduce the following. Recall the notation $\mathcal{R}(T)$ and $\mathcal{N}(T)$ for the range and null space of T (see Definition 8.7).

10.12 Lemma. *The following statements are equivalent:*

(1) $\{C, A^{-1}(b)\}$ *has the strong CHIP;*

(2) *For every* $x_0 \in C \cap A^{-1}(b)$,

(10.12.1) $$[C \cap A^{-1}(b) - x_0]^\circ = (C - x_0)^\circ + \mathcal{R}(A^*);$$

(3) *For every* $x_0 \in C \cap A^{-1}(b)$,

$$[C \cap A^{-1}(b) - x_0]^\circ \subset (C - x_0)^\circ + \mathcal{R}(A^*);$$

(4) *For every* $x_0 \in C \cap A^{-1}(b)$,

(10.12.3) $$\overline{\mathrm{con}}\,(C - x_0) \cap \mathcal{N}(A) \subset \overline{\mathrm{con}}\,[(C - x_0) \cap \mathcal{N}(A)]$$

and

(10.12.4) $$(C - x_0)^\circ + \mathcal{R}(A^*) \qquad \text{is closed.}$$

Proof. First observe that for every $x_0 \in C \cap A^{-1}(b)$, we have

$$C \cap A^{-1}(b) - x_0 = (C - x_0) \cap [A^{-1}(b) - x_0] = (C - x_0) \cap \mathcal{N}(A).$$

Next note that by Lemma 8.33,

$$\mathcal{N}(A)^\circ = \mathcal{N}(A)^\perp = \overline{\mathcal{R}(A^*)} = \mathcal{R}(A^*),$$

where the last equality holds because $\mathcal{R}(A^*)$ is finite-dimensional, hence closed by Theorem 3.7. Using these facts, an application of Lemma 10.3 yields the result. ∎

Now we can prove our main characterization theorem. It shows that the strong CHIP for the sets $\{C, A^{-1}(b)\}$ is the *precise* condition that allows us always to replace the approximation of any $x \in X$ from the set $K(b)$ by approximating a perturbation of x from the set C.

10.13 Theorem. *The following statements are equivalent:*

(1) $\{C, A^{-1}(b)\}$ *has the strong CHIP;*

(2) *For every* $x \in X$, *there exists* $y \in \ell_2(m)$ *such that*

(10.13.1) $$A[P_C(x + A^*y)] = b;$$

(3) *For every* $x \in X$, *there exists* $y \in \ell_2(m)$ *such that*

(10.13.2) $$P_{K(b)}(x) = P_C(x + A^*y).$$

In fact, for any given $y \in \ell_2(m)$, (10.13.1) *holds if and only if* (10.13.2) *holds.*

Proof. First note that if (10.13.2) holds, then $P_C(x + A^*y) \in K$ and hence (10.13.1) holds. Conversely, suppose (10.13.1) holds. Then $x_0 := P_C(x + A^*y) \in K$ and by Theorem 4.3,

$$(10.13.3) \qquad\qquad x + A^*y - x_0 \in (C - x_0)^\circ.$$

Hence $x - x_0 \in (C - x_0)^\circ + \mathcal{R}(A^*) \subset (K - x_0)^\circ$. By Theorem 4.3 again, $x_0 = P_K(x)$. That is, (10.13.2) holds. This proves the equivalence of (2) and (3) as well as the last statement of the theorem.

(1) $\implies$ (3). If (1) holds and $x \in X$, let $x_0 := P_K(x)$. Then Theorem 4.3 implies $x - x_0 \in (K - x_0)^\circ$. Using the strong CHIP,

$$(K - x_0)^\circ = (C - x_0)^\circ + \mathcal{R}(A^*).$$

Thus $x - x_0 + A^*y \in (C - x_0)^\circ$ for some $y \in \ell_2(m)$. By Theorem 4.3 again, $x_0 = P_C(x + A^*y)$. That is, $P_K(x) = P_C(x + A^*y)$ and (3) holds.

(3) $\implies$ (1). Suppose (3) holds and let $x_0 \in K$. Choose any $z \in (K - x_0)^\circ$ and set $x := z + x_0$. Note that $x - x_0 = z \in (K - x_0)^\circ$ so $x_0 = P_K(x)$ by Theorem 4.3. Since (3) holds, there exists $y \in \ell_2(m)$ such that $x_0 = P_C(x + A^*y)$. Hence Theorem 4.3 implies that

$$z = x - x_0 = x - P_C(x + A^*y) = x + A^*y - P_C(x + A^*y) - A^*y$$
$$\in [C - P_C(x + A^*y)]^\circ - A^*y \subset (C - x_0)^\circ + \mathcal{R}(A^*).$$

Since z was arbitrary in $(K - x_0)^\circ$, we see that $(K - x_0)^\circ \subset (C - x_0)^\circ + \mathcal{R}(A^*)$. Since $x_0 \in K$ was arbitrary, it follows by Lemma 10.12 that $\{C, A^{-1}(b)\}$ has the strong CHIP. ∎

Remarks. (1) This theorem allows us to determine the best approximation in $K(b)$ to any $x \in X$ by instead determining the best approximation in C to a *perturbation* of x. The usefulness of this is that it is usually much easier to determine best approximations from C than from the intersection $K(b)$. The price we pay for this simplicity is that now we must determine precisely just *which* perturbation A^*y of x works! However, this is determined by the (generally nonlinear) equation (10.13.1) for the unknown vector y. Since $y \in \ell_2(m)$ depends only on m parameters (the coordinates of y), (10.13.1) is an equation involving only a *finite* number of parameters and is often amenable to standard algorithms (e.g., descent methods) for their solution.

(2) Of course, to apply this theorem, one must first determine whether the pair of sets $\{C, A^{-1}(b)\}$ have the strong CHIP. Fortunately, some of the more interesting pairs that arise in practice do have this property (see Corollary 10.14 below).

But even if $\{C, A^{-1}(b)\}$ does not have the strong CHIP, we will show below that there exists a certain convex (extremal) subset C_b of C such that $K(b) = C_b \cap A^{-1}(b)$ and $\{C_b, A^{-1}(b)\}$ *does* have the strong CHIP! This means that we can still apply Theorem 10.13 but with C replaced by C_b.

(3) As an alternative to applying Theorem 10.13 for the numerical computation of $P_{K(b)}(x)$, we will show below that *Dykstra's algorithm* (see Chapter 9) is also quite suitable for computing $P_{K(b)}(x)$. Moreover, to use Dykstra's algorithm, it is not necessary to know whether $\{C, A^{-1}(b)\}$ has the strong CHIP. All that is needed to apply Dykstra's algorithm is to be able to compute $P_C(x)$ for any x. For example, in the particular important case when $C = \{x \in L_2 \mid x \geq 0\}$, we have the formula $P_C(x) = x^+$ (see Theorem 4.8).

10.14 Corollary. *If C is a polyhedral set, then $\{C, A^{-1}(b)\}$ has the strong CHIP. Hence for each $x \in X$, there exists $y \in \ell_2(m)$ that satisfies*

$$(10.14.1) \qquad A[P_C(x + A^*y)] = b.$$

Moreover, for any $y \in \ell_2(m)$ that satisfies (10.14.1), we have

$$(10.14.2) \qquad P_{K(b)}(x) = P_C(x + A^*y).$$

Proof. Note that $A^{-1}(b) = \cap_1^m H_i$, where each H_i is the hyperplane

$$H_i := \{z \in X \mid \langle z, x_i \rangle = b_i\}.$$

It follows by Example 10.10 that $\{C, A^{-1}(b)\}$ has the strong CHIP. The result now follows from Theorem 10.13. ∎

10.15 An Application of Corollary 10.14. Let

$$C_1 = \{x \in \ell_2 \mid x(1) \le 0\}, \quad C_2 = \left\{x \in \ell_2 \mid \sum_1^\infty \frac{x(n)}{2^{n/2}} = -2\right\},$$

and $K = C_1 \cap C_2$. What is the element of K having minimal norm? That is, what is $P_K(0)$?

Letting $h_1 = (1, 0, 0, \dots)$ and $h_2 = \left(1/2^{\frac{1}{2}}, 1/2^{\frac{2}{2}}, 1/2^{\frac{3}{2}}, \dots, 1/2^{\frac{n}{2}}, \dots\right)$, we see that $h_i \in \ell_2$ for $i = 1, 2$,

$$C_1 = \{x \in \ell_2 \mid \langle x, h_1 \rangle \le 0\} =: C, \text{ and } C_2 = \{x \in \ell_2 \mid \langle x, h_2 \rangle = -2\} = A^{-1}(-2),$$

where $A : \ell_2 \to \mathbb{R}$ is defined by $Ax = \langle x, h_2 \rangle$. Thus $K = C \cap A^{-1}(-2)$, and we want to compute $P_K(0)$.

Since C is a half-space, hence polyhedral, it follows by Corollary 10.14 that

$$(10.15.1) \qquad P_K(0) = P_C(A^*y)$$

for any $y \in \mathbb{R}$ with

$$(10.15.2) \qquad A[P_C(A^*y)] = -2.$$

But by Theorem 6.31,

$$P_C(A^*y) = P_C(yh_2) = yh_2 - \frac{1}{\|h_1\|^2}\{\langle yh_2, h_1 \rangle\}^+ h_1$$

$$(10.15.3) \qquad = yh_2 - \{y\langle h_2, h_1 \rangle\}^+ h_1$$

$$= yh_2 - \left\{\frac{1}{\sqrt{2}}y\right\}^+ h_1 = yh_2 - \frac{1}{\sqrt{2}}y^+ h_1.$$

Substituting this expression into (10.15.2), we see that we want $y \in \mathbb{R}$ such that $\langle yh_2 - \frac{1}{\sqrt{2}}y^+ h_1, h_2 \rangle = -2$, or

$$(10.15.4) \qquad y - \frac{1}{2}y^+ = -2.$$

If $y \ge 0$, then $y - \frac{1}{2}y^+ = \frac{1}{2}y \ge 0$ and (10.15.4) cannot hold. Thus we must have $y < 0$. Then we deduce from (10.15.4) that $y = -2$, and hence that

$$(10.15.5) \qquad P_K(0) = P_C(A^*y) = -2h_2$$

using (10.15.3). (Note also that $\|P_K(0)\| = 2$.)

Affine Sets

In general, if C is not polyhedral, it may not be obvious that $\{C, A^{-1}(b)\}$ has the strong CHIP, and hence not obvious whether Theorem 10.13 can be applied. What we seek next is a fairly simple sufficient condition that guarantees that $\{CA^{-1}(b)\}$ has the strong CHIP. Such a condition is that $b \in \text{ri } A(C)$, where ri (S) denotes the "relative interior" of S (see Definition 10.21 below). Before we prove this, we must develop the necessary machinery.

10.16 Definition. *A set V in X is called* **affine** *if $\alpha x + (1 - \alpha)y \in V$ whenever $x, y \in V$ and $\alpha \in \mathbb{R}$.*

In other words, V is affine if it contains the whole line through each pair of its points. Affine sets are also called *linear varieties* or *flats*. Obviously, every affine set is convex.

10.17 Affine Sets. *Let V be a nonempty subset of the inner product space X. Then:*

(1) *V is affine if and only if $\sum_1^n \alpha_i v_i \in V$ whenever $v_i \in V$ and $\alpha_i \in \mathbb{R}$ satisfy $\sum_1^n \alpha_i = 1$.*

(2) *V is affine if and only if $V = M + v$ for some (uniquely determined) subspace M and any $v \in V$. In fact, $M = V - V$.*

(3) *If $\{V_i\}$ is any collection of affine sets, then $\cap_i V_i$ is affine.*

Proof. (1) The "if" part is clear, since we use only the case where $n = 2$. For the converse, assume that V is affine. Let $v_i \in V$ $(i = 1, 2, \ldots, n)$ and $\alpha_i \in \mathbb{R}$ satisfy $\sum_1^n \alpha_i = 1$. We must show that $\sum_1^n \alpha_i v_i \in V$. Proceed by induction on n. For $n = 1$, the result is a tautology. Assume the result holds for some $n \geq 1$. Suppose that $v_1, v_2, \ldots, v_{n+1}$ are in V and $\sum_1^{n+1} \alpha_i = 1$. Clearly, not all $\alpha_i = 1$, since $\sum_1^{n+1} \alpha_i = 1$. By reindexing, we may assume that $\alpha_{n+1} \neq 1$. Then

$$\sum_1^{n+1} \alpha_i v_i = \sum_1^n \alpha_i v_i + \alpha_{n+1} v_{n+1} = (1 - \alpha_{n+1}) \sum_{i=1}^n \frac{\alpha_i}{1 - \alpha_{n+1}} v_i + \alpha_{n+1} v_{n+1}$$

$$= (1 - \alpha_{n+1})v + \alpha_{n+1} v_{n+1},$$

where $v := \sum_{i=1}^n \alpha_i/(1 - \alpha_{n+1})v_i \in V$ by the induction hypothesis (because $\sum_{i=1}^n \alpha_i/(1 - \alpha_{n+1}) = 1$). Then

$$\sum_1^{n+1} \alpha_i v_i = (1 - \alpha_{n+1})v + \alpha_{n+1} v_{n+1} \in V,$$

since V is affine. This completes the induction.

(2) Let V be affine and $v \in V$. Set $M := V - v$. We first show that M is a subspace. Let $x, y \in M$ and $\alpha \in \mathbb{R}$. Then $x = v_1 - v$, $y = v_2 - v$ for some $v_i \in V$ $(i = 1, 2)$. Further, $v_1 + v_2 - v \in V$ by part (1). Thus

$$x + y = (v_1 + v_2 - v) - v \in V - v = M$$

and

$$\alpha x = \alpha v_1 - \alpha v = [\alpha v_1 + (1 - \alpha)v] - v \in V - v = M.$$

Thus M is a subspace and $V = M + v$.

Next we verify the uniqueness of M by showing that $M = V - V$. Let $V = M + v$ for some $v \in V$. Then $M = V - v \subset V - V$. Conversely, if $x \in V - V$, then $x = v_1 - v_2$ for some $v_i \in V$, so $x = (v_1 - v_2 + v) - v \in V - v = M$ using part (1). Thus $V - V \subset M$, and so $M = V - V$ as claimed.

Conversely, let M be a subspace, $v \in V$, and $V = M + v$. To show that V is affine, let $x, y \in V$ and $\alpha \in \mathbb{R}$. Then $x = m_1 + v$ and $y = m_2 + v$ for some $m_i \in M$ $(i = 1, 2)$ implies that

$$\alpha x + (1 - \alpha)y = \alpha(m_1 + v) + (1 - \alpha)(m_2 + v) = [\alpha m_1 + (1 - \alpha)m_2] + v \in M + v = V.$$

Thus V is affine.

(3) Let $x, y \in \cap_i V_i$ and $\alpha \in \mathbb{R}$. Then $x, y \in V_i$ for each i and thus $\alpha x + (1-\alpha)y \in V_i$ for each i, since V_i is affine. Thus $\alpha x + (1 - \alpha)y \in \cap_i V_i$, and so $\cap_i V_i$ is affine. ∎

Part (2) states that a set is affine if and only if it is the translate of a subspace.

10.18 Definition. *The **affine hull** of a subset S in X, denoted by* aff(S), *is the intersection of all affine sets that contain S.*

10.19 Affine Hull. *Let S be a subset of the inner product space X. Then:*

(1) aff(S) *is the smallest affine set that contains S.*
(2) aff$(S) = \{\sum_1^n \alpha_i s_i \mid s_i \in S, \sum_1^n \alpha_i = 1, n \in \mathbb{N}\}$
(3) S *is affine if and only if $S = $ aff(S).*
(4) *If $0 \in S$, then* aff$(S) = $ span(S).

We leave the proof of this as an easy exercise (see Exercise 12 at the end of the chapter).

10.20 Definition. *The **dimension** of an affine set V, denoted by* dim V, *is -1 if V is empty, and is the dimension of the unique subspace M such that $V = M + v$ if V is nonempty. In general, the **dimension** of any subset S of X, denoted by* dim S, *is the dimension of the affine hull of S.*

For example, a single point has dimension 0, the line segment joining two distinct points has dimension 1, and the unit ball $B(0, 1)$ in X has the same dimension as X.

Using Lemma 1.15(7), we see that the interior of a set S is the set of all $x \in S$ such that $B(x, \epsilon) \subset S$ for some $\epsilon > 0$. Thus

$$\text{int } S = \{x \in S \mid B(x, \epsilon) \subset S \text{ for some } \epsilon > 0\}.$$

10.21 Definition. *The **relative interior** of a set S, denoted by* ri S, *is the interior of S relative to its affine hull. More precisely,*

$$\text{ri } S := \{x \in S \mid B(x, \epsilon) \cap \text{aff } S \subset S \text{ for some } \epsilon > 0\}.$$

Note that int $S \subset$ ri S, but the inclusion is strict in general. Indeed, if $X = \ell_2(2)$ is the Euclidean plane and $S = \{x \in X \mid x(2) = 0\}$ is the "horizontal axis," then int $S = \emptyset$ but ri $S = S$. Moreover, if $S_1 \subset S_2$, then it is obvious that int $S_1 \subset$ int S_2.

However, it is *not* true in general that $\operatorname{ri} S_1 \subset \operatorname{ri} S_2$. For example, if D is a cube in Euclidean 3-space, S_1 is an edge of D, and S_2 is a face of D that contains S_1, then the relative interiors of S_1 and S_2 are nonempty and disjoint.

There is another important distinction between interior and relative interior. While it is easy to give examples of nonempty convex sets in Euclidean n-space ($n \geq 1$) with empty interior (e.g., a point in 1-space), we will see below that the relative interior of any finite-dimensional nonempty convex set is always *nonempty*!

The next lemma often helps us to reduce problems concerning relative interiors to ones involving (ordinary) interiors.

10.22 Lemma. *Let S be a nonempty subset of X. Then*

$$(10.22.1) \qquad \operatorname{aff} S + y = \operatorname{aff}(S + y)$$

for each $y \in X$.

Moreover, for any $x \in S$, the following statements are equivalent:

(1) $x \in \operatorname{ri} S$;
(2) $0 \in \operatorname{ri}(S - x)$;
(3) $0 \in \operatorname{int}(S - x)$ *relative to the space $X_0 := \operatorname{span}(S - x)$.*

Proof. Let $y \in X$. If $x \in \operatorname{aff} S + y$, then $x = \sum_1^n \alpha_i s_i + y$ for some $s_i \in S$ and $\alpha_i \in \mathbb{R}$ with $\sum_1^n \alpha_i = 1$ (by 10.19(2)). Thus $x = \sum_1^n \alpha_i(s_i + y) \in \operatorname{aff}(S + y)$. This proves that $\operatorname{aff} S + y \subset \operatorname{aff}(S + y)$. Since S and y were arbitrary, it follows that $\operatorname{aff}(S + y) - y \subset \operatorname{aff}(S + y - y) = \operatorname{aff} S$ implies $\operatorname{aff}(S + y) \subset \operatorname{aff} S + y$, and so (10.22.1) holds.

(1) $\Longrightarrow$ (2). If $x \in \operatorname{ri} S$, then $B(x, \epsilon) \cap \operatorname{aff} S \subset S$ for some $\epsilon > 0$. Since $B(x, \epsilon) = x + B(0, \epsilon)$, it follows from (10.22.1) that

$$(10.22.2) \qquad B(0, \epsilon) \cap \operatorname{aff}(S - x) \subset S - x,$$

which is precisely statement (2).

(2) $\Longrightarrow$ (3). If (2) holds, then so does (10.22.2), so that

$$B(0, \epsilon) \cap X_0 \subset S - x,$$

which is precisely statement (3).

(3) $\Longrightarrow$ (1). If (3) holds, then by retracing the above steps we obtain (1). ∎

Our next result is the analogue of Theorem 2.8 for relative interiors.

10.23 Theorem. *Let K be a convex subset of an inner product space. Then for each $x \in \operatorname{ri} K$, $y \in \overline{K}$, and $0 < \lambda \leq 1$, it follows that*

$$(10.23.1) \qquad \lambda x + (1 - \lambda)y \in \operatorname{ri} K.$$

In particular, $\operatorname{ri} K$ is convex.

The proof is similar to that of Theorem 2.8 using Lemma 10.22.

The next theorem states in particular that the relative interior of a nonempty finite-dimensional convex set is nonempty.

10.24 Theorem. *Let K be a finite-dimensional convex subset in an inner product space. Then $\overline{K}$ and $\mathrm{ri}\, K$ are convex sets that have the same affine hull, and hence the same dimension, as K. In particular, $\mathrm{ri}\, K \neq \emptyset$ if $K \neq \emptyset$, and $\overline{\mathrm{ri}\, K} = \overline{K}$.*

Proof. $\overline{K}$ (respectively $\mathrm{ri}\, K$) is convex by Theorem 2.8 (respectively Theorem 10.23). Moreover, if $\mathrm{ri}\, K \neq \emptyset$, then $\overline{\mathrm{ri}\, K} = \overline{K}$ by Theorem 10.23.

Next suppose that C and D are nonempty finite-dimensional convex sets with $\overline{C} = \overline{D}$. We first show that

$$(10.24.1) \qquad \mathrm{aff}\, C \subset \mathrm{aff}\, \overline{C} = \mathrm{aff}\, \overline{D} \subset \overline{\mathrm{aff}\, D} = \mathrm{aff}\, D.$$

To see this, note that the first inclusion is obvious, since $\overline{C} \supset C$, while the last equality holds because $\mathrm{aff}\, D$ is finite-dimensional, hence closed by Theorem 3.7. It remains to show that $\mathrm{aff}\, \overline{D} \subset \overline{\mathrm{aff}\, D}$. For if $\overline{x} \in \mathrm{aff}\, \overline{D}$ and $\epsilon > 0$, choose $\overline{d}_1, \ldots, \overline{d}_n$ in $\overline{D}$ and $\alpha_i \in \mathbb{R}$ with $\sum_1^n \alpha_i = 1$ such that $\overline{x} = \sum_1^n \alpha_i \overline{d}_i$. Next choose $d_i \in D$ such that $\|d_i - \overline{d}_i\| < \epsilon / \sum_j |\alpha_j|$ for each i. Then $x := \sum_1^n \alpha_i d_i \in \mathrm{aff}\, D$ and

$$\|x - \overline{x}\| = \left\| \sum_1^n \alpha_i (d_i - \overline{d}_i) \right\| \leq \sum_1^n |\alpha_i| \|d_i - \overline{d}_i\| < \epsilon.$$

Hence $\overline{x} \in \overline{\mathrm{aff}\, D}$ and $\mathrm{aff}\, \overline{D} \subset \overline{\mathrm{aff}\, D}$. This verifies (10.24.1), and thus $\mathrm{aff}\, C \subset \mathrm{aff}\, D$. By interchanging the roles of C and D in (10.24.1), we obtain the reverse inclusion and, in particular,

$$(10.24.2) \qquad \mathrm{aff}\, C = \mathrm{aff}\, \overline{C} = \mathrm{aff}\, \overline{D} = \mathrm{aff}\, D.$$

This verifies that $C, \overline{C}, D$, and $\overline{D}$ all have the same affine hull and hence the same dimension. It follows that if $\mathrm{ri}\, K \neq \emptyset$, then $\mathrm{ri}\, K, K$, and $\overline{K}$ all have the same affine hulls, and hence the same dimensions.

To complete the proof, it remains to show that $\mathrm{ri}\, K \neq \emptyset$ if $K \neq \emptyset$.

Fix any $x_0 \in K$ and let $n := \dim K$. Then $n \geq 0$, and by Theorem 10.17, $\mathrm{aff}\, K = M + x_0$ for some n-dimensional subspace M. If $n = 0$, then $M = \{0\}$ and $\mathrm{aff}\, K = \{x_0\} = K$ implies that $\mathrm{ri}\, K = K$. Thus we may assume that $n \geq 1$.

Since

$$M = \mathrm{aff}\, K - x_0 = \mathrm{aff}\, (K - x_0) = \mathrm{span}\, (K - x_0)$$

is n-dimensional, it follows that there exist n vectors $x_1, \ldots, x_n$ in K such that the set $\{x_1 - x_0, x_2 - x_0, \ldots, x_n - x_0\}$ is a basis for M. Put $z = \frac{1}{n+1}(x_0 + x_1 + \cdots + x_n)$. We will show that $z \in \mathrm{ri}\, K$.

First note that

$$z = \frac{1}{n+1}(x_1 - x_0) + \frac{1}{n+1}(x_2 - x_0) + \cdots + \frac{1}{n+1}(x_n - x_0) + x_0.$$

Since $\mathrm{aff}\, K = M + x_0$ and $\{x_1 - x_0, \ldots, x_n - x_0\}$ is a basis for M, we see that for each $y \in \mathrm{aff}\, K$,

$$(10.24.3) \qquad y = \sum_1^n \alpha_i (x_i - x_0) + x_0$$

for some uniquely determined scalars α_i. Also,

$$\|y - z\| = \|(y - x_0) - (z - x_0)\| = \left\| \sum_{i=1}^{n} \left(\alpha_i - \frac{1}{n+1} \right) (x_i - x_0) \right\|.$$

By Theorem 3.7(4), we deduce that two points in a finite-dimensional space are close if and only if their corresponding coefficients (relative to a given basis) are close. Hence there exists $\epsilon > 0$ such that if $y \in B(z, \epsilon) \cap \text{aff}\, K$, then

$$(10.24.4) \qquad \left| \alpha_i - \frac{1}{n+1} \right| < \frac{1}{(n+1)^2} \quad \text{for} \quad i = 1, 2, \ldots, n.$$

In particular, (10.24.4) implies that $\alpha_i > 0$ for all i and

$$0 < \sum_{1}^{n} \alpha_i \leq \sum_{i=1}^{n} \left(\left| \alpha_i - \frac{1}{n+1} \right| + \frac{1}{n+1} \right) < \frac{n}{(n+1)^2} + \frac{n}{n+1}$$

$$= \frac{n + n(n+1)}{(n+1)^2} < \frac{(n+1)^2}{(n+1)^2} = 1.$$

But then (10.24.3) implies that

$$y = \sum_{1}^{n} \alpha_i (x_i - x_0) + x_0 = \sum_{1}^{n} \alpha_i x_i + \left(1 - \sum_{1}^{n} \alpha_i \right) x_0$$

is a convex combination of $x_0, x_1, \ldots, x_n$, and hence $y \in K$.

We have shown that $B(z, \epsilon) \cap \text{aff}\, K \subset K$, that is, $z \in \text{ri}\, K$. ∎

Relative Interiors and a Separation Theorem

Recall Definition 6.20 of two sets being separated by a hyperplane. The following separation theorem (which will be needed in a few places later) states that in a finite-dimensional Hilbert space, two convex sets having disjoint relative interiors can always be separated by a hyperplane.

10.25 Finite-Dimensional Separation Theorem. *If C, D are nonempty convex subsets of the finite-dimensional Hilbert space X such that $\text{ri}\, C \cap \text{ri}\, D = \emptyset$, then there exists $x_0 \in X \backslash \{0\}$ such that*

$$(10.25.1) \qquad \sup_{c \in C} \langle x_0, c \rangle \leq \inf_{d \in D} \langle x_0, d \rangle.$$

In particular, if α is any constant satisfying $\sup_{c \in C} \langle x_0, c \rangle \leq \alpha \leq \inf_{d \in D} \langle x_0, d \rangle$, then the hyperplane $\{x \in X \mid \langle x_0, x \rangle = \alpha\}$ separates C and D.

Proof. We have $\overline{\text{ri}\, C} = \overline{C}$ and $\overline{\text{ri}\, D} = \overline{D}$ by Theorem 10.24. By continuity,

$$\sup_{c \in \text{ri}\, C} \langle x_0, c \rangle = \sup_{c \in \overline{C}} \langle x_0, c \rangle \qquad \text{and} \qquad \inf_{d \in \text{ri}\, D} \langle x_0, d \rangle = \inf_{d \in \overline{D}} \langle x_0, d \rangle,$$

so that by replacing C and D by their respective relative interiors, we may assume that $C \cap D = \emptyset$.

Let $E = C - D$. Then E is convex by Proposition 2.3(5), and $0 \notin E$. We consider two cases.

Case 1: $0 \notin \overline{E}$.

In this case, set $x_0 := -P_{\overline{E}}(0) \neq 0$. Deduce from Theorem 4.1 that

$$\langle x_0, y + x_0 \rangle \leq 0 \quad \text{for all} \quad y \in \overline{E},$$

and so $\langle x_0, y \rangle \leq -\langle x_0, x_0 \rangle < 0$ for all $y \in \overline{E}$. Thus

$$(10.25.2) \qquad \langle x_0, c - d \rangle < 0 \qquad \text{for all} \quad c \in C, d \in D,$$

which proves (10.25.1).

Case 2: $0 \in \overline{E}$.

In this case, let $X_0 := \operatorname{span} \overline{E} = \operatorname{aff} \overline{E} = \operatorname{aff} E$. Working in X_0 rather than X, we see that

$$\operatorname{int} \overline{E} = \operatorname{ri} \overline{E} = \operatorname{ri} E = \operatorname{int} E.$$

Using Lemma 2.9, we deduce

$$0 \in \overline{E} \backslash E \subset \overline{E} \backslash \operatorname{int} E = \overline{E} \backslash \operatorname{int} \overline{E} = \operatorname{bd} \overline{E} = \operatorname{bd}(X_0 \backslash \overline{E}) \subset \overline{X_0 \backslash \overline{E}}.$$

Thus there exist $x_n \in X_0 \backslash \overline{E}$ such that $x_n \to 0$. Setting $z_n := P_{\overline{E}}(x_n)$, we obtain that $x_n - z_n \neq 0$ and, by Theorem 4.1,

$$(10.25.3) \qquad \langle x_n - z_n, y - z_n \rangle \leq 0 \quad \text{for all} \quad y \in \overline{E}, n \in \mathbb{N}.$$

Letting $e_n := (x_n - z_n)/\|x_n - z_n\|$ for each n, we see that $\|e_n\| = 1$ and

$$(10.25.4) \qquad \langle e_n, y \rangle \leq \langle e_n, z_n \rangle \qquad \text{for all} \quad y \in \overline{E}, \quad n \in \mathbb{N}.$$

Since $0 \in \overline{E}$, it follows that

$$(10.25.5) \qquad \|z_n\| \leq \|z_n - x_n\| + \|x_n\| \leq 2\|x_n\| \to 0.$$

That is, $z_n \to 0$.

Since $\{e_n\}$ is a sequence of norm-one elements, Theorem 3.7 implies that there exists a subsequence $\{e_{n_k}\}$ and a point $x_0 \in X_0$ such that $e_{n_k} \to x_0$. Since $\|e_{n_k}\| = 1$ for all k, $\|x_0\| = 1$. Moreover, in (10.25.4), we obtain

$$(10.25.6) \qquad \langle x_0, y \rangle \leq 0 \qquad \text{for all} \quad y \in \overline{E}.$$

Equivalently,

$$\langle x_0, c - d \rangle \leq 0 \quad \text{for all} \quad c \in C, d \in D,$$

and so (10.25.1) holds. ∎

Our next result gives three alternative characterizations of the relative interior of a convex set in a finite-dimensional space. Each characterization will prove useful in what follows.

10.26 Relative Interior. *Let D be a convex subset of a finite-dimensional inner product space, and let $x \in D$. Then the following statements are equivalent:*

(1) *$x \in \operatorname{ri} D$;*

(2) *For each $d \in D$, there exists $d' \in D$ and $0 < \lambda < 1$ such that $x = \lambda d + (1 - \lambda)d'$;*

(3) *For each $d \in D$, there exists $\mu > 0$ such that $x + \mu(x - d) \in D$;*

(4) *$(D - x)^\circ = (D - x)^\perp$.*

Proof. Since $x \in \operatorname{ri} D$ if and only if $0 \in \operatorname{ri}(D - x)$, by replacing D with $D - x$, we may assume $x = 0$.

(1) $\implies$ (2). Suppose $0 \in \operatorname{ri} D$. Then there exists $\epsilon > 0$ such that $B(0, \epsilon) \cap \operatorname{span} D \subset D$. Thus for each $d \in D \backslash \{0\}$, it follows that the element $-\epsilon(2\|d\|)^{-1}d$ is in D. Then $d' := -\epsilon(2\|d\|)^{-1}d \in D$ and $0 = \lambda d + (1 - \lambda)d'$, where

$$\lambda := \frac{\epsilon(2\|d\|)^{-1}}{1 + \epsilon(2\|d\|)^{-1}} \in (0, 1).$$

(2) $\implies$ (3). If (2) holds, let $d \in D$. Then there exist $d' \in D$ and $0 < \lambda < 1$ such that $0 = \lambda d + (1 - \lambda)d'$. Taking $\mu = \frac{\lambda}{1 - \lambda} > 0$, we see that $-\mu d = d' \in D$ and (3) holds.

(3) $\implies$ (4). Suppose (3) holds but (4) fails. Then we can choose $z \in D^\circ \backslash D^\perp$. Thus $\langle z, d \rangle \leq 0$ for all $d \in D$ and $\langle z, d_0 \rangle < 0$ for some $d_0 \in D$. Since (3) holds, there exists $\mu > 0$ such that $d_0' := -\mu d_0 \in D$. Hence

$$0 \geq \langle z, d_0' \rangle = -\mu \langle z, d_0 \rangle > 0,$$

which is absurd.

(4) $\implies$ (1). Suppose $D^\circ = D^\perp$. If $0 \notin \operatorname{ri} D$, then by setting $Y = \operatorname{span} D = \operatorname{aff} D$, we see that $0 \notin \operatorname{int} D$ relative to Y. By the separation theorem (Theorem 10.25) in the space Y, there exists $x_0 \in Y \backslash \{0\}$ such that $\sup \{\langle x_0, d \rangle \mid d \in D\} \leq 0$. Hence

$$x_0 \in D^\circ = D^\perp = [\operatorname{span} D]^\perp = Y^\perp = \{0\},$$

which contradicts $x_0 \neq 0$. ■

Remark. Statements (2) and (3) can be paraphrased geometrically to yield that $x \in \operatorname{ri} D$ if and only if each line segment in D having x for an endpoint can be extended beyond x and still remain in D.

In an analogous way, we can characterize the interior of a finite-dimensional convex set.

10.27 Interior and Relative Interior. *Let D be a convex subset of the finite-dimensional inner product space Y. Then*

(10.27.1) $$\operatorname{ri} D = \{x \in D \mid (D - x)^\circ = (D - x)^\perp\}$$

and

(10.27.2) $$\operatorname{int} D = \{x \in D \mid (D - x)^\circ = \{0\}\}.$$

Proof. Relation (10.27.1) is a consequence of Theorem 10.26. If $x \in \text{int } D$, then $B(x, \epsilon) \subset D$ for some $\epsilon > 0$. Hence $B(0, \epsilon) \subset D - x$ implies that

$$\{0\} = B(0, \epsilon)^\circ \supset (D - x)^\circ \supset \{0\},$$

so that $(D - x)^\circ = \{0\}$.

Conversely, if $x \in D$ and $(D - x)^\circ = \{0\}$, then

$$\{0\} = (D - x)^\circ \supset (D - x)^\perp \supset \{0\}$$

implies that $(D - x)^\circ = (D - x)^\perp = \{0\}$. By Theorem 10.26, $x \in \text{ri } D$, or $0 \in \text{ri } (D - x)$. Then

$$\{0\} = (D - x)^\perp = [\text{span} \, (D - x)]^\perp$$

implies that

$$Y = [\text{span} \, (D - x)]^{\perp\perp} = \overline{\text{span} \, (D - x)} = \text{span} \, (D - x)$$

using Theorem 4.5(9) and the fact that finite-dimensional subspaces are closed. Since $0 \in D - x$, $\text{span} \, (D - x) = \text{aff} \, (D - x)$, and hence $0 \in \text{int} \, (D - x)$, or $x \in \text{int } D$. This verifies (10.27.2). ∎

There are certain relationships between relative interiors of two convex sets that we record next.

10.28 Lemma. *Let K, K_1, and K_2 be nonempty convex subsets of a finite-dimensional Hilbert space. Then:*

(1) $\text{ri } \overline{K} = \text{ri } K$.
(2) $\overline{K}_1 = \overline{K}_2$ *if and only if* $\text{ri } K_1 = \text{ri } K_2$.
(3) $\text{ri } K_1 \cap \text{ri } K_2 \subset \text{ri} \, (\text{co} \, \{K_1 \cup K_2\})$.

Proof. (1) Since $K \subset \overline{K}$ and both K and $\overline{K}$ have the same affine hull by Theorem 10.24, we have $\text{ri } K \subset \text{ri } \overline{K}$. Conversely, let $x \in \text{ri } \overline{K}$. We must show that $x \in \text{ri } K$. Choose any $y \in \text{ri } K$. If $y = x$, we are done. Thus assume $y \neq x$. Then by Theorem 10.26,

$$z := x + \mu(x - y) \in \overline{K}$$

for $\mu > 0$ sufficiently small. Solve this equation for x to obtain that

$$x = \frac{1}{1 + \mu} z + \frac{\mu}{1 + \mu} y,$$

which is in $\text{ri } K$ by Theorem 10.23.

(2) If $\overline{K}_1 = \overline{K}_2$, then (1) implies that $\text{ri } K_1 = \text{ri } \overline{K}_1 = \text{ri } \overline{K}_2 = \text{ri } K_2$. Conversely, if $\text{ri } K_1 = \text{ri } K_2$, then Theorem 10.24 implies that $\overline{K}_1 = \overline{\text{ri } K_1} = \overline{\text{ri } K_2} = \overline{K}_2$.

(3) Let $x \in \text{ri } K_1 \cap \text{ri } K_2$ and let $y \in K_0 := \text{co} \, (K_1 \cup K_2)$. By Theorem 10.26, it suffices to show there exists $\mu > 0$ such that $x + \mu(x - y) \in K_0$. First note that $y = \lambda y_1 + (1 - \lambda) y_2$ for some $y_i \in K_i$ and $0 \leq \lambda \leq 1$. Since $x \in \text{ri } K_i$ $(i = 1, 2)$, Theorem 10.26 implies that there exist $\mu_i > 0$ such that

$$(10.28.1) \qquad x + \mu_i(x - y_i) \subset K_i \qquad (i = 1, 2).$$

By convexity of each K_i, (10.28.1) must also hold when μ_i is replaced by $\mu :=$ $\min\{\mu_1, \mu_2\}$:

(10.28.2) $x + \mu(x - y_i) \in K_i$ $(i = 1, 2)$.

It follows that $\mu > 0$ and, from (10.28.2), that

$$x + \mu(x - y) = x + \mu[x - \{\lambda y_1 + (1 - \lambda)y_2\}]$$
$$= \lambda[x + \mu(x - y_1)] + (1 - \lambda)[x + \mu(x - y_2)] \in K_0. \quad \blacksquare$$

10.29 Theorem. *Let X be a Hilbert space, C a closed convex cone, and M a finite-dimensional subspace such that $C \cap M$ is a subspace. Then $C + M$ is closed.*

Proof. Assume first that $C \cap M = \{0\}$. Let $x_n \in C + M$ and $x_n \to x$. We must show that $x \in C + M$. Write $x_n = c_n + y_n$, where $c_n \in C$ and $y_n \in M$. If some subsequence of $\{y_n\}$ is bounded, then by passing to a subsequence if necessary, we may assume that $y_n \to y \in M$. Then $c_n = x_n - y_n \to x - y$. Since C is closed, $c := x - y \in C$ and $x = c + y \in C + M$.

If no subsequence of $\{y_n\}$ is bounded, then $\|y_n\| \to \infty$. It follows that $\{y_n/\|y_n\|\}$ is bounded in M, so by passing to a subsequence if necessary, we may assume that $y_n/\|y_n\| \to y \in M$ and $\|y\| = 1$. Then $c_n/\|y_n\| \in C$ for all n and

$$\frac{c_n}{\|y_n\|} = \frac{x_n}{\|y_n\|} - \frac{y_n}{\|y_n\|} \to 0 - y \in C,$$

since C is closed. Thus $-y \in C \cap M = \{0\}$, which contradicts $\|y\| = 1$. This proves the theorem when $C \cap M = \{0\}$.

In general, $V = C \cap M$ is a closed subspace of M, so we can write $M = V + V^\perp \cap M$. Then

(10.29.1) $C + M = C + V + V^\perp \cap M = C + V^\perp \cap M$

and

$$C \cap (V^\perp \cap M) = V \cap V^\perp = \{0\}.$$

By the first part of the proof (with M replaced by $V^\perp \cap M$), we see that $C + V^\perp \cap M$ is closed. Thus $C + M$ is closed by (10.29.1). $\blacksquare$

The next result shows that the unit ball in a finite-dimensional space is contained in the convex hull of a *finite* set. This result is obviously false in an infinite-dimensional space.

10.30 Theorem. *Let $\{e_1, e_2, \ldots, e_n\}$ be an orthonormal basis for the inner product space Y. Then*

(10.30.1) $B\left(0, \dfrac{1}{\sqrt{n}}\right) \subset \mathrm{co}\,\{\pm e_1, \pm e_2, \ldots, \pm e_n\}.$

In other words, if $y \in Y$ and $\|y\| < 1/\sqrt{n}$, then there exist $2n$ scalars $\lambda_i \geq 0$ with $\sum_1^{2n} \lambda_i = 1$ such that $y = \sum_{i=1}^n (\lambda_i - \lambda_{n+i})e_i$.

Proof. If $y \in B\left(0, \frac{1}{\sqrt{n}}\right) \setminus \{0\}$, then $y = \sum_1^n \alpha_k e_k$, where $\alpha_k = \langle y, e_k \rangle$, and

$$\left(\sum_1^n |\alpha_k|^2\right)^{1/2} = \|y\| < \frac{1}{\sqrt{n}}.$$

Using Schwarz's inequality in $\ell_2(n)$, we obtain

$$0 < \sum_1^n |\alpha_k| \leq \left(\sum_1^n |\alpha_k|^2\right)^{1/2} \left(\sum_1^n 1\right)^{1/2} < \frac{1}{\sqrt{n}}\sqrt{n} = 1.$$

Moreover, since $\alpha_k = \alpha_k^+ - \alpha_k^-$, we set

$$\lambda_i := \begin{cases} \dfrac{\alpha_i^+}{\sum_1^n |\alpha_j|} & \text{if } 1 \leq i \leq n, \\[3mm] \dfrac{\alpha_{i-n}^-}{\sum_1^n |\alpha_j|} & \text{if } n < i \leq 2n, \end{cases}$$

and deduce that $\lambda_i \geq 0$, $\sum_1^{2n} \lambda_i = 1$, and

$$\sum_{i=1}^n \lambda_i e_i + \sum_{i=n+1}^{2n} \lambda_i(-e_{i-n}) \in \text{co}\,\{\pm e_1, \pm e_2, \ldots, \pm e_n\}.$$

Further, $0 \in \text{co}\,\{\pm e_1, \pm e_2, \ldots, \pm e_n\}$ implies that

$$y = \sum_1^n \alpha_i^+ e_i + \sum_1^n \alpha_i^-(-e_i)$$

$$= \sum_1^n |\alpha_j| \left[\sum_1^n \lambda_i e_i + \sum_{n+1}^{2n} \lambda_i(-e_{i-n})\right] + \left(1 - \sum_1^n |\alpha_j|\right) \cdot 0$$

$$\in \text{co}\,\{\pm e_1, \pm e_2, \ldots, \pm e_n\}. \quad \blacksquare$$

It turns out that the constant $1/\sqrt{n}$ in the inclusion (10.30.1) is "best possible" in the sense that (10.30.1) fails if $1/\sqrt{n}$ is replaced by any larger number (see Exercise 17 at the end of the chapter).

We need one more result of a general nature concerning the dual cones of images of linear mappings.

10.31 Dual Cone of Images. *Let A be a bounded linear mapping from a Hilbert space X to an inner product space Y. If S is any nonempty subset of X, then*

(10.31.1) $$[A(S)]^\circ = (A^*)^{-1}(S^\circ)$$

and

(10.31.2) $$[A(S)]^\perp = (A^*)^{-1}(S^\perp).$$

Proof. We have that $y \in [A(S)]^\circ \Longleftrightarrow \langle y, As \rangle \le 0$ for all $s \in S \Longleftrightarrow \langle A^*y, s \rangle \le 0$ for all $s \in S \Longleftrightarrow A^*y \in S^\circ \Longleftrightarrow y \in (A^*)^{-1}(S^\circ)$. This proves (10.31.1). The proof of (10.31.2) is strictly analogous. ∎

Now we can prove one of our main results. It gives an important sufficient condition that guarantees the strong CHIP.

For the remainder of this chapter, unless explicitly stated otherwise, we assume the following hypothesis: X **is a Hilbert space,** C **is a closed convex subset,** $b \in \ell_2(m)$, $A : X \to \ell_2(m)$ **is defined by**

$$Ax := (\langle x, x_1 \rangle, \langle x, x_2 \rangle, \dots, \langle x, x_m \rangle),$$

where $\{x_1, x_2, \dots, x_m\} \subset X \backslash \{0\}$ **is given, and** $b \in \ell_2(m)$. **Moreover, we assume that**

$$K = K(b) := C \cap A^{-1}(b) = \{x \in C \mid \langle x, x_i \rangle = b_i \quad (i = 1, \dots, m)\}$$

is not empty. That is, $b \in A(C)$.

10.32 Theorem. *If* $b \in \mathrm{ri}\, A(C)$, *then* $\{C, A^{-1}(b)\}$ *has the strong CHIP.*
   Proof. By Lemma 10.12, it suffices to show that for each fixed $x \in C \cap A^{-1}(b)$,

(10.32.1) $$(C - x)^\circ + \mathcal{R}(A^*)$$

is closed and

(10.32.2) $$\overline{\mathrm{con}}\,(C - x) \cap \mathcal{N}(A) \subset \overline{\mathrm{con}}\,[(C - x) \cap \mathcal{N}(A)].$$

Since $b \in \mathrm{ri}\, A(C)$, Theorem 10.26 implies that

(10.32.3) $$[A(C) - b]^\circ = [A(C) - b]^\perp.$$

Since $Ax = b$, we deduce that

$$[A(C - x)]^\circ = [A(C - x)]^\perp.$$

Using Theorem 10.31, it follows that

$$(A^*)^{-1}[(C - x)^\circ] = (A^*)^{-1}[(C - x)^\perp],$$

or, equivalently,

(10.32.4) $$\mathcal{R}(A^*) \cap (C - x)^\circ = \mathcal{R}(A^*) \cap (C - x)^\perp.$$

Since $\mathcal{R}(A^*)$ and $(C-x)^\perp$ are both subspaces, (10.32.4) shows that $\mathcal{R}(A^*) \cap (C-x)^\circ$ is a subspace. Using Theorem 10.29 (with C and M replaced by $(C-x)^\circ$ and $\mathcal{R}(A^*)$, respectively), we deduce that (10.32.1) holds.
   It remains to verify (10.32.2). Let $z \in \overline{\mathrm{con}}\,(C - x) \cap \mathcal{N}(A)$. To complete the proof, we will show that there exists a sequence $\{z_n\}$ in $\mathrm{con}\,[(C - x) \cap \mathcal{N}(A)]$ such

that $z_n \to z$, and hence $z \in \overline{\mathrm{con}}\,[(C - x) \cap \mathcal{N}(A)]$. If $z = 0$, we can take $z_n = 0$ for each n. Thus we may assume that $z \neq 0$.

Since $z \in \overline{\mathrm{con}}\,(C - x)$, there exists a sequence $\{z'_n\}$ in $\mathrm{con}\,(C - x)$ such that $z'_n \to z$. By Lemma 4.4(5), there exist $c_n \in C$ and $\rho_n > 0$ such that $z'_n = \rho_n(c_n - x)$. If $z'_n \in \mathcal{N}(A)$ for infinitely many n, say $z'_{n_k} \in \mathcal{N}(A)$ for $k = 1, 2, \ldots$, then

$$\frac{1}{\rho_{n_k}} z'_{n_k} = c_{n_k} - x \in (C - x) \cap \mathcal{N}(A),$$

so $z'_{n_k} \in \mathrm{con}\,[(C - x) \cap \mathcal{N}(A)]$ for all k. Setting $z_k = z'_{n_k}$, we see that

$$z_k \in \mathrm{con}\,[(C - x) \cap \mathcal{N}(A)] \quad \text{and} \quad z_k \to z.$$

Thus, by passing to a subsequence if necessary, we may assume that $z'_n \notin \mathcal{N}(A)$ for each n. If $\|c_n - x\| > 1$ for some n, then since $C - x$ is convex and $0 \in C - x$, we see that

$$\frac{c_n - x}{\|c_n - x\|} = \frac{1}{\|c_n - x\|}(c_n - x) + \left(1 - \frac{1}{\|c_n - x\|}\right) 0 \in C - x,$$

so $(c_n - x)/\|c_n - x\| = c'_n - x$ for some $c'_n \in C$ and $\|c'_n - x\| = 1$. Hence

$$z'_n = \rho_n(c_n - x) = \rho'_n(c'_n - x),$$

where $\rho'_n = \rho_n\|c_n - x\| > 0$ and $\|c'_n - x\| = 1$. In short, we may assume that each z'_n has the representation

$$z'_n = \rho_n(c_n - x),$$

where $\rho_n > 0$, $c_n \in C$, and $\|c_n - x\| \leq 1$.

Since $z'_n \to z$, by passing to a subsequence, we may assume that

$$(10.32.4) \qquad \frac{1}{2}\|z\| < \|z'_n\| = \rho_n\|c_n - x\| \leq \rho_n$$

for each n. Since $Az = 0$, we see that

$$(10.32.5) \qquad y_n := Az'_n \to Az = 0.$$

Then $y_n \neq 0$ for each n; and

$$y_n = \rho_n A(c_n - x) = \rho_n[Ac_n - b],$$

so

$$(10.32.6) \qquad \frac{1}{\rho_n} y_n + b = Ac_n \in A(C) \quad \text{for each } n.$$

Hence, using (10.32.4), we see that

$$(10.32.7) \qquad \gamma_n := \left\|\frac{y_n}{\rho_n}\right\| = \frac{1}{\rho_n}\|y_n\| < \frac{2}{\|z\|}\|y_n\| \to 0.$$

Since $b \in \operatorname{ri} A(C)$, there exists $\delta > 0$ such that
$$B(b, 2\delta) \cap \operatorname{aff} A(C) \subset A(C).$$
Subtract $Ax = b$ from both sides of this inclusion to obtain
$$(10.32.8) \qquad B(0, 2\delta) \cap \operatorname{span}[A(C - x)] \subset A(C - x).$$
Next we choose an orthogonal basis $\{e_1, \ldots, e_N\}$ of $\operatorname{span}[A(C-x)]$ such that $\|e_i\| = \delta$ for each i. Then $\{\pm e_1, \pm e_2, \ldots, \pm e_N\} \subset A(C - x)$ by (10.32.8), and by convexity of $A(C - x)$, there even follows $\operatorname{co}\{\pm e_1, \pm e_2, \ldots, \pm e_N\} \subset A(C - x)$. Now choose $u_i \in C$ such that
$$e_i = A(u_i - x), \qquad -e_i = A(u_{N+i} - x) \qquad (i = 1, 2, \ldots, N),$$
and set
$$\mu := \max\{\|u_i - x\| + 2\delta \mid i = 1, 2, \ldots, 2N\}.$$
Using Theorem 10.30, we deduce that
$$A[(C - x) \cap B(0, \mu)] \supset \operatorname{co}\{\pm e_1, \pm e_2, \ldots, \pm e_N\} = \delta \operatorname{co}\left\{\pm\frac{e_1}{\delta}, \pm\frac{e_2}{\delta}, \ldots, \pm\frac{e_N}{\delta}\right\}$$
$$\supset \delta B\left(0, \frac{1}{\sqrt{N}}\right) = B\left(0, \frac{\delta}{\sqrt{N}}\right).$$
Since $C - x$ is convex and $0 \in C - x$, it follows that $\lambda(C - x) \subset C - x$ for every $0 \leq \lambda \leq 1$. Thus, for each $0 < \lambda \leq 1$, we have
$$A[(C - x) \cap B(0, \lambda\mu)] \supset A[\lambda(C - x) \cap \lambda B(0, \mu)]$$
$$(10.32.9) \qquad\qquad = \lambda A[(C - x) \cap B(0, \mu)]$$
$$\supset \lambda B\left(0, \frac{\delta}{\sqrt{N}}\right) = B\left(0, \lambda\frac{\delta}{\sqrt{N}}\right).$$
Setting $\gamma := \lambda\delta/\sqrt{N}$, it follows from (10.32.9) that $\lambda\mu = (\gamma\sqrt{N}/\delta)\mu = \gamma\nu$, where $\nu := (\sqrt{N}/\delta)\mu > 0$, and hence
$$(10.32.10) \qquad A[(C - x) \cap B(0, \gamma\nu)] \supset B(0, \gamma)$$
for all $0 < \gamma \leq \delta/\sqrt{N}$. Add $Ax = b$ to both sides of (10.32.10) to get
$$(10.32.11) \qquad A[C \cap B(x, \gamma\nu)] \supset B(b, \gamma)$$
for all $0 < \gamma \leq \delta/\sqrt{N}$. By (10.32.7), $\gamma_n = \|y_n/\rho_n\| \to 0$ so by omitting finitely many n, we may assume that
$$(10.32.12) \qquad \gamma_n = \left\|\frac{y_n}{\rho_n}\right\| < \frac{\delta}{\sqrt{N}} \qquad \text{for all } n.$$
Note that $\gamma_n > 0$, $\gamma_n \to 0$, and $\gamma_n\rho_n = \|y_n\| \to 0$. By (10.32.11), we can choose $x_n \in C \cap B(x, \gamma_n\nu)$ such that
$$Ax_n = b - \frac{y_n}{\rho_n} \qquad \text{for all } n.$$
Thus $\|x_n - x\| < \gamma_n\nu \to 0$ and $A(x_n - x) = -y_n/\rho_n$ implies that $A[\rho_n(x_n - x)] = -y_n$. Hence the element $z_n := z_n' + \rho_n(x_n - x)$ is in $\operatorname{con}(C - x)$ and $Az_n = 0$. That is,
$$z_n \in \operatorname{con}(C - x) \cap \mathcal{N}(A) = \operatorname{con}[(C - x) \cap \mathcal{N}(A)],$$
where we used Lemma 10.6 for the last equality, and
$$\|z_n - z\| \leq \|z_n - z_n'\| + \|z_n' - z\| \leq \rho_n\|x_n - x\| + \|z_n' - z\|$$
$$< \rho_n\gamma_n\nu + \|z_n' - z\| = \|y_n\|\nu + \|z_n' - z\| \to 0. \qquad \blacksquare$$

10.33 Corollary. *Suppose $b \in \text{ri}\, A(C)$. Then, for each $x \in X$, there exists $y \in \ell_2(m)$ such that*

$$(10.33.1) \qquad \left\langle P_C\left(x + \sum_1^m y(i)x_i\right), x_j \right\rangle = b(j) \qquad (j = 1, 2, \ldots, m).$$

Moreover, for any $y \in \ell_2(m)$ that satisfies (10.33.1), *we have*

$$(10.33.2) \qquad P_K(x) = P_C\left(x + \sum_1^m y(i)x_i\right).$$

Proof. By Theorem 10.32, $\{C, A^{-1}(b)\}$ has the strong CHIP. By Theorem 10.13, $P_K(x) = P_C(x + A^*y)$ for any $y \in \ell_2(m)$ that satisfies $A[P_C(x + A^*y)] = b$. By virtue of (10.11.2) and (10.11.3), the result follows. ∎

10.34 An Application. Let T be any set that contains at least m points and let $X = \ell_2(T)$. Fix any m points $t_1, t_2, \ldots, t_m$ in T, define $x_i \in X$ by

$$x_i(t) = \begin{cases} 0 & \text{if } t \neq t_i, \\ 1 & \text{if } t = t_i, \end{cases}$$

for $i = 1, 2, \ldots, m$, put $C = \{x \in X \mid x(t) \geq 0 \text{ for all } t \in T\}$, and define $A : X \to \ell_2(m)$ by

$$Ax := (\langle x, x_1\rangle, \langle x, x_2\rangle, \ldots, \langle x, x_m\rangle) = (x(t_1), x(t_2), \ldots, x(t_m)).$$

Clearly,

$$A(C) = \{y \in \ell_2(m) \mid y(i) \geq 0 \quad \text{for } i = 1, 2, \ldots, m\}.$$

Claim 1: $\text{ri}\, A(C) = \text{int}\, A(C) = \{y \in \ell_2(m) \mid y(i) > 0 \ (i = 1, 2, \ldots, m)\}$.
To verify this claim, let

$$D = \{y \in \ell_2(m) \mid y(i) > 0 \quad \text{for} \quad i = 1, 2, \ldots, m\}.$$

Note that $D \subset A(C)$. Now let $y_0 \in D$ and set $\epsilon := \frac{1}{2} \min\{y_0(i) \mid i = 1, 2, \ldots, m\}$. Thus $\epsilon > 0$ and if $\|y - y_0\| < \epsilon$, then $\sum_{i=1}^m |y(i) - y_0(i)|^2 < \epsilon^2$ implies that $|y(i) - y_0(i)| < \epsilon$ for each i, so that

$$y(i) > y_0(i) - \epsilon \geq 2\epsilon - \epsilon = \epsilon > 0$$

for each i. That is, $y \in D$. Thus

$$(10.34.1) \qquad D \subset \text{int}\, A(C) \subset \text{ri}\, A(C).$$

Conversely, let $y \in \text{ri}\, A(C)$. Then $y(i) \geq 0$ for each i. If $y(i_0) = 0$ for some $i_0 \in \{1, 2, \ldots, m\}$, define $d \in \ell_2(m)$ by setting $d(i_0) = 1$ and $d(i) = 0$ if $i \neq i_0$. Then $d \in A(C)$, so that by Theorem 10.26, there exist $d' \in A(C)$ and $0 < \lambda < 1$ such that $y = \lambda d + (1 - \lambda)d'$. Then

$$0 = y(i_0) = \lambda d(i_0) + (1 - \lambda)d'(i_0) = \lambda + (1 - \lambda)d'(i_0) \geq \lambda > 0,$$

which is absurd. Thus we must have $y(i) > 0$ for all i. This proves that ri $A(C) \subset D$ and, combined with (10.34.1), verifies Claim 1.

Now fix any $b \in \ell_2(m)$ with $b(i) > 0$ for $i = 1, 2, \ldots, m$, and let

$$K = K(b) := C \cap A^{-1}(b).$$

By Claim 1, $b \in$ ri $A(C)$, and hence $K \neq \emptyset$.

Claim 2: For each $x \in X$, $P_K(x)$ is the function defined by

$$P_K(x)(t) = \begin{cases} x^+(t) & \text{if } t \in T \backslash \{t_1, t_2, \ldots, t_m\}, \\ b(i) & \text{if } t = t_i. \end{cases}$$

To verify this claim, we note that by Corollary 10.33, there exists $y \in \ell_2(m)$ such that

$$(10.34.2) \qquad \left\langle P_C \left(x + \sum_{i=1}^{m} y(i)x_i \right), x_j \right\rangle = b(j) \qquad (j = 1, 2, \ldots, m)$$

and

$$(10.34.3) \qquad P_K(x) = P_C \left(x + \sum_{1}^{m} y(i)x_i \right)$$

for any $y \in \ell_2(m)$ that satisfies (10.34.2). By Application 4.8, we have that $P_C(z) = z^+$ for any $z \in X$. Using this and the fact that $\langle z, x_j \rangle = z(t_j)$ for any $z \in X$, we see that (10.34.2) may be reduced to

$$\left[x + \sum_{i=1}^{m} y(i)x_i \right]^+ (t_j) = b(j),$$

or, more simply,

$$(10.34.4) \qquad [x(t_j) + y(j)]^+ = b(j) \qquad \text{for } j = 1, 2, \ldots, m.$$

But $y(j) := b(j) - x(t_j)$ $(j = 1, 2, \ldots, m)$ clearly satisfies (10.34.4). Using this, we deduce from (10.34.3) that

$$P_K(x) = \left[x + \sum_{i=1}^{m} \{b(i) - x(t_i)\}x_i \right]^+.$$

Evaluating this at any $t \in T$, we see that Claim 2 is verified.

Extremal Subsets of C

The above argument would not be valid if $b \in A(C) \backslash$ ri $A(C)$. How would we proceed in this case? It turns out that there is a result analogous to Corollary 10.33 even if $b \notin$ ri $A(C)$! However, in this case, the set C in equations (10.33.1) and (10.33.2) must be replaced by a certain convex *extremal* subset of C. To develop the theory in this more general situation, we first must define the notion of an extremal set.

10.35 Definition. *A subset E of a convex set S is called an* **extremal subset** *of S if x, y in S, $0 < \lambda < 1$, and $\lambda x + (1 - \lambda)y \in E$ implies that $x, y \in E$. In other words, if an interior point of a line segment in S lies in E, the whole line segment must lie in E.*

An **extreme point** *of S is an extremal subset of S consisting of a single point. Thus $e \in S$ is an extreme point of S if $x, y \in S$, $0 < \lambda < 1$, and $e = \lambda x + (1 - \lambda)y$ implies that $x = y = e$.*

For example, the convex extremal subsets of a cube in 3-space are the whole cube, the six faces, the twelve edges, and the eight vertices. The set of extreme points are the eight vertices. Also, for the unit disk in the plane, the convex extremal subsets are the whole disk and each point on the boundary (see Figure 10.35.1).

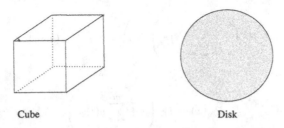

Cube   Disk

Figure 10.35.1

It is easy to verify that any intersection of extremal subsets of S is again an extremal subset of S (possibly empty).

Now we shall consider what can be said in the case where $\{C, A^{-1}(b)\}$ does *not* have the strong CHIP. The following proposition is the main result in this regard. It states that there exists a closed convex extremal subset C_b of C such that $C_b \cap A^{-1}(b) = C \cap A^{-1}(b)$, $b \in \text{ri}\, A(C_b)$, and $\{C_b, A^{-1}(b)\}$ has the strong CHIP! In other words, since the set $K(b) = C \cap A^{-1}(b)$ is unchanged if we replace C with C_b, we can apply Theorem 10.33 (with C replaced by C_b) to obtain a characterization of best approximations from the set $K(b)$ that does *not* require $b \in \text{ri}\, A(C)$.

10.36 Proposition. *There exists a unique convex subset C_b of C such that $b \in \text{ri}\, A(C_b)$ and C_b is larger than any other subset with this property. That is, if C' is a convex subset of C and $b \in \text{ri}\, A(C')$, then $C' \subset C_b$. Moreover,*

(1) C_b *is closed,*
(2) $C_b \cap A^{-1}(b) = C \cap A^{-1}(b) (= K(b))$,
(3) $A(C_b)$ *is an extremal subset of $A(C)$,*
(4) C_b *is an extremal subset of C,*
(5) $\{C_b, A^{-1}(b)\}$ *has the strong CHIP,*
(6) $C_b = C$ *if and only if $b \in \text{ri}\, A(C)$.*

Proof. Let

$$\mathcal{C} := \{C' \mid C' \subset C, \quad C' \text{ is convex}, \quad b \in \text{ri}\, A(C')\}.$$

The first statement of the proposition is that C has a unique largest element C_b. Let us assume this for the moment, and show how the remaining six statements follow.

(1) If C_b were not closed, then $C_0 := \overline{C_b}$ would be a convex subset of C that is strictly larger than C_b. Moreover,

$$A(C_0) = A(\overline{C_b}) \subset \overline{A(C_b)} \subset \overline{A(\overline{C_b})} = \overline{A(C_0)},$$

where the first inclusion follows from the continuity of A. It follows that

$$\overline{A(C_0)} = \overline{A(\overline{C_b})} = \overline{A(C_b)}.$$

By Lemma 10.28, ri $A(C_b) =$ ri $A(C_0)$. Since $b \in$ ri $A(C_b)$, we have that $b \in$ ri $A(C_0)$ and thus $C_0 \in C$. But this contradicts the maximality of C_b.

(2) If $C_b \cap A^{-1}(b) \neq K$, choose $x \in K \setminus C_b \cap A^{-1}(b)$. Then $x \in C \setminus C_b$ and $Ax = b$. Consider the set $C_0 := \text{co}(C_b \cup \{x\})$. Then C_0 is a convex subset of C that is strictly larger than C_b. Also, since $Ax = b \in$ ri $A(C_b) \subset A(C_b)$, and since $A(C_b)$ is convex, we deduce that

$$b \in \text{ri}\, A(C_b) = \text{ri}\,\{\text{co}\, A(C_b)\} = \text{ri}\,\{\text{co}\,[A(C_b \cup \{x\})]\} = \text{ri}\, A[\text{co}\,(C_b \cup \{x\})] = \text{ri}\, A(C_0).$$

Thus $C_0 \in C$, and this contradicts the maximality of C_b.

(3) If $A(C_b)$ were not extremal in $A(C)$, there would exist y_1, y_2 in $A(C)$, $y_1 \notin A(C_b)$, and $0 < \lambda < 1$ such that

$$y := \lambda y_1 + (1 - \lambda) y_2 \in A(C_b).$$

Choose $x_1 \in C \setminus C_b$ and $x_2 \in C$ such that $Ax_i = y_i$ $(i = 1, 2)$. Then

$$\lambda A x_1 + (1 - \lambda) A x_2 = \lambda y_1 + (1 - \lambda) y_2 = y \in A(C_b).$$

Since $b \in$ ri $A(C_b)$, Theorem 10.26 implies that there exists $\mu > 0$ such that

(10.36.1) $$y_3 := b + \mu(b - y) \in A(C_b).$$

Choose $x_3 \in C_b$ such that $Ax_3 = y_3$. Solve (10.36.1) for b to obtain

$$b = \frac{1}{1+\mu} y_3 + \frac{\mu}{1+\mu} y = \frac{1}{1+\mu} y_3 + \frac{\lambda \mu}{1+\mu} y_1 + \frac{\mu(1-\lambda)}{1+\mu} y_2.$$

Thus

$$b \in \text{ri co}\,\{y_1, y_2, y_3\} = \text{ri co}\,\{Ax_1, Ax_2, Ax_3\} = \text{ri}\, A(\text{co}\,\{x_1, x_2, x_3\}).$$

This proves that $C' := \text{co}\,\{x_1, x_2, x_3\} \in C$, and hence $C' \subset C_b$. But this contradicts $x_1 \notin C_b$.

(4) Let $x_1, x_2 \in C$, $0 < \lambda < 1$, and $x = \lambda x_1 + (1 - \lambda) x_2 \in C_b$. Then

$$\lambda A x_1 + (1 - \lambda) A x_2 = Ax \in A(C_b).$$

By (3), $A(C_b)$ is extremal in $A(C)$, and hence $Ax_i \in A(C_b)$ for $i = 1, 2$. If $x_1 \notin C_b$, then $C_0 := \mathrm{co}\,(C_b \cup \{x_1\})$ is a convex subset of C strictly larger than C_b and aff $A(C_0) = $ aff $A(C_b)$. It follows that

$$b \in \mathrm{ri}\, A(C_b) \subset \mathrm{ri}\, A(C_0).$$

Hence $C_0 \in \mathcal{C}$, which contradicts the maximality of C_b. This proves that $x_1 \in C_b$. Similarly, $x_2 \in C_b$. Hence C_b is extremal in C.

(5) Since $b \in \mathrm{ri}\, A(C_b)$ and $K = C_b \cap A^{-1}(b)$ by (2), it follows by Theorem 10.32 (with C_b in place of C) that $\{C_b, A^{-1}(b)\}$ has the strong CHIP.

(6) If $C = C_b$, then $b \in \mathrm{ri}\, A(C_b) = \mathrm{ri}\, A(C)$. Conversely, if $b \in \mathrm{ri}\, A(C)$, then since C_b is the largest convex subset C' of C with $b \in \mathrm{ri}\, A(C')$, we have that $C_b = C$.

Thus, to complete the proof, we must show that $\mathcal{C}$ has a unique largest element. For this, we use Zorn's lemma (see Appendix 1). Order $\mathcal{C}$ by containment: $C_1 \succ C_2$ if and only if $C_1 \supset C_2$. Let $\mathcal{T}$ be a totally ordered subset of $\mathcal{C}$ and set $C_0 := \cup\{C' \mid C' \in \mathcal{T}\}$.

Claim 1: $C_0 \in \mathcal{C}$.

First we show that C_0 is convex. Let $x, y \in C_0$ and $0 < \lambda < 1$. Then $x \in C_1$ and $y \in C_2$ for some $C_i \in \mathcal{T}$ $(i = 1, 2)$. Since $\mathcal{T}$ is totally ordered, either $C_1 \subset C_2$ or $C_2 \subset C_1$. We may assume $C_1 \subset C_2$. Then $x, y \in C_2$, and since C_2 is convex, $\lambda x + (1 - \lambda)y \in C_2 \subset C_0$, and C_0 is convex.

Next we show that $b \in \mathrm{ri}\, A(C_0)$. Let $y_0 \in A(C_0)$. Then $y_0 \in A(C')$ for some $C' \in \mathcal{T}$, and since $b \in \mathrm{ri}\, A(C')$, Theorem 10.26 implies that there exists $\mu > 0$ such that $b + \mu(b - y_0) \in A(C') \subset A(C_0)$. By Theorem 10.26 again, $b \in \mathrm{ri}\, A(C_0)$. Thus $C_0 \in \mathcal{C}$ and Claim 1 is proved.

Moreover, it is clear that $C_0 \supset C'$ for every $C' \in \mathcal{T}$. Thus C_0 is an upper bound for $\mathcal{T}$. By Zorn's lemma, $\mathcal{C}$ has a maximal element C_b.

Claim 2: $C_b \supset C'$ for all $C' \in \mathcal{C}$.

For otherwise $C_1 \not\subset C_b$ for some $C_1 \in \mathcal{C}$. Then the set $C_2 := \mathrm{co}\,\{C_1 \cup C_b\}$ is a convex subset of C that is *strictly larger* than C_b. Also, since $b \in \mathrm{ri}\, A(C_1) \cap \mathrm{ri}\, A(C_b)$, it follows from Lemma 10.28(3) that

$$b \in \mathrm{ri}\,\{\mathrm{co}\,[A(C_1) \cup A(C_b)]\} = \mathrm{ri}\,\{\mathrm{co}\,[A(C_1 \cup C_b)]\} = \mathrm{ri}\, A[\mathrm{co}\,(C_1 \cup C_b)] = \mathrm{ri}\, A(C_2).$$

Thus $C_2 \in \mathcal{C}$, $C_2 \supset C_b$, and $C_2 \neq C_b$. This contradicts the maximality of C_b, and proves Claim 2.

Finally, if C_b and C_b' are both maximal elements for $\mathcal{C}$, then Claim 2 shows that $C_b \supset C_b'$ and $C_b' \supset C_b$. That is $C_b' = C_b$, and the maximal element is unique. ∎

There is an alternative description for the set C_b that is often more useful for applications. We first observe the following lemma.

10.37 Lemma. *If S is a nonempty subset of a convex set C, then there exists a (unique) smallest convex extremal subset of C that contains S, namely,*

$$(10.37.1) \qquad \bigcap\{E \mid E \text{ is convex, extremal, and } S \subset E \subset C\}.$$

Proof. Since the intersection of extremal sets is extremal, we see that the set in (10.37.1) is extremal, convex, and is the smallest such set that contains S. ∎

10.38 Proposition. *Let E_b denote the smallest convex extremal subset of C that contains $K(b) = C \cap A^{-1}(b)$. Then $C_b = E_b$.*

In particular, C_b is the largest convex subset of C with the property that $b \in \operatorname{ri} A(C_b)$, and C_b is also the smallest convex extremal subset of C that contains $C \cap A^{-1}(b)$.

Proof. Assume first that the following three properties hold:

(10.38.1) $$b \in \operatorname{ri} A(E_b),$$

(10.38.2) $$A(E_b) \text{ is extremal in } A(C),$$

(10.38.3) $$A^{-1}(A(E_b)) \cap C = E_b.$$

Under these conditions, we will prove that $C_b = E_b$. Indeed, from (10.38.1) and the definition of C_b, we see that $C_b \supset E_b$. For the reverse inclusion, let $y \in A(C_b)$. Since $b \in \operatorname{ri} A(C_b)$ by Theorem 10.36, there exist (by Theorem 10.26) $y' \in A(C_b)$ and $0 < \lambda < 1$ such that $\lambda y + (1 - \lambda)y' = b \in A(E_b)$. Since $A(E_b)$ is extremal in $A(C)$, y and y' are in $A(E_b)$. That is, $A(C_b) \subset A(E_b)$ and hence

$$C_b \subset A^{-1}[A(E_b)] \cap C = E_b$$

by (10.38.3). Thus $C_b = E_b$, and the result is verified.

Hence to complete the proof, it suffices to verify (10.38.1)–(10.38.3). To this end, let F_e denote the smallest convex extremal subset of $A(C)$ that contains b, and set

(10.38.4) $$C_{F_e} := C \cap A^{-1}(F_e).$$

That is, F_e is the intersection of all convex extremal subsets of $A(C)$ that contain b.

Clearly, $C_{F_e} \supset C \cap A^{-1}(b) = K$. Now, C_{F_e} is convex, since F_e is convex and A is linear. Next let x, y in C, $0 < \lambda < 1$, and $z := \lambda x + (1 - \lambda)y \in C_{F_e}$. Then $z \in C$ and

$$\lambda Ax + (1 - \lambda)Ay = Az \in F_e.$$

Since F_e is extremal in $A(C)$, we have that Ax and Ay are in F_e. It follows that x and y are in $C \cap A^{-1}(F_e) = C_{F_e}$. This proves that C_{F_e} is extremal in C. Since $C_{F_e} \supset K$, we deduce $C_{F_e} \supset E_b$.

The mapping $A_C := A|_C$ from C to $A(C)$ is surjective, so that

$$A(C_{F_e}) = A_C(C_{F_e}) = A_C[C \cap A^{-1}(F_e)] = A_C[A_C^{-1}(F_e)] = F_e.$$

That is,

(10.38.5) $$A(C_{F_e}) = F_e.$$

Next we verify that $b \in \operatorname{ri} F_e$. This is equivalent to $0 \in \operatorname{ri}(F_e - b)$. Since $0 \in F_e - b$, we see that $\operatorname{aff}(F_e - b) = \operatorname{span}(F_e - b)$. By working in the space $Y_0 := \operatorname{span}(F_e - b)$ rather than $\ell_2(m)$, it suffices to show that $0 \in \operatorname{int}(F_e - b)$. If $0 \notin \operatorname{int}(F_e - b)$, then the separation theorem (Theorem 10.25) implies that there exists $y \in Y_0 \setminus \{0\}$ such that

(10.38.6) $$\langle y, f - b \rangle \le 0 \quad \text{for all} \quad f \in F_e.$$

If $y \in (F_e - b)^{\perp}$, then

$$y \in [\text{span}\,(F_e - b)]^{\perp} = Y_0^{\perp} = \{0\},$$

which contradicts $y \neq 0$. Thus there exists $f_0 \in F_e$ such that $\langle y, f_0 - b \rangle < 0$. Set

$$(10.38.7) \qquad\qquad H = \{z \in Y_0 \mid \langle y, z \rangle = 0\}.$$

Then the hyperplane H supports $F_e - b$ at 0, and $E := H \cap (F_e - b)$ is a convex extremal subset of $F_e - b$ such that $E \neq F_e - b$ (since $f_0 - b \notin E$). [To see that E is extremal, let $f_i \in F_e$ $(i = 1, 2)$, $0 < \lambda < 1$, and suppose $e := \lambda(f_1 - b) + (1 - \lambda)(f_2 - b) \in E$. Then $e \in H$, so

$$0 = \langle e, y \rangle = \lambda \langle f_1 - b, y \rangle + (1 - \lambda)\langle f_2 - b, y \rangle \leq 0$$

by (10.38.6). Thus equality must hold, so $\langle f_i - b, y \rangle = 0$ for $i = 1, 2$, and therefore $f_i - b \in E$ $(i = 1, 2)$.] It follows that $E + b$ is a convex extremal subset of F_e with $b \in E + b$ and $E + b \neq F_e$. Since F_e is extremal in $A(C)$, $E + b$ is also extremal in $A(C)$. But this contradicts the minimality of F_e. Hence $0 \in \text{int}\,(F_e - b)$ and thus

$$(10.38.8) \qquad\qquad b \in \text{ri}\, F_e = \text{ri}\, A(C_{F_e}).$$

We next show that $C_{F_e} \subset E_b$ and hence, since $C_{F_e} \supset E_b$, that $C_{F_e} = E_b$. Let $x_0 \in C_{F_e}$. Then $x_0 \in C$ and $Ax_0 \in F_e$. Since $b \in \text{ri}\, F_e$ by (10.38.8), Theorem 10.26 implies that there exist $y' \in F_e$ and $0 < \lambda < 1$ such that $b = \lambda Ax_0 + (1 - \lambda)y' \in F_e$. By (10.38.5), we can choose $x_0' \in C_{F_e}$ such that $y' = Ax_0'$. Then

$$b = \lambda Ax_0 + (1 - \lambda)Ax_0' = A[\lambda x_0 + (1 - \lambda)x_0'].$$

Thus the element $x := \lambda x_0 + (1 - \lambda)x_0'$ is in C and $Ax = b$. That is, $x \in K \subset E_b$. Since E_b is extremal in C, it follows that $x_0, x_0' \in E_b$. Thus $C_{F_e} \subset E_b$, and hence

$$(10.38.9) \qquad\qquad C_{F_e} = E_b.$$

Using (10.38.9), (10.38.5), and the fact that F_e is extremal, we obtain (10.38.2). From (10.38.8) and (10.38.9) we obtain (10.38.1).

Finally, using (10.38.9) and (10.38.5), we obtain

$$A^{-1}(A(E_b)) \cap C = A^{-1}(A(C_{F_e})) \cap C = A^{-1}(F_e) \cap C = C_{F_e} = E_b.$$

This verifies (10.38.3) and completes the proof. ∎

For the actual computation of C_b, the following result is sometimes helpful.

10.39 Proposition. *Let E be a convex extremal subset of C that contains $C \cap A^{-1}(b)$. Then the following statements are equivalent:*

(1) $E = C_b$;
(2) $b \in \text{ri}\, A(E)$;
(3) $[A(E) - b]^{\circ} = [A(E) - b]^{\perp}$.

Proof. The equivalence of (2) and (3) is from Theorem 10.27. The implication (1) $\Longrightarrow$ (2) is a consequence of Proposition 10.36. Finally, if (2) holds, then $C_b \supset E$ follows from Proposition 10.36. But Proposition 10.38 implies that $E \supset C_b$. Thus $E = C_b$. ∎

As an immediate consequence of Proposition 10.36 and Corollary 10.33, we obtain the main result of this chapter.

10.40 Perturbation Theorem . *For each $x \in X$, there exists $y \in \ell_2(m)$ such that*

$$(10.40.1) \qquad A\left[P_{C_b}\left(x + \sum_1^m y(i)x_i\right)\right] = b.$$

For any $y \in \ell_2(m)$ that satisfies (10.40.1), we have

$$(10.40.2) \qquad P_{K(b)}(x) = P_{C_b}\left(x + \sum_1^m y(i)x_i\right).$$

Moreover, $C_b = C$ if and only if $b \in \operatorname{ri} A(C)$.

Thus the problem of computing $P_{K(b)}(x)$ reduces to that of solving the (generally nonlinear) m equations (10.40.1) for the m unknowns $y = (y(1), y(2), \ldots, y(m))$. This in turn requires being able to compute $P_{C_b}(z)$ for any $z \in Z$. If $b \in \operatorname{ri} A(C)$, then $C_b = C$, and this requires being able to compute $P_C(z)$ for any $z \in X$. If $b \notin \operatorname{ri} A(C)$, then $C_b \neq C$, and we first have to determine the set C_b. In this regard, it is important to know how to recognize when $C_b = C$, that is, when $b \in \operatorname{ri} A(C)$. The following result is helpful in this regard.

10.41 Lemma. *The following six statements are equivalent:*

(1) $C = C_b$;
(2) $b \in \operatorname{ri} A(C)$;
(3) $[A(C) - b]^\circ = [A(C) - b]^\perp$;
(4) $\mathcal{R}(A^*) \cap [C - K(b)]^\circ \subset [C - K(b)]^\perp$;
(5) $\mathcal{R}(A^*) \cap (C - x)^\circ \subset (C - x)^\perp$ *for each $x \in K(b)$*;
(6) $\mathcal{R}(A^*) \cap (C - x)^\circ \subset (C - x)^\perp$ *for some $x \in K(b)$.*

Proof. The equivalence of the first three statements follows from Proposition 10.39.

Now fix any $x \in K(b)$. Then

$$(10.41.1) \qquad A(C - K(b)) = A(C) - A(K(b)) = A(C) - b = A(C - x).$$

Using (10.41.1), we see that (3) holds if and only if

$$(10.41.2) \qquad [A(C - K(b))]^\circ = [A(C - K(b))]^\perp$$

if and only if

$$(10.41.3) \qquad [A(C - x)]^\circ = [A(C - x)]^\perp.$$

Using Proposition 10.31, we deduce that (3) holds if and only if

$$(10.41.4) \qquad (A^*)^{-1}(C - x)^\circ = (A^*)^{-1}(C - x)^\perp$$

if and only if

$$(10.41.5) \qquad (A^*)^{-1}(C - K(b))^\circ = (A^*)^{-1}(C - K(b))^\perp.$$

Since it is always true that $S^\perp \subset S^\circ$ for any set S, the equal signs in (10.41.4) and (10.41.5) may be replaced by the inclusion symbol "$\subset$." From this, it is now easy to verify the equivalence of (3)–(6), and we omit the details. ∎

Before giving some applications of these results in the most important case (i.e., when C is the cone of nonnegative functions), we shall establish some improvements of the previous results in the particular case where C is a convex *cone*.

10.42 Definition. *A convex subset C of X is said to be* **generating** *for X if its span is dense in X, i.e., $\overline{\operatorname{span} C} = X$.*

10.43 Example. Let $X = \ell_2(T)$ and let C be the *positive cone* in X:

$$C = \{x \in X \mid x(t) \geq 0 \text{ for all } t \in T\}.$$

Then it is easy to see that for each $x \in X$, $x = x^+ - x^-$, and since both x^+ and x^- are in C, $X = C - C$. In particular, $X = \operatorname{span}(C)$, so C is generating.

Similarly, the positive cone is generating in $C_2[a, b]$ and in $L_2[a, b]$.

10.44 Lemma. *A convex subset C of a Hilbert space X is generating if and only if $C^\perp = \{0\}$.*

Proof. If C is generating, then $\overline{\operatorname{span} C} = X$. For any $x \in C^\perp$, choose a sequence $\{x_n\}$ in span C such that $x_n \to x$. Thus $x_n = \sum_{i=1}^{N(n)} \alpha_{ni} c_{ni}$ for some $\alpha_{ni} \in \mathbb{R}$ and $c_{ni} \in C$. Then $\langle x, c_{ni} \rangle = 0$ for all n and i, so that

$$\|x\|^2 = \langle x, x \rangle = \lim_n \langle x_n, x \rangle = \lim_n \sum_{i=1}^{N(n)} \alpha_{ni} \langle c_{ni}, x \rangle = 0.$$

Hence $x = 0$, and thus $C^\perp = \{0\}$.

Conversely, if $C^\perp = \{0\}$, then

$$X = \{0\}^\perp = C^{\perp\perp} = \{\overline{\operatorname{span} C}\}^{\perp\perp} = \overline{\operatorname{span} C}$$

using Theorem 4.5(3) and (9). Thus C is generating. ∎

When C is a convex cone, there is a strengthening of Lemma 10.41.

10.45 Lemma. *Let C be a closed convex cone in the Hilbert space X. Then the following five statements are equivalent:*

(1) $C = C_b$;
(2) $b \in \operatorname{ri} A(C)$;
(3) $[A(C)]^\circ \cap b^\perp \subset [A(C)]^\perp$;
(4) $\mathcal{R}(A^*) \cap C^\circ \cap [K(b)]^\perp \subset C^\perp$;
(5) $\mathcal{R}(A^*) \cap C^\circ \cap \{x\}^\perp \subset C^\perp$ *for some $x \in K(b)$.*

In addition, if C is generating, then these five statements are equivalent to each of the following three:

(6) $[A(C)]^\circ \cap b^\perp \subset \mathcal{N}(A^*)$;
(7) $\mathcal{R}(A^*) \cap C^\circ \cap [K(b)]^\perp = \{0\}$;
(8) $\mathcal{R}(A^*) \cap C^\circ \cap \{x\}^\perp = \{0\}$ *for some $x \in K(b)$.*

Finally, if C is a generating closed convex cone and A is surjective, these eight statements are equivalent to each of the following two:

(9) $[A(C)]^\circ \cap b^\perp = \{0\}$;
(10) $b \in \operatorname{int} A(C)$.

Proof. The equivalence of the first four statements follows from Lemma 10.41, the fact that $A(C)$ is a convex cone, and Theorem 4.5(5). The equivalence of (4) and (5) follows from the fact that

$$(10.45.1) \qquad \mathcal{R}(A^*) \cap C^\circ \cap [K(b)]^\perp = \mathcal{R}(A^*) \cap C^\circ \cap \{x\}^\perp$$

is an identity for any $x \in K(b)$. To verify this, it clearly suffices to show the inclusion

$$(10.45.2) \qquad \mathcal{R}(A^*) \cap C^\circ \cap [K(b)]^\perp \supset \mathcal{R}(A^*) \cap C^\circ \cap \{x\}^\perp.$$

But if $z \in \mathcal{R}(A^*) \cap C^\circ \cap \{x\}^\perp$, then $z = A^*y$ for some $y \in \ell_2(m)$, $z \in C^\circ$, and $\langle z, x \rangle = 0$. Then for any $x' \in K(b)$, we have that

$$\langle z, x' \rangle = \langle A^*y, x' \rangle = \langle y, Ax' \rangle = \langle y, b \rangle = \langle y, Ax \rangle = \langle A^*y, x \rangle = \langle z, x \rangle = 0.$$

Thus $z \in [K(b)]^\perp$ and (10.45.2) is verified.

If C is also generating, then $C^\perp = \{0\}$ by Lemma 10.44 and

$$[A(C)]^\perp = (A^*)^{-1}(C^\perp) = (A^*)^{-1}(\{0\}) = \mathcal{N}(A^*)$$

by Theorem 10.31. This proves the equivalence of (3) and (6), and of (4) and (7). The equivalence of (7) and (8) follows from (10.45.1).

If, moreover, A is surjective, then Lemma 8.33 implies that

$$\mathcal{N}(A^*) = \mathcal{R}(A)^\perp = \ell_2(m)^\perp = \{0\}.$$

Thus (6) and (9) are equivalent. But using 4.5(5), we can rewrite (9) as

$$\{0\} = [A(C) - b]^\circ.$$

Since $\{0\} \subset [A(C) - b]^\perp \subset [A(C) - b]^\circ$ is always true, it follows that

$$\{0\} = [A(C) - b]^\circ = [A(C) - b]^\perp.$$

Using Theorem 10.27 applied to $D = A(C)$, we deduce that (9) is equivalent to (10). ∎

Next we observe that if C is a convex cone, then any convex *extremal* subset of C must be a cone. In particular, C_b is a convex cone.

10.46 Extremal Subsets of Cones are Cones. *If E is a convex extremal subset of a convex cone C, then E is a convex cone.*

Proof. Let $x \in E$. If $\rho > 1$, then since 0 and ρx are in C and since

$$\frac{1}{\rho}(\rho x) + \left(1 - \frac{1}{\rho}\right)0 = x \in E,$$

it follows by extremality that ρx and 0 are in E. In particular, $\rho x \in E$ for all $\rho \geq 1$. If $0 \leq \lambda \leq 1$, then by convexity of E, $\lambda x = \lambda x + (1 - \lambda)0 \in E$. This proves that $\rho x \in E$ for all $\rho \geq 0$. Finally, if $x, y \in E$ and $\rho, \mu > 0$, then by convexity of E,

$$z := \frac{\rho}{\rho + \mu}x + \frac{\mu}{\rho + \mu}y \in E,$$

and $\rho x + \mu y = (\rho + \mu)z \in E$ by the statement above. Thus E is a convex cone. ∎

Constrained Interpolation by Positive Functions

Now we will exploit these results for what is perhaps the most important case: where C is the convex cone of nonnegative functions. In this case, C is a generating cone.

Let T be a nonempty set, $X = \ell_2(T)$,

$$C = \{x \in X \mid x(t) \geq 0 \text{ for all } t \in T\},$$

$\{x_1, x_2, \ldots, x_m\} \subset X$, and define $A : X \to \ell_2(m)$ by

$$Ax := (\langle x, x_1 \rangle, \langle x, x_2 \rangle, \ldots, \langle x, x_m \rangle).$$

By Theorem 8.28, the adjoint mapping $A^* : \ell_2(m) \to X$ is given by

$$A^*y = \sum_1^m y(i)x_i, \quad y \in \ell_2(m).$$

For any given $b \in \ell_2(m)$, let

$$(10.46.1) \qquad \begin{aligned} K(b) &:= C \cap A^{-1}(b) \\ &= \{x \in X \mid x \geq 0, \langle x, x_i \rangle = b(i) \text{ for } i = 1, 2, \ldots, m\}. \end{aligned}$$

We assume that $K(b) \neq \emptyset$, i.e., $b \in A(C)$.

We will give a useful formula for $P_{K(b)}(x)$ for any $x \in \ell_2(T)$.

For any $x \in \ell_2(T)$, the **support** of x is defined by

$$(10.46.2) \qquad \operatorname{supp} x := \{t \in T \mid x(t) \neq 0\}.$$

Also, for any subset S of T, we define the subset $C(S)$ of C by

$$(10.46.3) \qquad C(S) := \{x \in C \mid \operatorname{supp} x \subset S\}.$$

The importance of the set $C(S)$ is that it is *equal* to the minimal convex extremal subset C_b of C for a particular choice of S! Before we verify this, it is convenient to establish some basic properties of $C(S)$.

10.47 Properties of $C(S)$. *Fix any subset S of T. Then:*

(1) $C(S)$ *is a closed convex cone that is extremal in* C.
(2) $C(T) = C$ *and* $C(\emptyset) = \{0\}$.
(3) $C(S)^\circ = \{x \in \ell_2(T) \mid x(t) \leq 0 \quad \text{for all } t \in S\}$.
(4) $C(S)^\perp = \{x \in \ell_2(T) \mid x(t) = 0 \quad \text{for all } t \in S\}$.

Proof. (1) Let (x_n) be a sequence in $C(S)$ and $x_n \to x$. Then $x_n(t) \to x(t)$ for all t, and $\operatorname{supp} x_n \subset S$ for all n. Since $x_n(t) \geq 0$ for all t, it follows that $x(t) \geq 0$ for all t and $x \in C$. If $\operatorname{supp} x \not\subset S$, choose $t_0 \in T \backslash S$ such that $x(t_0) > 0$. Thus $x_n(t_0) > 0$ for n sufficiently large implies that $t_0 \in \operatorname{supp} x_n$ for large n, a contradiction to $\operatorname{supp} x_n \subset S$. This proves $\operatorname{supp} x \subset S$, and thus $C(S)$ is closed.

If $x, y \in C(S)$ and $\alpha, \beta \geq 0$, then $z := \alpha x + \beta y$ satisfies $z \geq 0$, and $\operatorname{supp} z \subset S$. Thus $z \in C(S)$, so $C(S)$ is a convex cone.

Finally, let $x, y \in C$, $0 < \lambda < 1$, and $z := \lambda x + (1-\lambda)y \in C(S)$. Then $\operatorname{supp} z \subset S$. Since $x, y \geq 0$ and λ, $1 - \lambda > 0$, it follows that $\operatorname{supp} x \subset S$, $\operatorname{supp} y \subset S$ and hence $x, y \in C(S)$. This proves that $C(S)$ is extremal in C.

(2) This is obvious.

(3) Let $D = \{z \in \ell_2(T) \mid z(t) \leq 0 \text{ for all } t \in S\}$. If $z \in D$, then for every $x \in C(S)$, $x = 0$ off S, $x \geq 0$ on S, and hence

$$\langle z, x \rangle = \sum_{t \in T} z(t)x(t) = \sum_{t \in S} z(t)x(t) \leq 0.$$

Thus $z \in C(S)^\circ$ and hence $D \subset C(S)^\circ$.

Conversely, let $z \in \ell_2(T)\backslash D$. Then $z(t_0) > 0$ for some $t_0 \in S$. The element $x_0 \in \ell_2(T)$ defined by $x_0(t_0) = 1$ and $x_0(t) = 0$ for $t \neq t_0$ is in $C(S)$, but $\langle z, x_0 \rangle = z(t_0) > 0$. Thus $z \notin C(S)^\circ$. This proves $C(S)^\circ \subset D$ and verifies (3).

(4) The proof is similar to (3), or it can be deduced from (3) and the fact that $C(S)^\perp = C(S)^\circ \cap [-C(S)^\circ]$. ∎

There is a simple formula for the best approximation to any $x \in \ell_2(T)$ from the set $C(S)$. Recall that the **characteristic function** of a set S is defined by

$$\chi_S(t) = \begin{cases} 1 & \text{if } t \in S, \\ 0 & \text{otherwise.} \end{cases}$$

10.48 Best Approximation from $C(S)$. *Let S be a given subset of T. Then $C(S)$ is a Chebyshev convex cone in $\ell_2(T)$, and*

$$(10.48.1) \qquad P_{C(S)}(x) = \chi_S \cdot x^+ \qquad \text{for each } x \in \ell_2(T).$$

In particular,

$$(10.48.2) \qquad P_C(x) = x^+ \qquad \text{for each } x \in \ell_2(T).$$

Proof. By Theorem 10.47, $C(S)$ is a closed convex cone in the Hilbert space $\ell_2(T)$, so $C(S)$ is Chebyshev by Theorem 3.5. Given any $x \in \ell_2(T)$, let $x_0 := \chi_S \cdot x^+$. Clearly, $x_0 \geq 0$ and $\operatorname{supp} x_0 \subset S$, so $x_0 \in C(S)$. Further, if $t \in S$, then $x(t) - x_0(t) = x(t) - x^+(t) \leq 0$ and $x - x_0 \in C(S)^\circ$ by Theorem 10.47(3). Moreover,

$$\langle x - x_0, x_0 \rangle = \sum_{t \in T}[x(t) - x_0(t)]x_0(t) = \sum_{t \in S}[x(t) - x_0(t)]x_0(t)$$

$$= \sum_{t \in S}[x(t) - x^+(t)]x^+(t) = 0,$$

since $[x(t) - x^+(t)]x^+(t) = 0$ for every $t \in T$. We have shown that $x - x_0 \in C(S)^\circ \cap x_0^\perp$. By Theorem 4.7, $x_0 = P_{C(S)}(x)$.

The last statement follows from the first, since $C = C(T)$ by Theorem 10.47(2). ∎

The next proposition shows that the important extremal subset C_b of C is just $C(S)$ for a prescribed subset S of T.

10.49 Proposition. *Let $S_b := \cup_{x \in K(b)} \text{supp} \, x$. Then*

$$(10.49.1) \qquad\qquad C_b = C(S_b).$$

Proof. Since $S_b \supset \text{supp} \, x$ for each $x \in K(b)$, it follows that $K(b) \subset C(S_b)$. Since $C(S_b)$ is a convex extremal subset of C by Theorem 10.47(1), it follows by Proposition 10.38 that $C(S_b) \supset C_b$. For the reverse inclusion, it suffices by Proposition 10.36 to show that $b \in \text{ri} \, A(C(S_b))$. But by Theorem 10.27, this is equivalent to showing that

$$(10.49.2) \qquad\qquad [A(C(S_b)) - b]^\circ = [A(C(S_b)) - b]^\perp.$$

Since $C(S_b)$ is a convex cone and A is linear, $A(C(S_b))$ is a convex cone and $b \in A(K(b)) \subset A(C(S_b))$. By Theorem 4.5(5), we get

$$[A(C(S_b)) - b]^\circ = [A(C(S_b))]^\circ \cap b^\perp.$$

Similarly,

$$[A(C(S_b)) - b]^\perp = [A(C(S_b))]^\perp \cap b^\perp.$$

Thus (10.49.2) may be rewritten as

$$(10.49.3) \qquad\qquad [A(C(S_b))]^\circ \cap b^\perp \subset [A(C(S_b))]^\perp.$$

To verify (10.49.3), let $y \in [A(C(S_b))]^\circ \cap b^\perp$. Then, using Theorem 10.31, we have that

$$(10.49.4) \qquad\qquad \langle y, b \rangle = 0$$

and

$$(10.49.5) \qquad\qquad A^* y \in C(S_b)^\circ.$$

By Proposition 10.47(3), $(A^* y)(t) \leq 0$ for all $t \in S_b$. Also, for any $x \in K(b)$, $Ax = b$, so that from (10.49.4) we obtain

$$(10.49.6) \quad 0 = \langle y, b \rangle = \langle y, Ax \rangle = \langle A^* y, x \rangle = \sum_{t \in T} (A^* y)(t) x(t) = \sum_{t \in S_b} (A^* y)(t) x(t),$$

where we used $\text{supp} \, x \subset S_b$ for the last equality. Since $(A^* y)(t) \leq 0$ and $x(t) \geq 0$ for all $t \in S_b$, (10.49.6) implies that

$$(10.49.7) \qquad\qquad (A^* y)(t) x(t) = 0 \qquad \text{for all } t \in S_b.$$

If $(A^* y)(t_0) < 0$ for some $t_0 \in S_b$, choose $x_0 \in K(b)$ such that $x_0(t_0) > 0$. Then $(A^* y)(t_0) x_0(t_0) < 0$, which contradicts (10.49.7). Thus $(A^* y)(t) = 0$ for all $t \in S_b$, and using 10.47(4), it follows that $A^* y \in C(S_b)^\perp$, or $y \in (A^*)^{-1}(C(S_b)^\perp)$. By (10.31.2), it follows that $y \in [A(C(S_b))]^\perp$, which verifies (10.49.3). ∎

Now we are in position to state and prove the main result of this section.

10.50 Interpolation Constrained by the Positive Cone. Let $K(b)$ be defined as in (10.46.1) and $S_b := \cup_{x \in K(b)} \text{supp}\, x$. Then for each $x \in \ell_2(T)$, there exists $y \in \ell_2(m)$ that satisfies

$$(10.50.1) \qquad \left\langle \left[x + \sum_1^m y(i) x_i \right]^+ \chi_{S_b}, x_j \right\rangle = b(j) \quad \text{for} \quad j = 1, 2, \ldots, m.$$

Moreover,

$$(10.50.2) \qquad P_{K(b)}(x) = \left(x + \sum_1^m y(i) x_i \right)^+ \chi_{S_b}$$

for any $y \in \ell_2(m)$ that satisfies (10.50.1).

Finally, the characteristic function χ_{S_b} may be deleted from both (10.50.1) and (10.50.2) if $\{x_1 \chi_{S_b}, x_2 \chi_{S_b}, \ldots, x_m \chi_{S_b}\}$ is linearly independent.

Proof. Combining Theorem 10.40, Proposition 10.49, and Theorem 10.48, we obtain the first paragraph of the theorem. To prove the last sentence of the theorem, suppose $\{x_1 \chi_{S_b}, x_2 \chi_{S_b}, \ldots, x_m \chi_{S_b}\}$ is linearly independent. We will show that $b \in \text{ri}\, A(C)$. It then follows that $C_b = C = C(T)$ and Theorems 10.40 and 10.48 show that the characteristic function χ_{S_b} may be dropped from (10.50.2).

Since C is generating (Example 10.43), it follows by Lemma 10.45 that it suffices to show that

$$(10.50.3) \qquad [A(C)]^\circ \cap b^\perp \subset \mathcal{N}(A^*).$$

To this end, let $y \in [A(C)]^\circ \cap b^\perp$. Using Theorem 10.31, we see that $A^* y \in C^\circ$. Then $A^* y \le 0$ by Proposition 10.47(3), so that for any $x \in K(b)$,

$$0 = \langle y, b \rangle = \langle y, Ax \rangle = \langle A^* y, x \rangle = \sum_{t \in T} (A^* y)(t) x(t) = \sum_{t \in T} (A^* y)(t) \chi_{S_b}(t) x(t),$$

since $\text{supp}\, x \subset S_b$ for every $x \in K(b)$. Since $A^* y \le 0$ and $x \ge 0$, it follows that $(A^* y)(t) \chi_{S_b}(t) = 0$ for all t. But $A^* y = \sum_{i=1}^m y(i) x_i$ implies that $\sum_{i=1}^m y(i) x_i \chi_{S_b} = 0$. Since $\{x_1 \chi_{S_b}, \ldots, x_m \chi_{S_b}\}$ is linearly independent, we must have $y(i) = 0$ for all i. That is, $y = 0$. Hence $y \in \mathcal{N}(A^*)$ and (10.50.3) holds. ∎

Remark. The proof actually has shown that if $\{x_1 \chi_{S_b}, \ldots, x_m \chi_{S_b}\}$ is linearly independent, then $b \in \text{int}\, A(C)$. To see this, we need only note that $b \in \text{int}\, A(C)$ if and only if $[A(C) - b]^\circ = \{0\}$ (by Theorem 10.27(2)) if and only if $[A(C)]^\circ \cap b^\perp = \{0\}$ (by Theorem 4.5(5)). And it was this latter condition that was established in verifying (10.50.3).

10.51 Example. Let $X = \ell_2$, let e_i denote the canonical unit basis vectors

$$e_i(j) = \delta_{ij}, \quad x_1 = e_1 + e_2, \quad x_2 = \sum_{n=3}^{\infty} \frac{1}{n} e_n,$$

and let

$$K := \{x \in \ell_2 \mid x \ge 0, \quad \langle x, x_1 \rangle = 1, \quad \langle x, x_2 \rangle = 0\}.$$

What is the minimal norm element of K? That is, what is $P_K(0)$?

Letting $b = (1, 0)$, we see that $K = K(b)$, and Theorem 10.50 applies. We first must compute $S_b = \cup_{x \in K(b)} \text{supp}\, x$. If $x \in K(b)$, then

$$(10.51.1) \qquad\qquad x(t) \geq 0 \text{ for all } t \in \mathbb{N},$$

$$(10.51.2) \qquad\qquad x(1) + x(2) = 1,$$

and

$$(10.51.3) \qquad\qquad \sum_{n=3}^{\infty} \frac{1}{n} x(n) = 0.$$

Now (10.51.1) and (10.51.3) imply that

$$(10.51.4) \qquad\qquad x(n) = 0 \quad \text{for all } n \geq 3.$$

Properties (10.51.1) and (10.51.2) imply that

$$(10.51.5) \qquad\qquad x(2) = 1 - x(1), \quad 0 \leq x(1) \leq 1.$$

In particular, $x = e_1$ and $x = e_2$ are both in $K(b)$. It follows from (10.51.4)–(10.51.5) that

$$(10.51.6) \qquad\qquad S_b = \bigcup_{x \in K(b)} \text{supp}\, x = \{1, 2\}.$$

It follows from Theorem 10.50 that

$$(10.51.7) \qquad\qquad P_K(0) = (y(1)x_1 + y(2)x_2)^+ \chi_{\{1,2\}}$$

for any choice of $y = (y(1), y(2)) \in \ell_2(2)$ such that

$$(10.51.8) \qquad\qquad \langle (y(1)x_1 + y(2)x_2)^+ \chi_{\{1,2\}}, \, x_1 \rangle = 1,$$

$$(10.51.9) \qquad\qquad \langle (y(1)x_1 + y(2)x_2)^+ \chi_{\{1,2\}}, \, x_2 \rangle = 0.$$

Now, (10.51.9) is automatically satisfied for any choice of y, since $x_2 \chi_{\{1,2\}} = 0$. Thus we need only satisfy (10.51.8). For the same reason, (10.51.8) reduces to

$$(10.51.10) \qquad\qquad \langle (y(1)x_1)^+, \, x_1 \rangle = 1.$$

Clearly, $y(1) = \frac{1}{2}$ satisfies (10.51.10) since $\langle x_1, x_1 \rangle = 2$. Thus

$$(10.51.11) \qquad\qquad P_K(0) = \frac{1}{2} x_1 = \frac{1}{2} (e_1 + e_2).$$

We should observe that although $P_K(0) = \frac{1}{2} x_1$ is the unique solution to this problem, there are *infinitely* many solutions y to equations (10.51.8)–(10.51.9), namely, $y = \left(\frac{1}{2}, y(2)\right)$ for any $y(2) \in \mathbb{R}$.

There is an analogous result that holds in the space $L_2[c, d]$, and it can be proved in a similar way.

Let $\{x_1, x_2, \ldots, x_m\} \subset L_2[c, d]$, $b \in \ell_2(m)$, and

$$(10.51.12) \qquad \begin{aligned} K(b) := \{ x \in L_2[c, d] \mid & x(t) \geq 0 \text{ for almost all } t \in [c, d], \\ & \text{and } \langle x, x_i \rangle = b(i) \text{ for } i = 1, 2, \ldots, m \}. \end{aligned}$$

For any $x \in L_2[c, d]$, its *support* is defined by

$$(10.51.13) \qquad\qquad \text{supp}\, x := \overline{\{t \in [c, d] \mid x(t) \neq 0\}}.$$

Of course, $\text{supp}\, x$ is defined only up to a set of measure zero.

★**10.52 Theorem.** Let $K(b)$ be as defined in (10.51.2) and let $S_b := \cup_{x \in K(b)} \operatorname{supp} x$. Then for each $x \in L_2[c,d]$, there exists $y \in \ell_2(m)$ that satisfies

$$(10.52.1) \qquad \left\langle \left[x + \sum_1^m y(i)x_i \right]^+ \chi_{S_b}, \, x_j \right\rangle = b(j) \text{ for } j = 1, 2, \ldots, m.$$

Moreover,

$$(10.52.2) \qquad P_{K(b)}(x) = \left[x + \sum_1^m y(i)x_i \right]^+ \chi_{S_b}$$

for any $y \in \ell_2(m)$ that satisfies (10.52.1).

Finally, the characteristic function χ_{S_b} may be deleted from both (10.52.1) and (10.52.2) if $\{x_1 \chi_{S_b}, x_2 \chi_{S_b}, \ldots, x_m \chi_{S_b}\}$ is linearly independent.

★**10.53 Example.** In the Hilbert space $L_2[0,3]$, let

$$x_i(t) := (1 - |t - i|)^+ \qquad (i = 1, 2)$$

and

$$K := \{x \in L_2[0,3] \mid x \geq 0, \quad \langle x, x_1 \rangle = 1, \quad \langle x, x_2 \rangle = 0\}.$$

What is the minimal norm element $P_K(0)$ of K?

Using the fact that $2\chi_{[0,1]} \in K$, it is easy to verify that

$$\bigcup \{\operatorname{supp} x \mid x \in K\} = [0,1].$$

By Theorem 10.52,

$$(10.53.1) \qquad P_K(0) = [y(1)x_1 + y(2)x_2]^+ \chi_{[0,1]}$$

for any choice of $y = (y(1), y(2)) \in \ell_2(2)$ that satisfies

$$(10.53.2) \qquad \langle [y(1)x_1 + y(2)x_2]^+ \chi_{[0,1]}, x_1 \rangle = 1,$$
$$(10.53.3) \qquad \langle [y(1)x_1 + y(2)x_2]^+ \chi_{[0,1]}, x_2 \rangle = 0.$$

Since $x_2 \chi_{[0,1]} = 0$, it follows that (10.53.3) is satisfied for any choice of y. Thus it suffices to determine a y that satisfies (10.53.2). Since

$$[y(1)x_1 + y(2)x_2]^+ \chi_{[0,1]} = [y(1)x_1\chi_{[0,1]} + y(2)x_2\chi_{[0,1]}]^+ = [y(1)x_1\chi_{[0,1]}]^+$$
$$= y(1)^+ x_1 \chi_{[0,1]},$$

it follows that (10.53.2) implies that

$$1 = \int_0^3 y(1)^+ x_1 \chi_{[0,1]} x_1 = y(1)^+ \int_0^1 x_1^2 = \frac{1}{3} y(1)^+,$$

so $y(1)^+ = 3 = y(1)$. Substituting this into (10.53.1), we obtain that

(10.53.4) $P_K(0) = 3x_1\chi_{[0,1]}.$

By the uniqueness of $P_K(0)$, it is now clear that it is *not* possible to drop the characteristic function $\chi_{[0,1]}$ from the expression (10.53.1) for $P_K(0)$.

There is an alternative approach to computing $P_{K(b)}(x)$ as described in Theorem 10.50. It uses *Dykstra's algorithm* (see Chapter 9). We describe this approach now.

Recall our setup. We are working in the Hilbert space $X = \ell_2(T)$ (or, more generally, in $X = L_2(T, \mu)$, where μ is any measure on T). Let $\{x_1, x_2, \ldots, x_m\} \subset X \backslash \{0\}$, $C = \{x \in X \mid x(t) \geq 0 \text{ for all } t \in T\}$, and define $A : X \to \ell_2(m)$ by

$$Ax := (\langle x, x_1 \rangle, \langle x, x_2 \rangle, \ldots, \langle x, x_m \rangle), \quad x \in X.$$

For any $b \in A(C)$, set

$$K(b) := C \cap A^{-1}(b) = \{x \in C \mid \langle x, x_i \rangle = b(i) \text{ for } i = 1, 2, \ldots, m\} = \bigcap_1^{m+1} K_i,$$

where

$$K_i := \{x \in X \mid \langle x, x_i \rangle = b(i)\}$$

is a hyperplane for $i = 1, 2, \ldots, m$, and $K_{m+1} = C$.

As we saw in Example 9.44, Dykstra's algorithm substantially simplifies in this application. Indeed, for any $x \in X$, we have that

(10.53.5) $P_{K_{m+1}}(x) = x^+$

by (10.48.2), and

(10.53.6) $P_{K_i}(x) = x - \dfrac{1}{\|x_i\|^2} \left[\langle x, x_i \rangle - b(i) \right] x_i$

for $i = 1, 2, \ldots, m$ by (6.17.1). Fix any $z \in X$ and set

$$[n] := \{1, 2, \ldots, m+1\} \cap \{n - k(m+1) \mid k = 0, 1, 2, \ldots\},$$

$z_0 = z$, $e_0 = 0$,

(10.53.7) $z_n = P_{K_{[n]}}(z_{n-1}) \qquad \text{if} \quad [n] \neq m+1,$

and

(10.53.8) $z_n = P_{K_{m+1}}(z_{n-1} + e_{n-(m+1)})$

and

(10.53.9) $e_n = z_{n-1} + e_{n-(m+1)} - z_n$

if $[n] = m+1$ $(n = 1, 2, \ldots)$. Then the Boyle–Dykstra theorem (Theorem 9.24) implies that the sequence $\{z_n\}$ converges to $P_{K(b)}(z)$: $\|z_n - P_{K(b)}(z)\| \to 0$.

For the *practical* implementation of these equations, it is convenient to rewrite them in the following form. Set

(10.53.10)
$$z_0 = z, \quad e_0 = 0,$$

and for $n \geq 0$ compute
(10.53.11)
$$\begin{cases} z_{(m+1)n+1} = z_{(m+1)n} - \dfrac{1}{\|x_1\|^2} \left[\langle z_{(m+1)n}, x_1 \rangle - b(1) \right] x_1, \\[2mm] z_{(m+1)n+2} = z_{(m+1)n+1} - \dfrac{1}{\|x_2\|^2} \left[\langle z_{(m+1)n+1}, x_2 \rangle - b(2) \right] x_2, \\[2mm] \cdots \\[2mm] z_{(m+1)n+m} = z_{(m+1)n+m-1} - \dfrac{1}{\|x_m\|^2} \left[\langle z_{(m+1)n+m-1}, x_m \rangle - b(m) \right] x_m, \end{cases}$$

(10.53.12)
$$z_{(m+1)(n+1)} = \left[z_{(m+1)n+m} + e_{(m+1)n} \right]^+,$$

(10.53.13)
$$e_{(m+1)(n+1)} = z_{(m+1)n+m} - z_{(m+1)(n+1)} + e_{(m+1)n}.$$

Exercises

(1) 1. Let $S = \{ \rho(\alpha, \beta, 1) \mid \rho \geq 0,\ \alpha^2 + \beta^2 \leq 1 \}$ and $T = \{ (\alpha, -1, 1) \mid \alpha \in \mathbb{R} \}$. Verify the following statements.
   (a) S is a closed convex cone and T is a closed affine set.
   (b) $(S \cap T)^\circ \neq \overline{S^\circ + T^\circ}$.
   (c) $\overline{\mathrm{con}\,(S \cap T)} = \overline{\mathrm{con}\,(S)} \cap \overline{\mathrm{con}\,(T)}$. Why does this example not contradict Lemma 10.3?

2. Let X and Y be inner product spaces, $A : X \to Y$ be linear, and suppose that $K \subset Y$ is a convex set (respectively a convex cone, a subspace). Then the inverse image $A^{-1}(K)$ is a convex set (respectively a convex cone, a subspace).

3. If C and D are closed convex cones in the Hilbert space X, show that $C \subset D$ if and only if $C^\circ \supset D^\circ$. More generally, show that for any two sets S and T in X, $\overline{\mathrm{con}\,(S)} \subset \overline{\mathrm{con}\,(T)}$ if and only if $S^\circ \supset T^\circ$.

4. (a) If $\{C_i\}$ is any collection of convex cones, show that

$$\mathrm{con}\left(\bigcup_i C_i \right) = \mathrm{co}\left(\bigcup_i C_i \right).$$

In other words, the *conical hull* of the union is the same as the *convex hull* of the union.

   (b) Show that if $\{C_i\}$ is any collection of complete convex cones, then

$$\left(\bigcap_i C_i \right)^\circ = \overline{\mathrm{co}\left(\bigcup_i C_i^\circ \right)}.$$

[Hint: Part (a).]

5. Let K be a nonempty convex set in X and $x_0 \in X$. Prove the equivalence of the following three statements.

(a) $x_0 \in \overline{K}$;

(b) $\langle x, x_0 \rangle \leq \sup_{y \in K} \langle x, y \rangle$;

(c) $(K - x_0)^\circ = \left\{ x \in X \mid \sup_{y \in K} \langle x, y \rangle = \langle x, x_0 \rangle \right\}$.

6. Let S and T be convex subsets of the Hilbert space X with $0 \in S \cap T$. Show that the following four statements are equivalent.

(a) $(S \cap T)^\circ = \overline{(S^\circ + T^\circ)}$;

(b) $(S \cap T)^\circ \subset \overline{(S^\circ + T^\circ)}$;

(c) $\overline{\mathrm{con}\,(S)} \cap \overline{\mathrm{con}\,(T)} \subset \overline{\mathrm{con}\,(S \cap T)}$;

(d) $\overline{\mathrm{con}\,(S)} \cap \overline{\mathrm{con}\,(T)} = \overline{\mathrm{con}\,(S \cap T)}$.

[Hint: Look at the proof of Lemma 10.3.]

7. One consequence of Theorem 4.6(4) is that if C_1 and C_2 are closed convex cones in the Hilbert space X, then

(7.1) $$(C_1 \cap C_2)^\circ = \overline{C_1^\circ + C_2^\circ}.$$

Show that (7.1) fails in general if the convex cones C_i are not closed. [Hint: Consider the sets $C_1 = \{(\alpha, \beta) \mid \alpha = \beta = 0 \text{ or } \alpha > 0 \text{ and } \beta \geq 0\}$ and $C_2 = \{(\alpha, \beta) \mid \alpha = \beta = 0 \text{ or } \alpha < 0 \text{ and } \beta \geq 0\}$ in $\ell_2(2)$.] This proves that the equivalent conditions of Exercise 6 do not always hold.

8. (a) If C is a convex set and α and β are positive real numbers, then

$$(\alpha + \beta)C = \alpha C + \beta C.$$

(b) If C_1 and C_2 are convex sets and $\alpha \in \mathbb{R}$, then

$$\alpha(C_1 + C_2) = \alpha C_1 + \alpha C_2.$$

9. If K is a convex subset of X, a function $f : K \to \mathbb{R} \cup \{\infty\}$ is called **convex** if

$$f(\lambda x + (1 - \lambda)y) \leq \lambda f(x) + (1 - \lambda)f(y)$$

for all $x, y \in K$ and $0 < \lambda < 1$.

(a) Show that $f(x) := \|x\|$ is convex on X.

(b) If $\{f_j \mid j \in J\}$ is an indexed collection of convex functions on K, show that $f := \sup_{j \in J} f_j$ is convex on K. (Here $f(x) := \sup\{f_j(x) \mid j \in J\}$, $x \in K$.)

(c) Show that $f(x) := \sup_{y \in K} \langle x, y \rangle$ is convex on X. (f is called the **support function** of K.)

(d) Every linear functional on X is a convex function.

(e) The function

$$f(x) = \begin{cases} 0 & \text{if } x \in K, \\ \infty & \text{if } x \notin K, \end{cases}$$

is convex on X. (f is called the **indicator function** of K.)

(f) If $f : K \to \mathbb{R} \cup \{\infty\}$ is convex on K, then the function $F : X \to \mathbb{R} \cup \{\infty\}$, defined by

$$F(x) := \begin{cases} f(x) & \text{if } x \in K, \\ \infty & \text{if } x \notin K, \end{cases}$$

is also convex. In other words, *every convex function has an extension to a convex function defined on the whole space.*

10. Let $\{C_1, C_2, \ldots, C_m\}$ be closed convex subsets of the Hilbert space X and suppose $\sum_1^n C_i \neq \emptyset$. Verify that $\{C_1, C_2, \ldots, C_m\}$ has the strong CHIP if and only if

(10.1) $$\partial \left(\sum_1^m I_{C_i} \right)(x) = \sum_1^m \partial I_{C_i}(x)$$

for every $x \in \cap_1^m C_i$, where I_C denotes the **indicator function** for C:

$$I_C(x) := \begin{cases} 0 & \text{if } x \in C, \\ \infty & \text{if } x \notin C, \end{cases}$$

and where ∂f denotes the **subdifferential** of f:

$$\partial f(x) := \{x^* \in X \mid \langle x^*, z - x \rangle + f(x) \leq f(z) \text{ for all } z \in X\}.$$

11. Let $X = \ell_2(T)$ (or, more generally, $X = L_2(T, \mu)$), $\{x_1, x_2, \ldots, x_m\} \subset X$, and define A on X by

$$Ax := (\langle x, x_1 \rangle, \langle x_2 x_2 \rangle, \ldots, \langle x, x_m \rangle), \qquad x \in X.$$

Prove that A is surjective (i.e., $A(X) = \ell_2(m)$) if and only if $\{x_1, x_2, \ldots, x_m\}$ is linearly independent.

[Hint: Use Lemma 8.33 to establish first that A is surjective if and only if A^* is injective.]

12. Prove Lemma 10.19 concerning affine hulls.

[Hint: Lemma 10.17.]

13. Prove Theorem 10.23, which shows, in particular, that the relative interior of a convex set is convex.

[Hint: Theorem 2.8 and Lemma 10.22.]

14. Prove that a nonempty set S has dimension n if and only if there exists a set of $n+1$ points $\{x_0, x_1, \ldots, x_n\}$ in S such that $\{x_1 - x_0, x_2 - x_0, \ldots, x_n - x_0\}$ is linearly independent, and no larger number of points in S has this property.

15. Let C_i be a convex subset of X with $\operatorname{ri} C_i \neq \emptyset$ for $i = 1, 2, \ldots, n$ (for example, if $\dim X < \infty$ and $C_i \neq \emptyset$ for each i). Prove that

$$\operatorname{ri} \left(\bigcap_1^n C_i \right) = \bigcap_1^n \operatorname{ri} C_i.$$

16. Let X and Y be inner product spaces, $A : X \to Y$ be linear, and $C \subset X$ be convex with $\operatorname{ri} C \neq \emptyset$. Show that

$$\operatorname{ri} A(C) = A(\operatorname{ri} C).$$

17. If $\{e_1, e_2, \ldots, e_n\}$ is an orthonormal basis for Y and $r > 1/\sqrt{n}$, prove that

$$B(0, r) \not\subset \text{co} \{\pm e_1, \pm e_2, \ldots, \pm e_n\}.$$

(This shows that the radius $1/\sqrt{n}$ is largest possible in Theorem 10.30.)
[Hint: Choose any ρ with $1/n < \rho < r/\sqrt{n}$ and show that the vector $x = \rho \sum_1^n e_i$ is in $B(0, r)$, but not in co $\{\pm e_1, \pm e_2, \ldots, \pm e_n\}$.]

18. Let C be a convex subset of X. Show that:
(a) C is extremal in C.
(b) The intersection of any collection of extremal subsets of C is either empty or extremal.
(c) The union of any collection of extremal subsets of C is an extremal subset of C.
(d) If E_1 is extremal in C and E_2 is extremal in E_1, then E_2 is extremal in C.

19. If E is a convex extremal subset of the convex cone C and $\text{ri}\, E \cap \text{ri}\, C \neq \emptyset$, then $E = C$.
[Hint: Proposition 10.46.]

20. Let K be a convex set and $x \in K$. Prove the equivalence of the following three statements.
(a) x is an extreme point of K;
(b) y, z in K and $x = \frac{1}{2}(y + z)$ implies that $y = z$;
(c) $K \backslash \{x\}$ is convex.

21. Let C be a convex cone in X. Prove that C is generating in X (i.e., $\overline{\text{span}}\, C = X$) if and only if $\overline{C - C} = X$.

22. Consider the mapping $A : \ell_2(2) \to \ell_2(2)$ defined by $A((\alpha, \beta)) = (0, \beta)$, and let

$$C := \left\{ (\alpha, \beta) \mid \alpha^2 + \beta^2 \leq 1 \right\}.$$

Verify the following statements.
(a) C is a closed convex set.
(b) A is a self-adjoint bounded linear operator with $\|A\| = 1$.
(c) $\mathcal{N}(A) = \{(\alpha, 0) \mid \alpha \in \mathbb{R}\}$.
(d) $A(C) = \{(0, \beta) \mid |\beta| \leq 1\}$.
(e) If $b = \pm(0, 1)$ and $K(b) := C \cap A^{-1}(b)$, then $K(b) = \{b\}$ and $C_b = \{b\}$.
(f) If $b = (0, \beta)$ with $-1 < \beta < 1$, then $C_b = C$.
(g) If $b = (0, 1)$, then
    (i) $(C - b) \cap \mathcal{N}(A) = \{0\}$.
    (ii) $\overline{\text{con}\,(C - b)} \cap \mathcal{N}(A) = \mathcal{N}(A)$.
    (iii) $\overline{\text{con}\,[(C - b) \cap \mathcal{N}(A)]} = \{0\} \neq \overline{\text{con}\,(C - b)} \cap \mathcal{N}(A)$.
(h) Theorem 10.40 fails in general if C_b is replaced by C in (10.40.1) and (10.40.2).
[Hint: Lemma 10.12 and part (g).]

★23. Let $X = L_2[0, 3]$, $x_i(t) = [1 - |t - i|]^+$ $(i = 1, 2)$, and

$$K = \left\{ x \in X \mid x \geq 0, \quad \langle x, x_1 \rangle = 1, \quad \langle x, x_2 \rangle = 0 \right\}.$$

Find $P_K(x)$, when $x(t) = \sin \left(\frac{\pi}{2} t \right)$.
[Hint: Theorem 10.52.]

★24. Let $X = L_2[0,1]$, $x_1 = \chi_{[0,1]}$, $x_2 = \sum_{n=1}^{\infty} n\chi_{[2^{-n},2^{-n+1}]}$, and

$$K = \{x \in X \mid x \geq 0, \quad \langle x, x_1 \rangle = 1, \quad \langle x, x_2 \rangle = 1\}.$$

Find $P_K(x)$, when $x(t) = \sin\left(\frac{\pi}{2}t\right)$.
[Hint: Theorem 10.52.]

★25. Fix any integer $m \geq 1$ and let $X = L_2[0, m+1]$. Consider the functions $\{x_1, x_2, \ldots, x_m\}$ in X defined by

$$x_i(t) = [1 - |t - i|]^+, \quad t \in [0, m+1]$$

for $i = 1, 2, \ldots, m$. Let $C := \{x \in X \mid x \geq 0\}$ and define $A : X \to \ell_2(m)$ by

$$Ax := (\langle x, x_1 \rangle, \langle x, x_2 \rangle, \ldots, \langle x, x_m \rangle), \quad x \in X.$$

Verify the following statements.
(a) $\{x_1, x_2, \ldots, x_m\}$ is a linearly independent set of nonnegative functions.
(b) $\overline{A(C)} = \{(\alpha_1, \alpha_2, \ldots, \alpha_m) \mid \alpha_i \geq 0 \text{ for all } i\}$.
(c) $\operatorname{ri} A(C) = \operatorname{int} A(C) = \{(\alpha_1, \alpha_2, \ldots, \alpha_m) \mid \alpha_i > 0 \text{ for all } i\}$.
(d) $\operatorname{bd} A(C) = \{(\alpha_1, \alpha_2, \ldots, \alpha_m) \mid \alpha_i \geq 0 \text{ for all } i \text{ and } \alpha_i = 0 \text{ for some } i\}$.

★26. With the notation of Exercise 25, $b = (1, \epsilon)$ for $0 < \epsilon < 1$, and $K(b) := C \cap A^{-1}(b)$, compute $P_{K(b)}(0)$, and compare this with $P_{K(b)}(0)$ when $\epsilon = 0$ or 1.

★27. Let $X = L_2[0,3]$, $C = \{x \in X \mid x \geq 0\}$, $x_i(t) = [1 - |t - i|]^+$ $(i = 1, 2)$, and define A on X by

$$Ax := (\langle x, x_1 \rangle, \langle x, x_2 \rangle), \quad x \in X.$$

Verify the following statements.
(a) A is a bounded linear operator from X onto $\ell_2(2)$.
(b) $A^* : \ell_2(2) \to X$ is given by

$$A^*y = y(1)x_1 + y(2)x_2, \quad y = (y(1), y(2)) \in \ell_2(2).$$

(c) $A(C) = \overline{A(C)} = \{(\alpha_1, \alpha_2) \mid \alpha_i \geq 0 \quad (i = 1, 2)\}$.
(d) $\operatorname{ri} A(C) = \operatorname{int} A(C) = \{(\alpha_1, \alpha_2) \mid \alpha_i > 0 \quad (i = 1, 2)\}$.
(e) $\operatorname{bd} A(C) = \{(\alpha_1, \alpha_2) \mid a_i \geq 0 \quad (i = 1, 2) \text{ and } \alpha_i = 0 \text{ for some } i\}$.

★28. Assume the hypothesis of Exercise 27, let $b = (1, 0)$, and let $K = K(b) = C \cap A^{-1}(b)$. Verify that if $x \in X$ satisfies $x \geq 0$ and $\int_0^1 tx(t)\, dt \leq 1$, then

$$P_K(x)(t) = [x(t) + \lambda_1 t]\chi_{[0,1]}(t),$$

where $\lambda_1 := 3\left[1 - \int_0^1 t\, x(t)\, dt\right]$.
[Hint: Theorem 10.50.]
    Using this fact, verify the following formulas:
(a) $P_K(0)(t) = 3t\chi_{[0,1]}(t)$.
(b) $P_K(t^\alpha)(t) = \left[t^\alpha + 3\left(\frac{\alpha+1}{\alpha+2}\right)t\right]\chi_{[0,1]}(t)$.

(c) $P_K(e^t)(t) = e^t \chi_{[0,1]}(t)$.

(d) $P_K(e^{t^2})(t) = \left[e^{t^2} + \dfrac{3}{2}(3 - e)t \right] \chi_{[0,1]}(t)$.

★29. Assume the hypothesis of Exercise 27, let $b = (1, 1)$, and $K = K(b) = C \cap A^{-1}(b)$. Verify that

$$P_K(0) = \frac{1}{2}(x_1 + x_2).$$

[Hint: Show that $b \in \operatorname{int} A(C)$ and apply Theorem 10.50.]

★30. Let $X = L_2[0, 1]$, $C = \{x \in X \mid x \geq 0\}$,

$$x_1 = \chi_{[0,1]}, \quad x_2 = \sum_{n=1}^{\infty} n \, \chi_{[2^{-n}, 2^{-n+1}]},$$

and define A on X by

$$Ax := (\langle x, x_1 \rangle, \langle x, x_2 \rangle), \quad x \in X.$$

Verify the following statements.

(a) A is a bounded linear operator from X onto $\ell_2(2)$.

(b) $A^* : \ell_2(s) \to X$ is given by

$$A^* y = y(1)x_1 + y(2)x_2, \quad y = (y(1), y(2)) \in \ell_2(2).$$

(c) $A(C) = \{(\alpha_1, \alpha_2) \mid \alpha_1 = \alpha_2 = 0 \text{ or } 0 < \alpha_1 \leq \alpha_2\}$.

(d) $A(C) = \{(\alpha_1, \alpha_2) \mid 0 < \alpha_1 \leq \alpha_2\}$. (In particular, $A(C)$ is not closed.)

(e) $\operatorname{ri} A(C) = \operatorname{int} A(C) = \{(\alpha_1, \alpha_2) \mid 0 < \alpha_1 < \alpha_2\}$.

(f) $\operatorname{bd} A(C) = \{(\alpha_1, \alpha_2) \mid 0 = \alpha_1 \leq \alpha_2 \text{ or } 0 < \alpha_1 = \alpha_2\}$.

★31. Assume the hypothesis of Exercise 30, let $b = (1, 1)$, and $K = C \cap A^{-1}(b)$. If $x \in X$ satisfies

$$x(t) + 2\left[1 - \int_{1/2}^{1} x(s)ds \right] \geq 0 \quad \text{for all} \quad t \in \left[\frac{1}{2}, 1 \right],$$

then show that

$$P_K(x) = [x + \lambda_0] \, \chi_{[\frac{1}{2}, 1]},$$

where

$$\lambda_0 := 2\left[1 - \int_{1/2}^{1} x(s)ds \right].$$

[Hint: Theorem 10.50.]

Using this fact, verify the following formulas.

(a) $P_K(0) = 2\chi_{[\frac{1}{2}, 1]}$.

(b) $P_K(t^\alpha)(t) = \left[t^\alpha + \frac{1 + \alpha 2^{\alpha+1}}{2^\alpha(1+\alpha)} \right] \chi_{[\frac{1}{2}, 1]}(t)$ for any $\alpha \geq 0$.

(c) $P_K(e^t)(t) = [e^t + 2(1 - e + \sqrt{e})]\chi_{[\frac{1}{2}, 1]}(t)$.

★32. In Example 10.53, it was shown that if $x_i(t) = [1 - |t - i|]^+$ $(i = 1, 2)$ and

$$K = \{x \in L_2[0,3] \mid x \geq 0, \ \langle x, x_1 \rangle = 1, \ \langle x, x_2 \rangle = 0\},$$

then

$$P_K(0)(t) = 3t \, \chi_{[0,1]}(t).$$

In this exercise you are asked to use Dykstra's algorithm (see the description following Example 10.53) to "compute" $P_K(0)$. Specifically, what is the result of applying Dykstra's algorithm to this problem after 10 iterations, 100 iterations, and 1000 iterations? (Here, of course, we are assuming that you have some familiarity with software like Mathematica or Maple to automate these computations.)

Historical Notes

The main problem studied in this chapter (see Example 10.1) evolved from the following "shape-preserving interpolation" problem. In the Sobolev space of real functions f on $[a, b]$ having absolutely continuous $(k - 1)$st derivative $f^{(k-1)}$ with $f^{(k)}$ in $L_2[a, b]$, find a solution of the problem to minimize

$$\left\{\int_a^b |f^{(k)}(t)|^2 \, dt \mid f^{(k)} \geq 0, \ f(t_i) = c_i \ (i = 1, 2, \ldots, n + k)\right\},$$

where $t_i \in [a, b]$ and $c_i \in \mathbb{R}$ are prescribed.

Numerous authors have studied this problem. For a survey of results up to 1986, see Ward (1986). Using an integration by parts technique, Favard (1940) and, more generally, de Boor (1976) showed that this probem has an equivalent reformulation as follows: Minimize

$$\{\|y\| \mid y \in L_2[a, b], \ y \geq 0, \ \langle y, x_i \rangle = b_i \ (i = 1, 2, \ldots, n)\},$$

where the functions $x_i \in L_2[a, b]$ and the numbers $b_i \in \mathbb{R}$ can be expressed in terms of the original data t_i and c_i.

The fundamental property, the strong CHIP, was introduced by Deutsch, Li, and Ward (1997) as a strengthening of the CHIP that had been studied earlier by Chui, Deutsch, and Ward (1990). Lemma 10.3, Examples 10.7, 10.9, and 10.10, Lemma 10.12, and Theorem 10.13 are from Deutsch, Li, and Ward (1997). Theorems 10.23 and 10.24 are taken from Rockafellar (1970; Theorems 6.1 and 6.2). The separation theorem (Theorem 10.25) is a special case of a result of Eidelheit (1936) who generalized a result of Mazur (1933; p. 73) (who had proved the special case where one of the convex sets is a singleton). The equivalence of (1)–(3) in Theorem 10.26 is from Rockafellar (1970; Theorem 6.4), while the equivalence of (1) and (4) is from Deutsch, Li, Ward (1997; Lemma 3.9). Lemma 10.28(1) (respectively Lemma 10.28(2)) is from Rockafellar (1970; Theorem 6.3) (respectively (1970; Corollary 6.3.1)). Theorem 10.29 is from Deutsch, Li, and Ward (1997; Theorem 3.11). Theorem 10.31 on dual cones of images can be found in Luenberger (1969). Theorem 10.32 is from Deutsch, Li, and Ward (1997; proof of Theorem 3.12). Special cases of Theorem 10.32 were known earlier. For example, when C is a closed convex cone,

$b \in \operatorname{int} A(C)$, and $x = 0$, see Micchelli and Utreras (1988). Also, when C is a closed convex cone, see Chui, Deutsch, and Ward (1992; Lemma 2.1).

The notion of an extreme point is due to Krein and Milman (1940), who proved the important result that any compact convex subset of a locally convex linear topological space is the closed convex hull of its set of extreme points. This result has many far-reaching ramifications.

Proposition 10.36 was suggested by Wu Li (1998; unpublished). The original definition of C_b given in Deutsch, Li, and Ward (1997) (and in Chui, Deutsch, and Ward (1992) for the case where C is a convex cone) was that C_b was the smallest convex extremal subset of C that contains $C \cap A^{-1}(b)$. Proposition 10.38 shows that the two definitions of C_b agree. The perturbation theorem (Theorem 10.40) (respectively Proposition 10.39) is from Deutsch, Li, and Ward (1997; Theorem 4.5 (respectively Proposition 4.6)).

The notion of a *generating set* in a Hilbert space, as well as Lemma 10.44, was given in Deutsch, Ubhaya, Ward, and Xu (1996).

The section on constrained interpolation by positive functions essentially follows Deutsch, Ubhaya, Ward, and Xu (1996) (who worked in the space $L_2[a, b]$). The main results of that section, Theorems 10.50 and 10.52, are both special cases of the following theorem.

10.54 Theorem. *Let $(T, \mathcal{S}, \mu)$ be a measure space and $L_2(\mu) = L_2(T, \mathcal{S}, \mu)$. Let $C = \{x \in L_2(\mu) \mid x \geq 0 \text{ a.e. } (\mu)\}$, $\{x_1, \ldots, x_m\} \subset L_2(\mu) \setminus \{0\}$, $b \in \ell_2(m)$, $Ax := (\langle x, x_1 \rangle, \ldots, \langle x, x_m \rangle)$ for all $x \in L_2(\mu)$, $K(b) := C \cap A^{-1}(b) \neq \emptyset$, and $S_b := \cup \{\operatorname{supp} x \mid x \in K(b)\}$. Then, for every $x \in L_2(\mu)$, there exists $y \in \ell_2(m)$ that satisfies*

$$(10.54.1) \qquad \left\langle \left[x + \sum_1^m y(i)x_i \right]^+ \chi_{S_b}, \; x_j \right\rangle = b(j) \quad (j = 1, \ldots, m).$$

Moreover,

$$(10.54.2) \qquad P_{K(b)}(x) = \left(x + \sum_1^m y(i)x_i \right)^+ \chi_{S_b}$$

for any $y \in \ell_2(m)$ that satisfies (10.54.1). Finally, the characteristic function χ_{S_b} may be deleted from both (10.54.1) and (10.54.2) if $\{x_1 \chi_{S_b}, \ldots, x_m \chi_{S_b}\}$ is linearly independent.

Theorem 10.52, excluding its last sentence, was first proved by Micchelli, Smith, Swetits, and Ward (1985) using variational methods. It was probably this paper, more than any other, that motivated the general study presented in this chapter. Other work related to the Micchelli, Smith, Swetits, and Ward paper (1985) include Illiev and Pollul (1984), (1984a), Dontchev, Illiev, and Konstantinov (1985), Irvine (1985), Irvine, Marin, and Smith (1986), Borwein and Wolkowicz (1986), Smith and Wolkowicz (1986), Anderson and Ivert (1987), Dontchev (1987), Smith and Ward (1988), Swetits, Weinstein, and Xu (1990), and Li, Pardalos, and Han (1992). A related question as to the *feasibility* of the problem, i.e., when is $C \cap A^{-1}(b) \neq \emptyset$, was studied by Mulansky and Neamtu (1998).

The problem of minimizing $\{\|x\| \mid x \in C, \; Ax = b\}$ can be easily cast as a certain optimization problem whose "optimal" solution can be studied via Fenchel duality. This was done by Rockafellar (1967), Massam (1979), Borwein and Wolkowicz

(1981), Gowda and Teboulle (1990), and Borwein and Lewis (1992), (1992a). Interpolation problems have also been considered from a control theory perspective, see Dontchev, Illiev, and Konstantinov (1985) and Dontchev (1987). A practical motivation for the study of such constrained problems arises in connection with L_2 spectral estimation (see Goodrich and Steinhardt (1986)).

Some of the results of this chapter have been extended and generalized. For example, results analogous to Theorem 10.52, in the case where C is the cone of increasing functions or the cone of convex functions, was given in Chui, Deutsch, and Ward (1992); more generally, results analogous to Theorem 10.52 when C is the cone of n-convex functions were established in Deutsch, Ubhaya, and Xu (1995) and Deutsch, Ubhaya, Ward, and Xu (1996). In this situation, it is necesary to know how to recognize best approximations from C. One noteworthy result along these lines is the following.

10.55 Theorem. Let C denote the set of nondecreasing functions on $[a, b]$, $f \in L_2[a, b]$, $f \notin C$, and $g \in C$. Then $g = P_C(f)$ if and only if g is equal (almost everywhere on $[a, b]$) to the derivative of the greatest convex minorant of the indefinite integral $s \mapsto \int_a^s f(t)\, dt$ of f.

Recall that the *greatest convex minorant* k of a function h is the largest convex function bounded above by h; that is,

$$k(t) := \sup\{\, c(t) \mid c \text{ convex on } [a, b], \ c \le h \,\} \quad \text{for all } t \in [a, b].$$

This result was established by Reid (1968) (in the case that f is bounded) using methods of optimal control, and by Deutsch, Ubhaya, and Xu (1995; Theorem 5.2) in the general case by the methods of this chapter.

Deutsch, Li, and Ward (2000) extended the main results of this chapter to include *inequality* constraints as well as equality ones. That is, they considered the approximating set to be of the form

$$K = C \cap \{x \in X \mid Ax \le b\}.$$

Li and Swetits (1997) developed an algorithm to handle inequality constraints in finite-dimensional spaces.

Deutsch (1998) showed that the strong CHIP is the *precise* condition that allowed a Karush–Kuhn–Tucker-type characterization of optimal solutions in convex optimization. Bauschke, Borwein, and Li (1999) studied the relationship between the strong CHIP, various kinds of "regularity," and Jameson's property (G).

It should be mentioned that while the Dykstra algorithm can be used for computing $P_{K(b)}(x)$, there are descent methods for directly solving the equation

$$A[P_C(x + A^*y)] = b$$

for $y \in \ell_2(m)$, and hence for computing $P_{K(b)}(x) = P_C(x + A^*y)$ when C is a polyhedron; see Deutsch, Li, and Ward (1997; section 6). Moreover, algorithms that apply to the problem considered here when C is a convex cone were developed in Irvine, Marin, and Smith (1986) and Micchelli and Utreras (1988).

CHAPTER 11

INTERPOLATION AND APPROXIMATION

Interpolation

In this chapter we first consider the general problem of finite interpolation. Then we study the related problems of simultaneous approximation and interpolation (SAI), simultaneous approximation and norm-preservation (SAN), simultaneous interpolation and norm-preservation (SIN), and simultaneous approximation and interpolation with norm-preservation (SAIN).

It is a well known basic fact in analysis that it is always possible to find a (unique) polynomial of degree $n - 1$ to interpolate any prescribed set of n real numbers at n prescribed points (see Corollary 11.4 below). That is, if $\{t_1, t_2, \ldots, t_n\}$ is a set of n distinct real numbers and $\{c_1, c_2, \ldots, c_n\}$ is any set of n real numbers, then there is a unique polynomial $y \in \mathcal{P}_{n-1}$ such that $y(t_i) = c_i$ $(i = 1, 2, \ldots, n)$.

Setting $a = \min_i t_i$, $b = \max_i t_i$, $X = C_2[a, b] \cap \mathcal{P}_{n-1}$, and defining the *point evaluation* functionals x_i^* on X by

$$x_i^*(x) := x(t_i), \qquad x \in X,$$

we see that X is n-dimensional, $x_i^* \in X^*$, and the problem of polynomial interpolation can be reworded as follows: For any given set of n real numbers c_i, there exists a unique $y \in X$ such that

$$x_i^*(y) = c_i \qquad (i = 1, 2, \ldots, n).$$

Thus polynomial interpolation is a special case of the **general problem of finite interpolation**: *Let M be an n-dimensional subspace of the inner product space X and let $\{x_1^*, x_2^*, \ldots, x_n^*\}$ be a set of n (not necessarily continuous) linear functionals on X. Given any set of n scalars $\{c_1, c_2, \ldots, c_n\}$ in $\mathbb{R}$, determine whether there exists an element $y \in M$ such that*

$$x_i^*(y) = c_i \qquad (i = 1, 2, \ldots, n).$$

The next theorem governs this situation. But first we state a definition.

11.1 Definition. *Let x, y in X and suppose Γ is a set of linear functionals on X. We say that y **interpolates** x relative to Γ if*

$$(11.1.1) \qquad x^*(y) = x^*(x) \quad \text{for every} \quad x^* \in \Gamma.$$

Note that if $\Gamma = \{x_1^*, x_2^*, \ldots, x_n^*\}$, then y interpolates x relative to Γ if and only if y interpolates x relative to span Γ. Now we describe when interpolation can always be carried out.

11.2 Characterization of When Finite Interpolation Is Possible. *Let M be an n-dimensional subspace of the inner product space X and let $x_1^*, x_2^*, \ldots, x_n^*$ be n (not necessarily bounded) linear functionals on X. Then the following statements are equivalent:*

(1) For each set of n scalars $\{c_1, c_2, \ldots, c_n\}$ in $\mathbb{R}$, there exists $y \in M$ such that

$$(11.2.1) \qquad\qquad x_i^*(y) = c_i \qquad (i = 1, 2, \ldots, n);$$

(2) For each set of n scalars $\{c_1, c_2, \ldots, c_n\}$ in $\mathbb{R}$, there exists a unique $y \in M$ such that (11.2.1) holds;

(3) For each $x \in X$, there exists a unique $y \in M$ that interpolates x relative to $\Gamma := \{ x_1^*, x_2^*, \ldots, x_n^* \}$; i.e.,

$$(11.2.2) \qquad\qquad x_i^*(y) = x_i^*(x) \qquad (i = 1, 2, \ldots, n);$$

(4) The only solution $y \in M$ to the equations

$$(11.2.3) \qquad\qquad x_i^*(y) = 0 \qquad (i = 1, 2, \ldots, n)$$

is $y = 0$;

(5) For every basis $\{x_1, x_2, \ldots, x_n\}$ of M,

$$(11.2.4) \qquad\qquad \det\,[x_i^*(x_j)] \neq 0;$$

(6) For some basis $\{x_1, x_2, \ldots, x_n\}$ of M, (11.2.4) holds;
(7) There exists a set $\{x_1, x_2, \ldots, x_n\}$ in M (necessarily a basis) such that

$$(11.2.5) \qquad\qquad x_i^*(x_j) = \delta_{ij} \qquad (i, j = 1, 2, \ldots, n);$$

(8) The set of restrictions $\{x_1^*|_M, x_2^*|_M, \ldots, x_n^*|_M\}$ is linearly independent (i.e., if $\sum_1^n \alpha_i x_i^*(y) = 0$ for all $y \in M$, then $\alpha_i = 0$ for $i = 1, 2, \ldots, n$).

Proof. (1) $\Longrightarrow$ (2). Suppose (1) holds but (2) fails. Then there exist scalars $c_1, c_2, \ldots, c_n$ and distinct elements y_1, y_2 in M such that $x_i^*(y_j) = c_i$ $(i = 1, 2, \ldots, n)$ for $j = 1, 2$. Then the nonzero vector $y = y_1 - y_2$ is in M and satisfies $x_i^*(y) = 0$ $(i = 1, 2, \ldots, n)$. Letting $\{x_1, x_2, \ldots, x_n\}$ be a basis for M, it follows that $y = \sum_1^n a_j x_j$ for some a_j, not all zero, and

$$\sum_{j=1}^n \alpha_j x_i^*(x_j) = 0 \qquad (i = 1, 2, \ldots, n).$$

It follows that $\det\,[x_i^*(x_j)] = 0$, and hence that there exist scalars $d_1, d_2, \ldots, d_n$ such that the system of equations

$$x_i^*\left(\sum_1^n \beta_j x_j\right) = \sum_{j=1}^n \beta_j x_i^*(x_j) = d_i \qquad (i = 1, 2, \ldots, n)$$

fails to have any solution $(\beta_1, \beta_2, \ldots, \beta_n)$. That is, there is *no* element $z \in M$ such that $x_i^*(z) = d_i$ $(i = 1, 2, \ldots, n)$. This contradicts (1), so that (2) must hold.

(2) $\Longrightarrow$ (3). This follows by taking $c_i = x_i^*(x)$ $(i = 1, 2, \ldots, n)$.

(3) $\implies$ (4). This follows by taking $x = 0$ in (3).

(4) $\implies$ (5). Let $\{x_1, x_2, \ldots, x_n\}$ be a basis of M. If (5) fails, then $\det [x_i^*(x_j)] = 0$, and the homogeneous system of equations $\sum_{j=1}^{m} \alpha_j x_i^*(x_j) = 0$ $(i = 1, 2, \ldots, n)$ has a nontrivial solution $(\alpha_1, \alpha_2, \ldots, \alpha_n)$. Thus the element $y = \sum_1^n \alpha_i x_i$ is in M, $y \neq 0$, and $x_i^*(y) = 0$ $(i = 1, 2, \ldots, n)$. This proves that the equations (11.2.3) have a nonzero solution in M, so that (4) fails.

(5) $\implies$ (6). This is obvious.

(6) $\implies$ (7). Let $\{x_1, x_2, \ldots, x_n\}$ be a basis for M such that (11.2.4) holds. Then, for each $k \in \{1, 2, \ldots, n\}$, the equations

$$\sum_{j=1}^{n} \alpha_{kj} x_i^*(x_j) = \delta_{ik} \qquad (i = 1, 2, \ldots, n)$$

have a unique solution $(\alpha_{k1}, \alpha_{k2}, \ldots, \alpha_{kn})$. Setting $y_k = \sum_{j=1}^n \alpha_{kj} x_j$, it follows that $y_k \in M$ and

$$x_i^*(y_k) = \delta_{ik} \qquad (i = 1, 2, \ldots, n).$$

To see that $\{y_1, y_2, \ldots, y_n\}$ is a basis for M, it suffices to show that $\{y_1, y_2, \ldots, y_n\}$ is linearly independent. If $\sum_1^n \alpha_i y_i = 0$, then for $j = 1, 2, \ldots, n$,

$$0 = x_j^* \left(\sum_1^n \alpha_i y_i \right) = \sum_{i=1}^{n} \alpha_i x_j^*(y_i) = \alpha_j.$$

(7) $\implies$ (8). Let $\{x_1, x_2, \ldots, x_n\}$ be a set in M such that (11.2.5) holds. If $\sum_1^n \alpha_i x_i^*(y) = 0$ for every $y \in M$, then in particular,

$$0 = \sum_{i=1}^{n} \alpha_i x_i^*(x_j) = \alpha_j \qquad (j = 1, 2, \ldots, n).$$

Thus $\{x_1^*|_M, x_2^*|_M, \ldots, x_n^*|_M\}$ is linearly independent.

(8) $\implies$ (1). Let $\{x_1, x_2, \ldots, x_n\}$ be a basis for M. If $\{x_1^*|_M, x_2^*|_M, \ldots, x_n^*|_M\}$ is linearly independent, then the only solution to the system of equations

$$\sum_{i=1}^{n} \alpha_i x_i^*(x_j) = 0 \qquad (j = 1, 2, \ldots, n)$$

is $(\alpha_1, \alpha_2, \ldots, \alpha_n) = (0, 0, \ldots, 0)$. Hence $\det [x_i^*(x_j)] \neq 0$. Thus if $\{c_1, c_2, \ldots, c_n\} \subset \mathbb{R}$, there is a (unique) solution $(\beta_1, \beta_2, \ldots, \beta_n)$ to the system of equations

$$\sum_{i=1}^{n} \beta_j x_i^*(x_j) = c_i \qquad (i = 1, 2, \ldots, n).$$

Then $y = \sum_1^n \beta_j x_j \in M$ satisfies (11.2.1). ∎

In particular, if X is n-dimensional, the following corollary is an immediate consequence of Theorem 11.2.

11.3 Corollary. *If X is an n-dimensional inner product space and $x_1^*, x_2^*, \ldots, x_n^*$ are n linear functionals on X, then the following statements are equivalent:*
(1) *For each set of n scalars $\{c_1, c_2, \ldots, c_n\}$ in $\mathbb{R}$, there exists an $x \in X$ such that*

$$(11.3.1) \qquad x_i^*(x) = c_i \qquad (i = 1, 2, \ldots, n);$$

(2) *For each set of n scalars $\{c_1, c_2, \ldots, c_n\}$ in $\mathbb{R}$, there exists a unique $x \in X$ such that (11.3.1) holds;*
(3) *The only solution $x \in X$ of the equations*

$$(11.3.2) \qquad x_i^*(x) = 0 \qquad (i = 1, 2, \ldots, n)$$

is $x = 0$;
(4) *For every basis $\{x_1, x_2, \ldots, x_n\}$ of X,*

$$(11.3.3) \qquad \det [x_i^*(x_j)] \neq 0;$$

(5) *For some basis $\{x_1, x_2, \ldots, x_n\}$ of X, (11.3.3) holds;*
(6) *There exists a set $\{x_1, x_2, \ldots, x_n\}$ in X (necessarily a basis) such that*

$$(11.3.4) \qquad x_i^*(x_j) = \delta_{ij} \qquad (i, j = 1, 2, \ldots, n);$$

(7) *$\{x_1^*, x_2^*, \ldots, x_n^*\}$ is linearly independent.*

As one application of this corollary, we prove the statement about polynomial interpolation made at the beginning of the chapter.

11.4 Polynomial Interpolation. *Let $t_1, t_2, \ldots, t_n$ be n distinct real numbers and $\{c_1, c_2, \ldots, c_n\}$ be any set of n real numbers. Then there is a unique polynomial $p \in \mathcal{P}_{n-1}$ of degree at most $n - 1$ such that*

$$(11.4.1) \qquad p(t_i) = c_i \qquad (i = 1, 2, \ldots, n).$$

In particular, if two polynomials in $\mathcal{P}_{n-1}$ coincide at n distinct points, they must be identical.

Proof. Let $[a, b]$ be any finite interval in $\mathbb{R}$ containing all the t_i's and let $X = C_2[a, b] \cap \mathcal{P}_{n-1}$. Then X is n-dimensional. Define n linear functionals x_i^* on X by

$$x_i^*(x) := x(t_i) \qquad (i = 1, 2, \ldots, n).$$

By Corollary 11.3, it suffices to show that X contains a set $\{x_1, x_2, \ldots, x_n\}$ such that $x_i(t_j) = \delta_{ij} \; (i, j = 1, 2, \ldots, n)$. Set

$$x_i(t) := \prod_{\substack{j=1 \\ j \neq i}}^{n} \frac{(t - t_j)}{(t_i - t_j)} \qquad (i = 1, 2, \ldots, n).$$

Clearly, $x_i \in \mathcal{P}_{n-1}$ and $x_i(t_j) = \delta_{ij}$. ∎

We should observe that interpolation can *fail* even if "the number of unknowns equals the number of conditions."

11.5 Example. Let $X = C_2[-1, 1]$ and $M = \text{span}\{x_1, x_2\}$, where $x_1(t) = 1$ and $x_2(t) = t^2$. Then, for any given pair of distinct points t_1, t_2 in $[-1, 1]$, define the linear functionals x_i^* on X by $x_i^*(x) = x(t_i)$ $(i = 1, 2)$. Clearly,

$$\det[x_i^*(x_j)] = \begin{vmatrix} 1 & 1 \\ t_1^2 & t_2^2 \end{vmatrix} = t_2^2 - t_1^2 = (t_2 - t_1)(t_2 + t_1).$$

In particular, $\det[x_i^*(x_j)] = 0$ if $t_2 = -t_1$, and so by Theorem 11.2, interpolation fails in general for this case. Specifically, if c_1 and c_2 are *distinct* real scalars, it is not possible to choose coefficients α_1, α_2 such that the polynomial $p(t) = \alpha_1 + \alpha_2 t^2$ satisfies $p(t_1) = c_1$ and $p(-t_1) = c_2$. This is also obvious from the fact that no matter how the coefficients α_i are chosen, $p(t) = \alpha_1 + \alpha_2 t^2$ is an *even* function and hence $p(-t) = p(t)$ for all t.

If any of the equivalent statements of Theorem 11.2 hold, then (by Theorem 11.2(8)) the set $\{x_1^*, x_2^*, \ldots, x_n^*\}$ must be linearly independent. There is an essential converse to this fact. Namely, if $\{x_1^*, x_2^*, \ldots, x_n^*\}$ is a linearly independent set of linear functionals on X, then there is an n-dimensional subspace M of X such that interpolation can be carried out in M. That is, all the statements of Theorem 11.2 are valid. This result was established earlier in Theorem 6.36, and we state it here again for convenience.

11.6 Interpolation Theorem. *Let $\{x_1^*, x_2^*, \ldots, x_n^*\}$ be a linearly independent set of linear functionals on X. Then there exists a set $\{x_1, \ldots, x_n\}$ in X such that*

$$(11.6.1) \qquad x_i^*(x_j) = \delta_{ij} \qquad (i, j = 1, 2, \ldots, n).$$

In particular, for each set of n real scalars $\{c_1, c_2, \ldots, c_n\}$, the element $y = \sum_1^n c_i x_i$ satisfies

$$(11.6.2) \qquad x_i^*(y) = c_i \qquad (i = 1, 2, \ldots, n).$$

11.7 Definition. *Let $x_1, \ldots, x_n$ be n elements in X and $x_1^*, \ldots, x_n^*$ be n linear functionals on X. The pair of sets $\{x_1, x_2, \ldots, x_n\}$ and $\{x_1^*, x_2^*, \ldots, x_n^*\}$ is called* **biorthogonal** *if*

$$(11.7.1) \qquad x_i^*(x_j) = \delta_{ij} \qquad (i, j = 1, 2, \ldots, n).$$

Theorem 11.6 shows that if $\{x_1^*, x_2^*, \ldots, x_n^*\}$ is linearly independent, there exists a set $\{x_1, x_2, \ldots, x_n\}$ in X such that the pair of sets is biorthogonal.

The usefulness of biorthogonal sets stems from the fact (Theorem 11.6) that one can easily construct solutions to interpolation problems. That is, if $\{x_1, \ldots, x_n\}$ and $\{x_1^*, \ldots, x_n^*\}$ are biorthogonal, then the linear combination $y = \sum_1^n c_i x_i$ solves the interpolation problem

$$x_i^*(y) = c_i \qquad (i = 1, 2, \ldots, n).$$

11.8 Examples of Biorthogonal Sets.

(1) (**Lagrange interpolation**) Let $a \leq t_1 < t_2 < \cdots < t_n \leq b$, $w(t) = (t - t_1)(t - t_2) \cdots (t - t_n)$, and

$$x_j(t) = \prod_{\substack{i=1 \\ i \neq j}}^{n} \frac{t - t_i}{t_j - t_i} = \frac{w(t)}{(t - t_j)w'(t_j)} \qquad (j = 1, 2, \ldots, n).$$

Then $x_j \in \mathcal{P}_{n-1}$, and defining x_i^* on $C_2[a, b]$ by

$$x_i^*(x) := x(t_i) \qquad (i = 1, 2, \ldots, n),$$

we see that

$$x_i^*(x_j) = x_j(t_i) = \delta_{ij} \qquad (i, j = 1, 2, \ldots, n).$$

That is, $\{x_1, x_2, \ldots, x_n\}$ and $\{x_1^*, x_2^*, \ldots, x_n^*\}$ are biorthogonal. Thus the polynomial $p = \sum_1^n c_i x_i \in \mathcal{P}_{n-1}$ satisfies $p(t_i) = c_i$ $(i = 1, 2, \ldots, n)$.

(2) (**Taylor interpolation**) In the inner product space $X = \{x \in C_2[-1, 1] \mid x^{(n-1)}$ exists on $[-1, 1]\}$, consider the polynomials $x_i(t) = t^{i-1}/(i - 1)!$ and linear functionals x_i^* on X defined by $x_i^*(x) := x^{(i-1)}(0)$ $(i = 1, 2, \ldots, n)$. It is easy to verify that the sets $\{x_1, x_2, \ldots, x_n\}$ and $\{x_1^*, x_2^*, \ldots, x_n^*\}$ are biorthogonal. Hence the polynomial $p = \sum_1^n c_i x_i \in \mathcal{P}_{n-1}$ satisfies

$$p^{(i-1)}(0) = c_i \qquad (i = 1, 2, \ldots, n).$$

Additional examples of biorthogonal sets are given in the exercises (namely, Exercises 11.14.5–11.14.7).

Simultaneous Approximation and Interpolation

If Y and Z are subsets of X, we recall that Y is **dense** in Z if for each $z \in Z$ and $\epsilon > 0$, there exists $y \in Y$ such that $\|z - y\| < \epsilon$. That is, $\overline{Y} \supset Z$.

In modeling the behavior of a system, we often try to approximate the system by a simpler one that preserves certain characteristics of the original system. For example, the next result shows that *it is always possible to simultaneously approximate and interpolate from a dense subspace.*

11.9 Simultaneous Approximation and Interpolation (SAI). *Let Y be a dense subspace of X and let $\{x_1^*, x_2^*, \ldots, x_n^*\} \subset X^*$. Then, for every $x \in X$ and $\epsilon > 0$, there exists $y \in Y$ such that*

$$\text{(11.9.1)} \qquad \qquad \|x - y\| < \epsilon$$

and

$$\text{(11.9.2)} \qquad x_i^*(y) = x_i^*(x) \qquad (i = 1, 2, \ldots, n).$$

Proof. It is no loss of generality to assume that $\{x_1^*, x_2^*, \ldots, x_n^*\}$ is linearly independent. Further, we may even assume that the set of restrictions $\{x_1^*|_Y, x_2^*|_Y, \ldots, x_n^*|_Y\}$ is linearly independent. [For if $\sum_1^n \alpha_i x_i^*(y) = 0$ for every

$y \in Y$, then since Y is dense in X, $\sum_1^n \alpha_i x_i^*(x) = 0$ for all $x \in X$; that is, $\sum_1^n \alpha_i x_i^* = 0$. By linear independence of the x_i^*'s, $\alpha_1 = \alpha_2 = \cdots = \alpha_n = 0$.] The linear functionals $y_i^* = x_i^*|_Y$ $(i = 1, 2, \ldots, n)$ thus form a linearly independent set in Y^*. By Theorem 11.6, there exists a set $\{x_1, x_2, \ldots, x_n\}$ in Y such that

$$x_i^*(x_j) = y_i^*(x_j) = \delta_{ij} \quad (i, j = 1, 2, \ldots, n).$$

Let $c = \sum_1^n \|x_j^*\| \, \|x_j\|$ and $\epsilon > 0$ be given. Choose $y_1 \in Y$ such that $\|x - y_1\| < \epsilon(1 + c)^{-1}$. Set $y_2 = \sum_1^n x_j^*(x - y_1)x_j$ and $y = y_1 + y_2$. Then $y \in Y$,

$$x_i^*(y) = x_i^*(y_1) + x_i^*(y_2) = x_i^*(y_1) + x_i^*(x - y_1) = x_i^*(x) \quad (i = 1, 2, \ldots, n),$$

and

$$\|x - y\| \le \|x - y_1\| + \|y_2\| < \epsilon(1 + c)^{-1} + \sum_1^n |x_j^*(x - y_1)| \, \|x_j\|$$

$$\le \epsilon(1 + c)^{-1} + \epsilon(1 + c)^{-1} \sum_1^n \|x_j^*\| \, \|x_j\| = \epsilon. \quad \blacksquare$$

This theorem implies that any result that concludes that a certain subspace is dense in X is actually a result about simultaneous approximation *and* interpolation, and not just approximation alone. As an application, we consider the following.

11.10 Corollary (Approximation by polynomials with matching moments). Let $x_1, x_2, \ldots, x_n$ be in $C_2[a, b]$. Then for each $x \in C_2[a, b]$ and $\epsilon > 0$, there exists a polynomial p such that

$$(11.10.1) \qquad\qquad \|x - p\| < \epsilon$$

and

$$(11.10.2) \qquad \int_a^b p(t)x_i(t)dt = \int_a^b x(t)x_i(t)dt \quad (i = 1, 2, \ldots, n).$$

Proof. Let $X = C_2[a, b]$, Y be the subspace of all polynomials, and define x_i^* on X by

$$x_i^*(z) := \int_a^b z(t)x_i(t)dt \quad (i = 1, 2, \ldots, n).$$

Then Y is dense by the Weierstrass approximation theorem (Corollary 7.20), and $x_i^* \in X^*$ by Theorem 5.18. The result now follows by Theorem 11.9. $\blacksquare$

In some problems (see, e.g., Problem 4 of Chapter 1), the energy of a system is proportional to $\int_a^b x^2(t)dt$ for a certain function x. Using this terminology, the next result shows that it is always possible to approximate from a dense subspace while simultaneously preserving energy.

11.11 Simultaneous Approximation and Norm-preservation (SAN).
Let Y be a dense subspace of X. Then, for each $x \in X$ and $\epsilon > 0$, there exists $y \in Y$ such that

(11.11.1) $\|x - y\| < \epsilon$

and

(11.11.2) $\|y\| = \|x\|$.

Proof. If $x = 0$, take $y = 0$. If $x \neq 0$, choose $z \in Y \backslash \{0\}$ such that $\|x - z\| < \epsilon/2$. Then $y = \|x\|(\|z\|^{-1})z \in Y$, $\|y\| = \|x\|$, and

$$\|x - y\| = \|x - \|x\|(\|z\|^{-1})z\| \leq \|x - z\| + \|z - \|x\|(\|z\|^{-1})z\|$$
$$\leq \|x - z\| + |\|z\| - \|x\|| \leq 2\|x - z\| < \epsilon. \quad \blacksquare$$

Simultaneous Approximation, Interpolation, and Norm-preservation

In contrast to Theorems 11.9 and 11.11, it is *not* always possible to simultaneously interpolate and preserve norm (SIN), and a fortiori, it is not always possible to simultaneously approximate, interpolate, and preserve norm (SAIN). These two latter properties turn out to be equivalent and can be characterized. First we state an essential lemma.

11.12 Lemma. Let Y be a dense subspace of X, $\{x_1^*, x_2^*, \ldots, x_n^*\}$ a subset of X^*, and

$$Z = \bigcap_1^n \{x \in X \mid x_i^*(x) = 0\}.$$

Then $Y \cap Z$ is dense in Z.

Proof. By Theorem 11.9, for each $z \in Z$ and $\epsilon > 0$, there exists $y \in Y$ such that $\|z - y\| < \epsilon$ and $x_i^*(y) = x_i^*(z) = 0$ for $i = 1, 2, \ldots, n$. Thus $y \in Y \cap Z$ and $\|z - y\| < \epsilon$. $\quad \blacksquare$

11.13 Theorem. Let Y be a dense subspace of the Hilbert space X and let $\{x_1^*, x_2^*, \ldots, x_n^*\}$ be a subset of X^*. Then the following four statements are equivalent:

(1) (SAIN) For each $x \in X$ and $\epsilon > 0$, there exists $y \in Y$ such that

(11.13.1) $\|x - y\| < \epsilon$,

(11.13.2) $x_i^*(y) = x_i^*(x) \quad (i = 1, 2, \ldots, n)$,

and

(11.13.3) $\|y\| = \|x\|$;

(2) (SIN) For each $x \in X$, there exists $y \in Y$ such that (11.13.2) and (11.13.3) hold;

(3) Each x_i^* attains its norm at a point in Y;

(4) Each x_i^* has its representer in Y.

Proof. First observe that since X is a Hilbert space, Theorem 6.10 implies that each x_i^* has a representer $x_i \in X$. Further, it is no loss of generality to assume that $\{x_1, x_2, \ldots, x_n\}$ is linearly independent.

The implication (1) $\implies$ (2) is trivial.

(2) $\implies$ (3). Assume (2) holds, fix any $i \in \{1, 2, \ldots, n\}$, and let $x = x_i / \|x_i\|$. Choose $y \in Y$ such that $x_i^*(y) = x_i^*(x)$ and $\|y\| = \|x\| = 1$. Then

$$x_i^*(y) = x_i^*(x) = \langle x, x_i \rangle = \|x_i\| = \|x_i^*\|,$$

so x_i^* attains its norm at y.

(3) $\iff$ (4). This is just Theorem 6.12.

(4) $\implies$ (1). If (4) holds, then the representers x_i are all in Y. We must show that for each $x \in X$ and $\epsilon > 0$, there is a $y \in Y$ such that $\|x - y\| < \epsilon$, $\langle y, x_i \rangle = \langle x, x_i \rangle$ $(i = 1, 2, \ldots, n)$, and $\|y\| = \|x\|$. Let $\{y_1, y_2, \ldots, y_n\}$ be an orthonormal basis for $M := \mathrm{span}\{x_1, x_2, \ldots, x_n\}$. Now,

$$M^\perp = \bigcap_1^n \{y \in X \mid \langle y, y_i \rangle = 0\}$$

and $X = M \oplus M^\perp$ by Theorem 5.9. Thus we can write

$$x = \sum_1^n \alpha_i y_i + z$$

for some $z \in M^\perp$ and $\alpha_i = \langle x, y_i \rangle$. Since $Y \cap M^\perp$ is dense in $M^\perp$ by Lemma 11.12, we can use Theorem 11.11 to obtain $w \in Y \cap M^\perp$ such that

$$\|w - z\| < \epsilon \text{ and } \|w\| = \|z\|.$$

Setting $y = \sum_1^n \alpha_i y_i + w$, we see that $y \in Y$, $\|x - y\| = \|z - w\| < \epsilon$,

$$\langle y, y_i \rangle = \alpha_i = \langle x, y_i \rangle \qquad (i = 1, 2, \ldots, n),$$

and

$$\|y\|^2 = \left\| \sum_1^n \alpha_i y_i \right\|^2 + \|w\|^2 = \left\| \sum_1^n \alpha_i y_i \right\|^2 + \|z\|^2 = \|x\|^2.$$

Thus (1) holds. ∎

Exercises

1. (a) Let X be n-dimensional, $\{x_1, x_2, \ldots, x_n\}$ be in X, and $\{x_1^* x_2^*, \ldots, x_n^*\}$ be in X^*. Show that any two of the following conditions implies the third:
   (i) $\{x_1, x_2, \ldots, x_n\}$ is linearly independent;
   (ii) $\{x_1^*, x_2^*, \ldots, x_n^*\}$ is linearly independent;
   (iii) $\det[x_i^*(x_j)] \neq 0$.

   (b) More generally than (a), let $\{x_1, x_2, \ldots, x_n\}$ be in X, let M denote span$\{x_1, x_2, \ldots, x_n\}$, and let $\{x_1^*, x_2^*, \ldots, x_n^*\}$ be in X^*. Show that any two of the following conditions implies the third:

(i) $\{x_1, x_2, \ldots, x_n\}$ is linearly independent;

(ii) $\{x_1^*|_M, x_2^*|_M, \ldots, x_n^*|_M\}$ is linearly independent;

(iii) $\det[x_i^*(x_j)] \neq 0$.

2. Exercise 1 can be used to construct a basis for X^* if a basis for the n-dimensional space X is given; it suffices to choose a set of linear functionals $\{x_1^*, x_2^*, \ldots, x_n^*\}$ on X such that $\det[x_i^*(x_j)] \neq 0$. As an application, let $X = $ span$\{x_1, x_2, \ldots, x_6\}$, where the functions x_i are defined on $[-1, 1] \times [-1, 1]$ by $x_1(s, t) = 1$, $x_2(s, t) = s$, $x_3(s, t) = t$, $x_4(s, t) = s^2$, $x_5(s, t) = st$, and $x_6(s, t) = t^2$. Find a basis for X^*.

3. Let X be the 6-dimensional space of functions defined as in Exercise 2. Show that it is *not* always possible to find an $x \in X$ that assumes arbitrarily prescribed values at six distinct points $z_i = (s_i, t_i)$ in $[-1, 1] \times [-1, 1]$.

4. True or False? For each of the statements below, determine whether it is possible to construct a polynomial $p \in \mathcal{P}_2$ satisfying the stated conditions. Determine also whether the solution is unique when it exists. Justify your answers.

(a) $p(0) = 1$, $p'(1) = 2$, $p''(2) = -1$.

(b) $p(0) = 0$, $\int_0^1 p(t)dt = 1$, $\int_0^1 t\, p(t)dt = 2$.

(c) $p(0) = 1$, $p''(0) = 0$, $p'''(0) = -1$.

(d) $p(0) = 1$, $p''(0) = 0$, $p'''(0) = 0$.

5. Show that if t_1 and t_2 are distinct real numbers, then the functions $x_1(t) := 1$, $x_2(t) := t$ and the linear functionals

$$x_1^*(x) := \frac{t_2 x(t_1) - t_1 x(t_2)}{t_2 - t_1}, \quad x_2^*(x) := \frac{x(t_2) - x(t_1)}{t_2 - t_1}$$

are biorthogonal.

6. (**Hermite or osculatory interpolation**). Let $X = \mathcal{P}_{2n-1}$ (regarded as a subspace of $C_2[a, b]$) and $a \leq t_1 \leq t_2 \leq \cdots \leq t_n \leq b$.

(a) Show that for any set of $2n$ real numbers $\{c_1, \ldots, c_n, d_1, \ldots, d_n\}$, there exists a unique $p \in X$ such that

$$p(t_i) = c_i, \; p'(t_i) = d_i \quad (i = 1, 2, \ldots, n).$$

(b) Verify that the sets of functions $\{x_1, x_2, \ldots, x_{2n}\}$ and functionals $\{x_1^*, x_2^*, \ldots, x_{2n}^*\}$ are biorthogonal, where

$$w(t) = \prod_1^n (t - t_i), \quad l_k(t) = \frac{w(t)}{(t - t_k)w'(t_k)},$$

$$x_k(t) = [1 - \frac{w''(t_k)}{w'(t_k)}(t - t_k)]l_k^2(t), \quad x_{n+k}(t) = (t - t_k)l_k^2(t),$$

$$x_k^*(x) := x(t_k), \quad x_{n+k}^*(x) := x'(t_k) \quad (k = 1, 2, \ldots, n).$$

(c) Construct the polynomial $p \in \mathcal{P}_{2n-1}$ that solves part (a). [Hint: Part (b).]

7. Let $[a, b]$ be any interval with $b - a < 2\pi$. The set of **trigonometric polynomials** of degree n is the subspace $\mathcal{T}_n$ of $C_2[a, b]$ defined by

$$\mathcal{T}_n = \operatorname{span}\{1, \cos t, \cos 2t, \ldots, \cos nt, \sin t, \sin 2t, \ldots, \sin nt\}.$$

(a) Show that if $a \le t_1 < t_2 < \cdots < t_{2n+1} \le b$, the functions $\{x_1, x_2, \ldots, x_{2n+1}\}$ and the functionals $\{x_1^*, x_2^*, \ldots, x_{2n+1}^*\}$ are biorthogonal, where

$$x_j(t) = \prod_{\substack{k=1 \\ k \ne j}}^{2n+1} \left[\frac{\sin \frac{1}{2}(t - t_k)}{\sin \frac{1}{2}(t_j - t_k)} \right] \quad \text{and} \quad x_j^*(x) := x(t_j).$$

Of course, it must also be verified that $x_j \in \mathcal{T}_n$. [Hint: Induction on n.]
(b) Verify that $\dim \mathcal{T}_n = 2n + 1$.
(c) Show that for any set of $2n + 1$ real scalars c_i, there exists a unique $x \in \mathcal{T}_n$ such that

$$x(t_i) = c_i \qquad (i = 1, 2, \ldots, 2n + 1).$$

Construct x explicitly.

8. Given any n real numbers t_i, define the n-th order **Vandermonde determinant** by

$$V_n(t_1, t_2, \ldots, t_n) := \begin{vmatrix} 1 & 1 & \ldots & 1 \\ t_1 & t_2 & \ldots & t_n \\ \ldots & & & \\ t_1^{n-1} & t_2^{n-1} & \ldots & t_n^{n-1} \end{vmatrix}.$$

(a) Prove that

$$V_n(t_1, t_2, \ldots, t_n) = \prod_{1 \le j < i \le n} (t_i - t_j).$$

[Hint: Induct on n using the fact that $V_{n+1}(t_1, t_2, \ldots, t_n, t)$ is a polynomial of degree n in t that vanishes for $t = t_i$ $(i = 1, 2, \ldots, n)$.]
(b) Give another proof of Theorem 11.4 based on Theorem 11.2 and part (a).

9. (a) Prove the following "converse" to Theorem 11.2. Let $\{x_1^*, x_2^*, \ldots, x_n^*\}$ be n linear functionals on X. Then the following statements are equivalent.
   (i) $\{x_1^*, x_2^*, \ldots, x_n^*\}$ is linearly independent;
   (ii) There exist n vectors $x_1, x_2, \ldots, x_n$ in X such that $x_i^*(x_j) = \delta_{ij}$ $(i, j = 1, 2, \ldots, n)$;
   (iii) For each set of n real scalars $\{c_1, c_2, \ldots, c_n\}$, there exists $y \in X$ such that $x_i^*(y) = c_i$ $(i = 1, 2, \ldots, n)$;
   (iv) There exists an n-dimensional subspace M of X such that $\{x_1^*|_M, x_2^*|_M, \ldots, x_n^*|_M\}$ is linearly independent.
(b) When is the element y of (iii) in part (a) unique?
[Hint: When does $\cap_1^n \{x \in X. \mid x_i^*(x) = 0\}$ consist of only the zero element?]

10. If $\{x_1, x_2, \ldots, x_n\}$ is linearly independent in X, then there exists $\{x_1^*, x_2^*, \ldots, x_n^*\}$ in X^* such that $x_i^*(x_j) = \delta_{ij}$ $(i, j = 1, 2, \ldots, n)$.
   [Hint: Regard the x_i as linear functionals $\tilde{x}_i$ on X^* via the definition $\tilde{x}_i(x^*) := x^*(x_i)$ for every $x^* \in X^*$. Apply Theorem 11.6.]

11. Let $\{x^*, x_1^*, \ldots, x_n^*\}$ be in X^*. Show that $x^* \in \operatorname{con}\{x_1^*, x_2^*, \ldots, x_n^*\}$ if and only if $x^*(x) \geq 0$ whenever $x_i^*(x) \geq 0$ for all i. (Recall that $\operatorname{con}(A)$ denotes the conical hull of A: the set of all nonnegative linear combinations of elements of A.)

12. Show that Theorem 11.13 is also valid in a *complex* Hilbert space.

13. There are several results in this chapter that are valid in *any* (real or complex) *normed linear space*. Verify that each of the following is of this type.
    (a) Theorem 11.2, where the scalars in (2) and (3) are in $\mathbb{C}$.
    (b) Corollary 11.3, where the scalars in (1) and (2) are in $\mathbb{C}$.
    (c) Theorem 11.6, where the scalars c_i are in $\mathbb{C}$.
    (d) Theorem 11.9.
    (e) Theorem 11.11.
    (f) Lemma 11.12.
    (g) Exercise 11.14.1.
    (h) Exercise 11.14.9.
    (i) Exercise 11.14.10.

Historical Notes

The classical results concerning polynomial interpolation, Taylor's theorem, divided differences, etc., can be found in the books of Davis (1963), Walsh (1956), and Steffensen (1950). For a more modern approach to polynomial interpolation of functions, with the emphasis on what happens when the number of interpolation points becomes infinitely large, see Szabados and Vértesi (1990). The equivalence of (1), (5), and (7) of Corollary 11.3 can be found in Davis (1963; p. 26). The first example of a "SAI" (simultaneous approximation and interpolation) theorem seems to be that of Walsh in 1928 (see Walsh (1956; p. 310)). He proved that a continuous function f on a compact subset of the complex plane can be uniformly approximated by polynomials that also interpolate f on a prescribed finite subset of the domain. The abstract SAI Theorem 11.9, under the more general normed linear space setting with Y dense and convex (and not necessarily a subspace), was established by Yamabe (1950). Singer (1959) further generalized Yamabe's theorem to a linear topological space setting. Unaware of Singer's result, Deutsch (1966) generalized Theorem 11.9 to the linear topological space setting.

Theorem 11.11 is a special case of a result of Deutsch and Morris (1969; Theorem 3.1). The first "SAIN" (simultaneous approximation and interpolation with norm-preservation) theorem was given by Wolibner (1951), who proved the following. Let $a \leq t_1 < t_2 < \cdots < t_n \leq b$ and let f be a real continuous function on $[a, b]$. Then for each $\epsilon > 0$, there exists a polynomial p such that $\|f - p\|_\infty < \epsilon$, $p(t_i) = f(t_i)$ $(i = 1, 2, \ldots, n)$, and $\|p\|_\infty = \|f\|_\infty$. (Here $\|g\|_\infty := \max\{|g(t)| \mid t \in [a, b]\}$.) Paszkowski (1957) proved the first "SIN" (simultaneous interpolation and norm-preservation) theorem when he proved the following. If f is a continuous function on an interval $[a, b]$ and $a \leq t_1 < t_2 < \cdots < t_n \leq b$, then there exists a polynomial p such that $p(t_i) = f(t_i)$ for $i = 1, 2, \ldots, n$ and $\|p\|_\infty = \|f\|_\infty$. (In fact, he even showed that the *degree* of the polynomial that works is *independent* of the function, but depends only on the points t_i!) The equivalence of (1), (3), and (4) of Theorem 11.13 was proved by Deutsch and Morris (1969). The equivalence of (1) and (2) in Theorem 11.13 was established by Deutsch (1985). Results of SAIN (respectively SIN) type, in a more general normed linear space setting, can be found in Deutsch and Morris (1969) (respectively Deutsch (1985)).

CHAPTER 12

CONVEXITY OF CHEBYSHEV SETS

Is Every Chebyshev Set Convex?

In Theorem 3.5 we saw that every closed convex subset of a Hilbert space is a Chebyshev set. In this chapter we will study the converse problem of whether or not every Chebyshev subset of a Hilbert space must be convex.

The main result is this: *Every boundedly compact Chebyshev set must be convex* (Theorem 12.6). In particular, *every Chebyshev subset of a finite-dimensional Hilbert space is convex*. Whether this result is also true when the Hilbert space is infinite-dimensional is perhaps the major unsolved problem in (abstract) approximation theory today.

It is convenient to introduce notation for line segments or intervals in X. The **line segment** or **interval** joining the two points x and y in X is the set

$$[x, y] := \{\lambda x + (1 - \lambda)y \mid 0 \le \lambda \le 1\}.$$

That is, $[x, y] = \text{co}\{x, y\}$ is the convex hull of x and y.

Our first observation is that every point on the line segment joining x to its best approximation $P_K(x)$ also has $P_K(x)$ as its best approximation (see Figure 12.0.1).

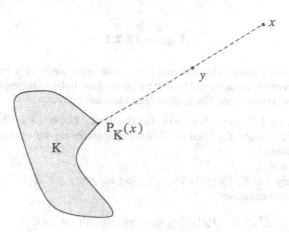

Figure 12.0.1

12.1 Lemma. *Let K be a Chebyshev set in X and $x \in X$. Then $P_K(y) = P_K(x)$ for every $y \in [x, P_K(x)]$.*

Proof. If the result were false, then there would exist $y \in [x, P_K(x)]$ and $z \in K$ such that $\|y - z\| < \|y - P_K(x)\|$. We can write $y = \lambda x + (1 - \lambda)P_K(x)$ for some

$0 < \lambda < 1$. Then

$$\|x - z\| \leq \|x - y\| + \|y - z\| < \|x - y\| + \|y - P_K(x)\|$$
$$= (1 - \lambda)\|x - P_K(x)\| + \lambda\|x - P_K(x)\| = \|x - P_K(x)\| = d(x, K),$$

which is absurd. ∎

Chebyshev Suns

A concept that will prove useful in studying Chebyshev sets is that of a "sun."

12.2 Definition. *A Chebyshev set K in X is called a* **sun** *if for each $x \in X$,*

$$P_K[x + \lambda(x - P_K(x))] = P_K(x) \qquad \text{for every } \lambda \geq 0.$$

In other words (using Lemma 12.1), K is a sun if and only if each point on the ray from $P_K(x)$ through x also has $P_K(x)$ as its best approximation in K (see Figure 12.2.1).

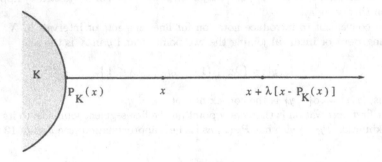

Figure 12.2.1

The next result shows that not only is a sun equivalent to a convex set, but that the contractive property of the metric projection onto a convex Chebyshev set actually characterizes those Chebyshev sets that are convex.

12.3 Suns and Convex Sets are Equivalent. *Let K be a Chebyshev set in an inner product space X. Then the following statements are equivalent:*
(1) *K is convex;*
(2) *K is a sun;*
(3) *For every $x \in X$, $P_K(x) = P_{[y, P_K(x)]}(x)$ for every $y \in K$;*
(4) *P_K is nonexpansive:*

(12.3.1) $\|P_K(x) - P_K(y)\| \leq \|x - y\|$ *for all* $x, y \in X$.

Proof. (1) $\Rightarrow$ (2). If K is not a sun, then there exist $x \in X$ and $\lambda > 0$ such that $z := x + \lambda[x - P_K(x)]$ satisfies $P_K(z) \neq P_K(x)$. Thus there exists $y \in K$ such that $\|z - y\| < \|z - P_K(x)\|$. Substituting for z and dividing both sides of this inequality by $1 + \lambda$, we obtain

$$\left\| x - \left[\frac{\lambda}{1 + \lambda} P_K(x) + \frac{1}{1 + \lambda} y \right] \right\| < \|x - P_K(x)\|.$$

This proves that K cannot be convex, since otherwise $\lambda(1+\lambda)^{-1}P_K(x)+(1+\lambda)^{-1}y$ would be an element of K closer to x than $P_K(x)$.

(2) $\Rightarrow$ (3). If (3) fails, there exist $x \in X$ and $y \in K$ such that $P_K(x) \neq P_{[y,P_K(x)]}(x)$. Thus there exists $0 < \lambda \le 1$ such that $z := \lambda y + (1-\lambda)P_K(x) = P_{[y,P_K(x)]}(x)$ and

$$\|x - z\| < \|x - P_K(x)\|.$$

Dividing by λ, we get

$$\left\| x + \frac{1-\lambda}{\lambda}[x - P_K(x)] - y \right\| < \left\| \frac{1}{\lambda}x - \frac{1}{\lambda}P_K(x) \right\|$$
$$= \left\| x + \frac{1-\lambda}{\lambda}[x - P_K(x)] - P_K(x) \right\|.$$

This proves that y is a better approximation to $x + (1-\lambda)\lambda^{-1}[x - P_K(x)]$ than $P_K(x)$. Hence K is not a sun.

(3) $\Rightarrow$ (4). Assume (3) holds and let x and y be in X. Since both $P_K(x) = P_{[P_K(x),P_K(y)]}(x)$ and $P_K(y) = P_{[P_K(x),P_K(y)]}(y)$, then by the characterization theorem (Theorem 4.1) (applied to the convex set $[P_K(x), P_K(y)]$), we deduce $\langle x - P_K(x), P_K(y) - P_K(x) \rangle \le 0$ and $\langle y - P_K(y), P_K(x) - P_K(y) \rangle \le 0$. Equivalently,

$$\langle x - P_K(x), P_K(x) - P_K(y) \rangle \ge 0 \quad \text{and} \quad \langle P_K(y) - y, P_K(x) - P_K(y) \rangle \ge 0.$$

Adding these, we deduce that

$$0 \le \langle x - y + P_K(y) - P_K(x), P_K(x) - P_K(y) \rangle$$
$$= \langle x - y, P_K(x) - P_K(y) \rangle - \|P_K(x) - P_K(y)\|^2$$
$$\le \|x - y\| \, \|P_K(x) - P_K(y)\| - \|P_K(x) - P_K(y)\|^2,$$

which proves (4).

(4) $\Rightarrow$ (1). If K is not convex, then there exist points x and y in K such that $z := \frac{1}{2}(x+y) \notin K$ (see Exercise 12.8.1). Next we observe that

$$(12.3.2) \qquad \max\{\|x - P_K(z)\|, \|y - P_K(z)\|\} > \|x - y\|/2.$$

For otherwise, $\|x - P_K(z)\| \le \|x - y\|/2$ and $\|y - P_K(z)\| \le \|x - y\|/2$. Then

$$\|x - y\| \le \|x - P_K(z)\| + \|P_K(z) - y\| \le \|x - y\|$$

implies that equality must hold for all these inequalities. In particular,

$$\|[x - P_K(z)] + [P_K(z) - y]\| = \|x - P_K(z)\| + \|P_K(z) - y\|.$$

By the condition of equality in the triangle inequality, we must have $x - P_K(z) = \rho[P_K(z) - y]$ for some $\rho \ge 0$. Since $\|x - P_K(z)\| = \|x - y\|/2 = \|y - P_K(z)\|$, it follows that $\rho = 1$ and

$$P_K(z) = \frac{1}{2}(x + y) = z \notin K,$$

which is absurd. Thus (12.3.2) holds.

Without loss of generality, we may assume that $\|x - P_K(z)\| > \|x - y\|/2$. But since $x \in K$, we have that $P_K(x) = x$ and $\|x - y\|/2 = \|x - z\|$ implies that

$$\|P_K(x) - P_K(z)\| > \|x - z\|.$$

That is, P_K is not nonexpansive. ∎

The metric projection onto an approximately compact Chebyshev set is always continuous.

12.4 Lemma. *If K is an approximatively compact Chebyshev set, then P_K is continuous.*

Proof. If not, there is a point $x \in X$ and a sequence $\{x_n\}$ converging to x such that

$$(12.4.1) \qquad \|P_K(x_n) - P_K(x)\| \geq \epsilon > 0 \quad \text{for every } n.$$

Now by Theorem 5.3,

$$d(x, K) \leq \|x - P_K(x_n)\| \leq \|x - x_n\| + \|x_n - P_K(x_n)\|$$
$$= \|x - x_n\| + d(x_n, K) \to d(x, K).$$

Thus $\{P_K(x_n)\}$ is a minimizing sequence. By approximative compactness, there is a subsequence $\{P_K(x_{n_i})\}$ that converges to a point $y \in K$. Then

$$\|x - y\| = \lim \|x - P_K(x_{n_i})\| = d(x, K)$$

implies that $y = P_K(x)$ and $P_K(x_{n_i}) \to P_K(x)$. But this contradicts (12.4.1). ∎

For proving the main result of this chapter we shall need the following well-known fixed-point theorem, which we state without proof.

12.5 Schauder Fixed-Point Theorem. *Let B be a closed convex subset of the inner product space X and let F be a continuous mapping from B into a compact subset of B. Then F has a "fixed-point"; that is, there is $x_0 \in B$ such that $F(x_0) = x_0$.*

Convexity of Boundedly Compact Chebyshev Sets

Recall that a subset K of X is *boundedly compact* if each bounded sequence in K has a subsequence converging to a point in K. The main result of this chapter can be stated as follows.

12.6 Boundedly Compact Chebyshev Sets are Convex. *If K is a boundedly compact Chebyshev set in an inner product space, then K is convex.*

Proof. By Theorem 12.3, it suffices to verify that K is a sun. We proceed to prove this by contradiction.

Suppose K were not a sun. Then there would exist $x \in X \backslash K$ such that *not* every element of the form $x(\lambda) := x + \lambda(x - P_K(x))$, $\lambda \geq 0$, has $P_K(x)$ as its best approximation in K. Let

$$\lambda_1 := \sup\{\, \lambda \geq 0 \mid P_K(x(\lambda)) = P_K(x) \,\}.$$

Using Lemma 12.1 we see that $P_K(x(\lambda)) = P_K(x)$ for every $0 \leq \lambda < \lambda_1$ and $P_K(x(\lambda)) \neq P_K(x)$ for every $\lambda > \lambda_1$. If $\lambda_1 > 0$, the continuity of the distance function (Theorem 5.3) implies that

$$d(x(\lambda_1), K) = \lim_{\lambda \to \lambda_1^-} d(x(\lambda), K) = \lim_{\lambda \to \lambda_1^-} \|x(\lambda) - P_K(x)\|$$
$$= \lim_{\lambda \to \lambda_1^-} (1 + \lambda)\|x - P_K(x)\| = (1 + \lambda_1)\|x - P_K(x)\|$$
$$= \|x(\lambda_1) - P_K(x)\|.$$

If $\lambda_1 = 0$, then $x(\lambda_1) = x$ and the above equality holds trivially. Thus $P_K(x(\lambda_1)) = P_K(x)$ and $x(\lambda_1)$ is the farthest point on the ray from $P_K(x)$ through x that has $P_K(x)$ as its best approximation. For simplicity, set $x_1 = x(\lambda_1)$.

Now choose any $r > 0$ such that the closed ball $B = \{y \in X \mid \|x_1 - y\| \le r\}$ is disjoint from K. Define a mapping F on B by

(12.6.1) $\qquad F(y) := x_1 + \dfrac{r}{\|x_1 - P_K(y)\|}[x_1 - P_K(y)], \quad y \in B.$

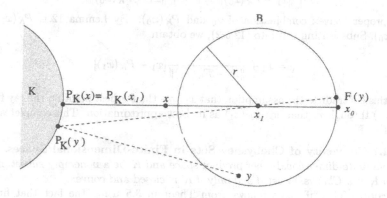

Figure 12.6.3

Clearly, $\|F(y) - x_1\| = r$, so F maps B into itself (see Figure 12.6.3).

Since K is boundedly compact, it is approximatively compact (by Lemma 3.11), so P_K is continuous by Lemma 12.4. Since the norm is continuous and $\|x_1 - P_K(y)\| > r$ for all $y \in B$, it follows that F is continuous.

Claim: *The range of F is contained in a compact subset of B.*

It suffices to show that $\overline{F(B)}$, the closure of the range of F, is compact. Let $\{z_n\}$ be any sequence in $\overline{F(B)}$ $(n \ge 2)$. For each $n \ge 2$, choose $y_n \in F(B)$ such that $\|z_n - y_n\| < 1/n$. Then choose $x_n \in B$ such that $y_n = F(x_n)$; that is,

$$y_n = x_1 + \frac{r}{\|x_1 - P_K(x_n)\|}[x_1 - P_K(x_n)].$$

By the nonexpansiveness of the distance function $d(x, K)$ (Theorem 5.3), it follows that for every $k_0 \in K$,

$$\|x_n - k_0\| \ge |d(x_n, K) - d(k_0, K)| = \|x_n - P_K(x_n)\|,$$

and so $\{P_K(x_n)\}$ is a bounded sequence in K. Hence there exists a subsequence $\{P_K(x_{n_i})\}$ converging to a point $z_0 \in K$. Thus

$$y_{n_i} \to x_1 + \frac{r}{\|x_1 - z_0\|}[x_1 - z_0] =: y_0.$$

Since $y_{n_i} \in F(B)$ for all i, $y_0 \in \overline{F(B)}$. Also,

$$\|z_{n_i} - y_0\| \le \|z_{n_i} - y_{n_i}\| + \|y_{n_i} - y_0\| \to 0.$$

Hence $\{z_n\}$ has a subsequence converging to a point in $\overline{F(B)}$. Thus $\overline{F(B)}$ is compact, and this proves the claim.

By the Schauder fixed-point theorem (Theorem 12.5), F has a fixed point $x_0 \in B$:

$$(12.6.2) \qquad x_0 = F(x_0) = x_1 + \frac{r}{\|x_1 - P_K(x_0)\|}[x_1 - P_K(x_0)].$$

Thus

$$x_1 = \frac{\|x_1 - P_K(x_0)\|}{r + \|x_1 - P_K(x_0)\|}x_0 + \frac{r}{r + \|x_1 - P_K(x_0)\|}P_K(x_0)$$

is a proper convex combination of x_0 and $P_K(x_0)$. By Lemma 12.1, $P_K(x_1) = P_K(x_0)$. Substituting this into (12.6.2), we obtain

$$x_0 = x_1 + \frac{r}{\|x_1 - P_K(x_1)\|}[x_1 - P_K(x_1)].$$

But this contradicts the assumption that x_1 was the farthest point on the ray from $P_K(x_1)$ through x_1 that has $P_K(x_1)$ as its best approximation. This completes the proof. ∎

12.7 Convexity of Chebyshev Sets in Finite-Dimensional Spaces. *Let X be a finite-dimensional inner product space and K be a nonempty subset of X. Then K is a Chebyshev set if and only if K is closed and convex.*

Proof. The "if" part follows from Theorem 3.5 using the fact that finite-dimensional spaces are complete (Theorem 3.7(3)). Conversely, if K is a Chebyshev set, then K is closed by Theorem 3.1, and K is boundedly compact by Lemma 3.7(1). The result now follows by Theorem 12.6. ∎

The main unanswered question concerning these results is the following:

Question. Must every Chebyshev set in (an infinite-dimensional) Hilbert space be convex?

We believe that the answer is no.

Exercises

1. A set K in X is called **midpoint convex** if $\frac{1}{2}(x+y) \in K$ whenever x, $y \in K$.

   (a) Show that every convex set is midpoint convex; and every midpoint convex set that is closed must be convex.

   (b) Give an example of a set in $\mathbb{R}$ that is midpoint convex, but not convex.

   (c) Show that if K is midpoint convex, then K is dense in its closed convex hull, $\overline{co(K)}$.

2. Let K be a convex Chebyshev set in the inner product space X. Show that for each pair $x, y \in X$, either $\|P_K(x) - P_K(y)\| < \|x-y\|$ or $P_K(x) - P_K(y) = x - y$.

3. Prove that for any pair of vectors x, z in X,

   $$(1) \qquad \{y \in X \mid \langle y - z, x - z \rangle > 0\} = \cup_{\lambda > 0} B(z + \lambda(x - z), \lambda\|x - z\|).$$

   In words, the open half-space containing x that is determined by the hyperplane through z and orthogonal to $x - z$ is the union of all open balls with centers on the ray from z through x whose radius is the distance from the center to z. [Hint: Let L and R denote the sets on the left and right sides of (1). Note that $y \in R$ if and only if $\|z + \lambda(x - z) - y\| < \lambda\|x - z\|$ for some $\lambda > 0$. Square both sides of this inequality and simplify to deduce that $y \in L$.]

4. Show that a Chebyshev set K is a sun if and only if for every $x \in X \setminus K$,

$$K \cap \{y \in X \mid \langle y - P_K(x), x - P_K(x) \rangle > 0\} = \emptyset.$$

[Hint: Exercise 3 with $z = P_K(x)$.]

5. Give a proof that a Chebyshev set in $l_2(2)$ is convex without appealing to any fixed-point theorem.

6. Let K be a Chebyshev set in any *normed linear space* X. For the statements below, prove that (a) $\Rightarrow$ (b) $\Rightarrow$ (c).
   (a) K is convex;
   (b) K is a sun;
   (c) For every $x \in X$, $P_K(x) = P_{[y, P_K(x)]}(x)$ for every $y \in K$.

7. Prove the following generalization of Theorem 12.6 to any normed linear space. *Every boundedly compact Chebyshev set in a normed linear space X is a sun.* [Hint: Schauder's fixed-point theorem is valid in any normed linear space.]

8. Let X denote the 2-dimensional space $\mathbb{R}^2$ with the norm $\|x\| = \max_{1 \leq i \leq 2}\{|x(i)|\}$. Show that the set $K = \{x \in X \mid x(2) \geq -x(1)/3 \text{ or } x(2) \geq -3x(1)\}$ is a *nonconvex* Chebyshev set.

Historical Notes

The idea of a sun was first developed and used by Klee (1953), but the terminology "sun" was first proposed a little later by Efimov and Stechkin (1958).

Phelps (1957) showed that (4) implies (1) in Theorem 12.3. That is, he showed that if P_K is nonexpansive, then K must be convex. The proof that a Chebyshev set in a finite-dimensional Hilbert space must be convex was first given in the Dutch doctoral thesis of Bunt (1934). Motzkin (1935) gave one proof of the convexity of Chebyshev sets in $l_2(2)$, and another one (1935a) for bounded Chebyshev sets in $l_2(2)$. Kritikos (1938), apparently unaware of the earlier proofs, gave another proof of the convexity of Chebyshev sets in $l_2(n)$. Jessen (1940), aware of Kritikos's proof, gave yet another one. Busemann (1947) and Valentine (1964) generalized these results by showing the convexity of Chebyshev sets in any finite-dimensional smooth and strictly convex normed linear space. Klee (1953) showed that strict convexity was not essential for the Busemann–Valentine result. Klee (1961) generalized this further by showing the convexity of a Chebyshev set in a smooth reflexive Banach space if its metric projection satisfies certain continuity criteria. Vlasov (1961) showed that in a smooth Banach space, every boundedly compact Chebyshev set is convex. In particular, this yields Theorem 12.6 as a corollary. Theorem 12.6 seems to be the strongest (positive) result known about the convexity of Chebyshev sets in a Hilbert space. However, every item of the following theorem is a stronger result in the sense that it actually characterizes those Chebyshev sets that are convex.

12.8 Theorem. *Let C be a Chebyshev set in the Hilbert space X. Each of the following statements is equivalent to C being convex.*

(1) *C is weakly closed (Klee (1961))*;

(2) C is approximatively compact (Efimov–Stechkin (1961));

(3) P_C is continuous (Vlasov (1967) and Asplund (1969));

(4) P_C is "radially" continuous (Vlasov (1967a)), i.e., for any $x \in X$, the restriction of P_C to the ray $\{x + \lambda[x - P_C(x)] \mid \lambda \geq 0\}$ is continuous at x;

(5)

$$\lim_{\epsilon \to 0^+} \frac{d(x_\epsilon, C) - d(x, C)}{\|x_\epsilon - x\|} = 1$$

for every $x \in X \setminus C$, where $x_\epsilon = x + \epsilon[x - P_C(x)]$ (Vlasov (1967a);

(6) P_C is continuous except possibly at a countable number of points (Balaganskii (1982));

(7) P_C is continuous except possibly on a subset of a hyperplane (Vlasov and Balaganskii (1996; Corollary 2.8));

(8) C is a sun (Efimov–Stechkin (1961));

(9) C intersects each closed half-space in a proximinal set (Asplund (1969)).

Asplund (1969) showed that if a nonconvex Chebyshev set exists in a Hilbert space X, then one also exists that has a bounded convex complement. Johnson (1987) gave an example of a nonconvex Chebyshev set in an *incomplete* inner product space. Jiang (1990) found some errors in Johnson's example, but showed that all were correctable. Johnson's example does not answer the question, since the space was not complete, but it tends to support the conjecture of Klee (1966) that nonconvex Chebyshev sets exist in (some) infinite-dimensional Hilbert space. By modifying his construction in (Johnson, 1987), Johnson (1998) showed that the nonconvex Chebyshev set could be constructed so as also to be bounded. Berens and Westphal (1978) have shown that the "maximal monotonicity" of P_C is equivalent to the convexity of C. See also Berens (1980) for an exposition of the monotonicity aspect of the metric projection in Hilbert space. An excellent (unpublished) exposition of the convexity of Chebyshev sets in finite-dimensional Hilbert spaces, along with proofs, was given by Kelly (1978). Surveys on the convexity of Chebyshev sets problem were given by Vlasov (1973), Narang (1977), Singer (1977), Deutsch (1993), and Balaganskii and Vlasov (1996). The latter survey also contains a geometric proof of an example of Vlasov, similar to Johnson's example. In addition, the Balaganskii and Vlasov survey (1996) contains a large amount of material relating the geometry of Banach spaces with the approximative properties of certain subsets.

Related to this circle of ideas is the problem of characterizing geometrically those finite-dimensional *normed* linear spaces X in which each Chebyshev set is convex. When $n := \dim X \leq 4$, it is known that each Chebyshev set in X is convex if and only if each "exposed point" of the unit sphere $S(X)$ is a "smooth point." (The case $n = 1$ is trivial; the case $n = 2$ follows because if each exposed point of $S(X)$ is a smooth point, then X is smooth; the proof when $n = 3$ is due to Brøndsted (1966); the proof when $n = 4$ was given by Brown (1980), who developed quite an elaborate topological machinery for this purpose.) Whether the same result holds true when $n > 4$ is still open, but I suspect it is. In particular, there are finite-dimensional normed linear spaces (e.g., the "smooth spaces") that are *not* Hilbert spaces, but all of whose Chebyshev subsets are convex (Klee (1961)).

However, by restricting the Chebyshev sets being considered, a characterization can be given. Tsar'kov (1984) showed that every *bounded* Chebyshev set in a finite-dimensional normed linear space X is convex if and only if the extreme points of $S(X^*)$ are dense in $S(X^*)$. See also Brown (1986) for related material.

Finally, we note that the question of the convexity of Chebyshev sets in an infinite-dimensional Hilbert space was first posed explicitly by Klee (1961) and implicitly by Efimov and Stechkin (1961), since they asked the question (which is equivalent by Theorem 12.8(2)) whether each Chebyshev set must be approximatively compact.

APPENDIX 1

ZORN'S LEMMA

Definition. A **partially ordered set** *is a pair* $(S, \prec)$, *where* S *is a nonempty set and* $\prec$ *is a relation on* S *that satisfies the following three properties:*

(1) $x \prec x$ *for all* $x \in S$.

(2) *If* $x \prec y$ *and* $y \prec z$, *then* $x \prec z$.

(3) *If* $x \prec y$ *and* $y \prec x$, *then* $x = y$.

The relation $\prec$ is called a **partial order** on S. In other words, a partial order is (1) "reflexive," (2) "transitive," and (3) "anti-symmetric." For example, the inclusion relation $\subset$ is a partial order on the collection $S = 2^X$ of all subsets of a set X, and the relation $\leq$ is a partial order on the set of real numbers $\mathbb{R}$.

We use terminology suggested by these two examples by saying that x is *greater* than y when we write $y \prec x$. A **totally ordered** (or linearly ordered) subset of the partially ordered set $(S, \prec)$ is a subset T of S such that *for each pair* of elements x and y in T, either $x \prec y$ or $y \prec x$. For example, any subset T of $\mathbb{R}$ is totally ordered in $(\mathbb{R}, \leq)$, but 2^X is not totally ordered in $(2^X, \subset)$.

Let $(S, \prec)$ be a partially ordered set and let $T \subset S$. An element $b \in S$ is called an **upper bound** for T if $t \prec b$ for every $t \in T$. An element $m \in S$ is called **maximal** if $m \prec x$ implies $x \prec m$. That is, the only element of S that is greater than m is m itself.

Zorn's Lemma. *A partially ordered set has a maximal element if each totally ordered subset has an upper bound.*

It is known that Zorn's lemma is logically equivalent to the axiom of choice, which asserts the possibility of forming a new set by choosing exactly one element from each set in an arbitrary collection of nonempty sets.

APPENDIX 2

EVERY HILBERT SPACE IS $\ell_2(I)$

In this appendix we prove that every Hilbert space is equivalent to the Hilbert space $\ell_2(I)$ for some index set I.

We recall (see Exercise 4 of Chapter 6) that two Hilbert spaces X and Y are called **equivalent** if there is a bijective mapping U from X onto Y that preserves linearity,

$$U(\alpha x + \beta y) = \alpha\, U(x) + \beta\, U(y) \qquad \text{for all } x, y \in X, \quad \alpha, \beta \in \mathbb{R},$$

preserves the inner product,

$$\langle U(x), U(y) \rangle = \langle x, y \rangle \qquad \text{for all } x, y \in X,$$

and preserves the norm,

$$\|U(x)\| = \|x\| \qquad \text{for all } x \in X.$$

That is, two equivalent Hilbert spaces are essentially the same. The mapping U amounts to nothing more than a relabeling of the elements.

In Exercise 4 of Chapter 6 it was outlined that X and Y are equivalent if and only if there exists a linear mapping U from X onto Y that is norm-preserving: $\|U(x)\| = \|x\|$ for all $x \in X$.

Theorem. *Every Hilbert space $X \neq \{0\}$ is equivalent to the Hilbert space $\ell_2(I)$ for some index set I.*

Proof. By Corollary 4.20 and Theorem 4.21, X has an orthonormal basis $\{e_i \mid i \in I\}$. Thus each $x \in X$ has the representation

$$x = \sum_{i \in I} \langle x, e_i \rangle e_i \quad \text{and} \quad \|x\| = \left[\sum_{i \in I} |\langle x, e_i \rangle|^2 \right]^{1/2}.$$

Define the map U on X into the space of functions on I by

$$(Ux)(i) := \langle x, e_i \rangle, \quad i \in I.$$

Note that since

$$\sum_{i \in I} |(Ux)(i)|^2 = \sum_{i \in I} |\langle x, e_i \rangle|^2 = \|x\|^2 < \infty$$

by Corollary 4.20, $Ux \in \ell_2(I)$. Thus $U : X \to \ell_2(I)$. It is easy to see that U is linear. Moreover, the last equation above shows that U is norm-preserving. To complete the proof, it suffices to show that U is *onto*.

First note that functions $\delta_i : I \to \mathbb{R}$ defined for each $i \in I$ by

$$\delta_i(j) = \delta_{ij}, \quad j \in I \qquad \text{(Kronecker's delta)},$$

form an orthonormal basis for $\ell_2(I)$. In fact, if $y \in \ell_2(I)$, then

$$y = \sum_{i \in I} y(i)\delta_i, \qquad \sum_{i \in I} |y(i)|^2 = \|y\|^2.$$

Given any $y \in \ell_2(I)$, define x by

$$x = \sum_{i \in I} y(i)e_i.$$

Then $x \in X$ and $(Ux)(i) = y(i)$ for each $i \in I$. That is, $Ux = y$, so U is onto. ∎

Implicit in the proof was the fact that if X is equivalent to $\ell_2(I)$, then the cardinality of I is equal to the cardinality of an orthonormal basis for X. In other words, the dimensions of X and $\ell_2(I)$ are the same. (The dimension of a Hilbert space is the cardinality of an arbitrary orthonormal basis.)

REFERENCES

N. I. Achieser
[1] **Theory of Approximation**, Ungar, New York, 1956.

N. I. Achieser and M. G. Krein
[1] **Some Questions in the Theory of Moments**, Kharkov, 1938 (Russian). [Translated in: Translations of Mathematical Monographs, Vol. 2, Amer. Math. Soc., Providence, RI, 1962.]

P. Alexandroff and H. Hopf
[1] **Topologie**, Springer, Berlin, 1935. (Reprinted by Chelsea Publ. Co., New York, 1965.)

D. Amir
[1] **Characterizations of Inner Product Spaces**, Birkhäuser Verlag, Basel, 1986.

D. Amir, F. Deutsch, and W. Li
[1] Characterizing the Chebyshev polyhedrons in an inner product space, preprint, 2000.

V. E. Anderson and P. A. Ivert
[1] Constrained interpolants with minimal $W^{k,p}$-norm, J. Approx. Theory, 49(1987), 283–288.

W. N. Anderson, Jr. and R. J. Duffin
[1] Series and parallel addition of matrices, J. Math. Anal. Appl., 26(1969), 576–594.

N. Aronszajn
[1] Theory of reproducing kernels, Trans. Amer. Math. Soc., 68(1950), 337–403.
[2] **Introduction to the Theory of Hilbert Spaces, Vol. 1**, Research Foundation of Oklahoma A & M College, Stillwater, OK, 1950a.

N. Aronszajn and K. T. Smith
[1] Invariant subspaces of completely continuous operators, Annals Math., 60(1954), 345–350.

G. Ascoli
[1] Sugli spazi lineari metrici e le loro varietà lineari, Ann. Mat. Pura Appl., (4) 10(1932), 33–81, 203–232.

E. Asplund
[1] Čebyšev sets in Hilbert space, Trans. Amer. Math. Soc., 144(1969), 235–240.

F. V. Atkinson
[1] The normal solubility of linear equations in normed spaces (Russian), Mat. Sbornik N.S., 28 (70) (1951), 3–14.
[2] On relatively regular operators, Acta Sci. Math. Szeged, 15(1953), 38–56.

B. Atlestam and F. Sullivan
[1] Iteration with best approximation operators, Rev. Roum. Math. Pures et Appl., 21(1976), 125–131.

J.-P. Aubin
[1] **Applied Functional Analysis**, Wiley-Interscience, New York, 1979.

R. Baire
[1] Sur les fonctions de variables réelles, Ann. Mat. Pura Appl., 3(1899), 1–123.

V. S. Balaganskii
[1] Approximative properties of sets in Hilbert space, Math. Notes, 31(1982), 397–404.

V. S. Balaganskii and L. P. Vlasov
[1] The problem of convexity of Chebyshev sets, Russian Math. Surveys, 51(1996), 1127–1190. [English translation of: Uspekhi Mat. Nauk, 51(1996), 125–188.]

S. Banach
  [1] Sur les fonctionnelles linéaires I, Studia Math., 1(1929a), 211–216.
  [2] Sur les fonctionnelles linéaires II, Studia Math., 1(1929b), 223–239.
  [3] Théorèmes sur les ensembles de premières catégorie, Fund. Math., 16(1930), 395–398.
  [4] **Théorie des opérations linéaires**, Monografje Matematyczne, Warsaw, 1932. (Reprinted by Chelsea Publ. Co., New York, 1955.)

S. Banach and H. Steinhaus
  [1] Sur le principe de la condensation de singularités, Fund. Math., 9(1927), 50–61.

V. Barbu and Th. Precupanu
  [1] **Convexity and Optimization in Banach Spaces**, Sijthoff & Noordhoff, the Netherlands, 1978.

H. H. Bauschke
  [1] Projection Algorithms and Monotone Operators, Ph.D. thesis, Simon Fraser University, 1996.

H. H. Bauschke and J. M. Borwein
  [1] On the convergence of von Neumann's alternating projection algorithm for two sets, Set-Valued Analysis, 1(1993), 185–212.
  [2] On projection algorithms for solving convex feasibility problems, SIAM Review, 38(1996) 367–426.

H. H. Bauschke, J. M. Borwein, and A. S. Lewis
  [1] The method of cyclic projections for closed convex sets in Hilbert space, in **Optimization and Nonlinear Analysis** (edited by Y. Censor and S. Reich), Contemporary Mathematics, Amer. Math. Soc., 1996.

H. H. Bauschke, J. M. Borwein, and W. Li
  [1] Strong conical hull intersection property, bounded linear regularity, Jameson's property (G), and error bounds in convex optimization, Math. Program., 86(1999), 135–160.

H. H. Bauschke, F. Deutsch, H. Hundal, and S.-H. Park
  [1] Accelerating the convergence of the method of alternating projections, preprint, July 23, 1999.

A. Ben-Israel and T. N. E. Greville
  [1] **Generalized Inverses: Theory and Applications**, John Wiley & Sons, New York, 1974.

H. Berens
  [1] **Vorlesung über Nichtlineare Approximation**, Universität Erlangen–Nürnberg Lecture Notes, 1977/78.
  [2] Best approximation in Hilbert space, in **Approximation Theory III** (edited by E. W. Cheney), Academic Press, New York, 1980, 1–20.

H. Berens and U. Westphal
  [1] Kodissipative metrische Projektionen in normierten linearen Räume, in **Linear Spaces and Approximation** (edited by P. L. Butzer and B. Sz.-Nagy), ISNM, 40(1978), 120–130.

S. N. Bernstein
  [1] Démonstration du théorème de Weierstrass fardée sur le calcul de probabilités, Communications of the Kharkov Mathematical Society, 13(1912), 1–2.

S. Bernstein and C. de la Vallée Poussin
  [1] **L'Approximation**, Chelsea Publ. Co., New York, 1970. [This is a reprint in one volume of two books "Leçons sur les Propriétés Extrémales et la Meilleure Approximation des Fonctions Analytiques d'une Variable Réele", by S. Bernstein, Paris, 1926, and "Leçons sur L'Approximation des Fonctions d'une Variable Réelle" by C. de la Vallée Poussin, Paris, 1919. Text was unaltered except for correction of errata.]

A. Bjerhammar
  [1] Rectangular reciprocal matrices, with special reference to geodetic calculations, Bull. Géodésique, 1951, 188–220.

A. Björck
  [1] **Numerical Methods for Least Squares Problems**, SIAM, Philadelphia, 1996.

H. Bohman
  [1] On approximation of continuous and analytic functions, Arkiv för Matematik, 2(1952), 43–56.

H. F. Bohnenblust and A. Sobczyk
  [1] Extensions of functionals on complex linear spaces, Bull. Amer. Math. Soc., 44(1938), 91–93.

B. P. J. N. Bolzano
  [1] Rein analytischer Beweis des Lehrsatzes ..., Prague, 1817; Abh. Gesell. Wiss. Prague, (3) 5 (1818), 1–60; Ostwald's Klassiker, 153 (edited by P. E. B. Jourdain), Leipzig, 1905, 3–43.

C. de Boor
  [1] On "best" interpolation, J. Approx. Theory, 16(1976), 28–42.

E. Borel
  [1] Thèse, Annales Scientifiques de l'Ecole Normal Supérieure, (3), 12(1895), 9–55.

J. M. Borwein and A. S. Lewis
  [1] Partially finite convex programming, Part I: Quasi relative interior and duality theory, Math. Programming, 57(1992), 15–48.
  [2] Partially finite convex programming, Part II: Explicit lattice models, Math. Programming, 57(1992a), 49–83.

J. M. Borwein and H. Wolkowicz
  [1] Facial reduction for a cone-convex programming problem, J. Austral. Math. Soc., Ser. A, 30(1981), 369–380.
  [2] A simple constraint qualification in infinite-dimensional programming, Math. Programming, 35(1986), 83–96.

T. L. Boullion and P. L. Odell
  [1] **Proceedings of the Symposium on Theory and Applications of Generalized Inverses of Matrices**, Texas Tech Press, Lubbock, Texas, 1968.
  [2] **Generalized Inverse Matrices**, Wiley-Interscience, New York, 1971.

J. P. Boyle and R. L. Dykstra
  [1] A method for finding projections onto the intersection of convex sets in Hilbert spaces, in **Advances in Order Restricted Statistical Inference**, Lecture Notes in Statistics, Springer-Verlag, 1985, 28–47.

A. P. Boznay
  [1] A remark on the alternating algorithm, Periodica Math. Hungarica, 17(1986), 241–244.

D. Braess
  [1] **Nonlinear Approximation Theory**, Springer-Verlag, New York, 1986.

A. Brøndsted
  [1] Convex sets and Chebyshev sets II, Math. Scand., 18(1966), 5–15.

B. Brosowski and F. Deutsch
  [1] An elementary proof of the Stone–Weierstrass theorem, Proc. Amer. Math. Soc., 81(1981), 89–92.

A. L. Brown
   [1] **Abstract Approximation Theory**, Seminar in Analysis 4, MATSCIENCE, Madras, 1969/70.
   [2] Chebyshev sets and facial systems of convex sets in finite-dimensional spaces, Proc. London Math. Soc., 41(1980), 297–339.
   [3] Chebyshev sets and the shapes of convex bodies, in **Methods of Functional Analysis in Approximation Theory** (edited by C. A. Micchelli, D. V. Pai, and B. V. Limaye), Birkhäuser Verlag, Boston, 1986, 97–121.

V. Buniakowsky
   [1] Sur quelques inégalités concernant les intégrales ordinaires et les intégrales aux différences finies, Mém. Acad. St. Petersburg (7)1, no. 9(1859).

L. N. H. Bunt
   [1] Bitdrage tot de theorie der konvekse puntverzamelingen, thesis, Univ. of Groningen, Amsterdam (Dutch), 1934.

R. B. Burckel and S. Saeki
   [1] An elementary proof of the Müntz–Szász theorem, Expo. Math., 4(1983), 335–341.

H. Busemann
   [1] Note on a theorem on convex sets, Mat. Tidsskrift B, 1947, 32–34.

R. Cakon
   [1] Alternative proofs of Weierstrass theorem of approximation: an expository paper, masters paper, The Pennsylvania State University, 1987.

A. L. Cauchy
   [1] Cours d'analyse algébrique, Oeuvres, (2) III, 1821.
   [2] Oeuvres, sér. II, t. 3, Gauthier–Villars, Paris, 1900.

E. W. Cheney
   [1] **Introduction to Approximation Theory**, McGraw–Hill, New York, 1966.

W. Cheney and W. Light
   [1] **A Course in Approximation Theory**, Brooks/Cole Pub. Co., Pacific Grove, CA, 2000.

E. W. Cheney and A. A. Goldstein
   [1] Proximity maps for convex sets, Proc. Amer. Math. Soc., 10(1959), 448–450.

C. K. Chui, F. Deutsch, and J. D. Ward
   [1] Constrained best approximation in Hilbert space, Constr. Approx., 6(1990), 35–64.
   [2] Constrained best approximation in Hilbert space II, J. Approx. Theory, 71(1992), 213–238.

J. B. Conway
   [1] **A Course in Functional Analysis**, Second ed., Springer-Verlag, New York, 1990.

P. Cousin
   [1] Sur les fonctions de n variables complexes, Acta Math., 19(1895), 1–61.

G. Davis, S. Mallat, and Z. Zhang
   [1] Adaptive time-frequency approximations with matching pursuits, in **Wavelets: Theory, Algorithms, and Applications** (edited by C. L. Chui, L. Montefusco, and L. Puccio), Academic Press, New York, 1994, pp. 271–293.

P. J. Davis
   [1] **Interpolation and Approximation**, Blaisdell, New York, 1963. (Reprinted by Dover, New York, 1975.)

M. M. Day
  [1] Reflexive Banach spaces not isomorphic to uniformly convex spaces, Bull. Amer. Math.
      Soc., 47(1941), 313–317.

L. Debnath and P. Mikusinski
  [1] **Introduction to Hilbert Spaces with Applications**, Academic Press, Inc., Boston,
      1990.

C. A. Desoer and B. H. Whalen
  [1] A note on pseudoinverses, J. Soc. Industr. Appl. Math., 11(1963), 442–447.

F. Deutsch
  [1] Applications of Functional Analysis to Approximation Theory, Ph.D. thesis, Brown
      University, 1966.
  [2] Simultaneous interpolation and approximation in topological linear spaces, J. SIAM
      Appl. Math., 14(1966), 1180–1190.
  [3] The alternating method of von Neumann, in **Multivariate Approximation Theory**
      (edited by W. Schempp and K. Zeller), ISNM 51, Birkhäuser Verlag, Basel, 1979, 83–96.
  [4] Representers of linear functionals, norm-attaining functionals, and best approximation
      by cones and linear varieties in inner product spaces, J. Approx. Theory, 36(1982),
      226–236.
  [5] Von Neumann's alternating method: the rate of convergence, in **Approximation
      Theory IV**, Academic Press, New York, 1983, 427–434.
  [6] Rate of convergence of the method of alternating projections, ISNM, Vol. 72, Birkhäuser
      Verlag, Basel, 1984, 96–107.
  [7] Simultaneous interpolation and norm-preservation, ISNM, Vol. 74, Birkhäuser Verlag,
      Basel, 1985, 122–132.
  [8] The method of alternating orthogonal projections, in **Approximation Theory, Spline
      Functions and Applications** (edited by S. .P. Singh), Kluwer Academic Publishers, the
      Netherlands, 1992, 105–121.
  [9] The convexity of Chebyshev sets in Hilbert space, in **Topics in Polynomials of One
      and Several Variables and their Applications** (edited by Th. M. Rassias,
      H. M. Srivastava, and A. Yanushauskas), World Scientific, 1993, 143–150.
  [10] Dykstra's cyclic projections algorithm: the rate of convergence, in **Approximation
      Theory, Wavelets and Applications** (edited by S. P. Singh), Kluwer Academic
      Publishers, the Netherlands, 1995, 87–94.
  [11] The role of the stong conical hull intersection property in convex optimization and
      approximation, in **Aproximation Theory IX** (edited by C. K. Chui and L. L.
      Schumaker), Vanderbilt University Press, Nashville, TN, 1998, 105–112.
  [12] Accelerating the convergence of the method of alternating projections via a line search:
      a brief survey, in the proceedings of the research workshop on parallel algorithms (edited
      by D. Butnariu, Y. Censor, and S. Reich), 2001, to appear.

F. Deutsch and H. Hundal
  [1] The rate of convergence of Dykstra's cyclic projections algorithm: the polyhedral case,
      Numer. Funct. Anal. and Optimiz., 15 (1994), 537–565.
  [2] The rate of convergence for the method of alternating projections II, J. Math. Anal.
      Appl., 205(1997), 381–405.

F. Deutsch, W. Li, and J. D. Ward
  [1] A dual approach to constrained interpolation from a convex subset of Hilbert space, J.
      Approx. Theory, 90(1997), 385–414.
  [2] Best approximation from the intersection of a closed convex set and a polyhedron in
      Hilbert space, weak Slater conditions, and the strong conical hull intersection property,
      SIAM J. Optimiz. 10(2000), 252–268.

F. R. Deutsch and P. H. Maserick
  [1] Applications of the Hahn–Banach theorem in approximation theory, SIAM Review,
      9(1967), 516–530.

F. Deutsch, J. H. McCabe, and G. M. Phillips
[1] Some algorithms for computing best approximations from convex cones, SIAM J. Numer. Anal., 12(1975), 390–403.

F. Deutsch and P. D. Morris
[1] On simultaneous approximation and interpolation which preserves the norm, J. Approx. Theory, 2(1969), 355–373.

F. Deutsch, V. A. Ubhaya, J. D. Ward, and Y. Xu
[1] Constrained best aproximation in Hilbert space III. Applications to n-convex functions, Constr. Approx., 12(1996), 361–384.

F. Deutsch, V. A. Ubhaya, and Y. Xu
[1] Dual cones, constrained n-convex L_p-approximation, and perfect splines, J. Approx. Theory, 80(1995), 180–203.

R. A. DeVore and G. G. Lorentz
[1] **Constructive Approximation**, Springer-Verlag, New York, 1993.

J. Dieudonné
[1] La dualité dans les espaces vectoriels topologiques, Ann. Sci. École Norm. Sup., 59(1942), 107–139.
[2] **History of Functional Analysis**, North Holland Math. Studies 49, Elsevier, New York, 1981.

J. Dixmier
[1] Étude sur les variétés et les opératerus de Julia avec quelques application, Bull. Soc. Math. France, 77(1949), 11–101.

A. L. Dontchev
[1] Duality methods for constrained best interpolation, Math. Balkanica, 1(1987), 96–105.

A. L. Dontchev, G. Illiev, and M. M. Konstantinov
[1] Constrained interpolation via optimal control, in Proceedings of the Fourteenth Spring Conference of the Union of Bulgarian Mathematicians, Sunny Beach, April 6–9, 1985, 385–392.

N. Dunford and J. T. Schwartz
[1] **Linear Operators Part I: General Theory**, Interscience Publ., New York, 1958.

R. L. Dykstra
[1] An algorithm for restricted least squares regression, J. Amer. Statist. Assoc., 78(1983), 837–842.

N. V. Efimov and S. B. Stechkin
[1] Some properties of Chebyshev sets, Dokl. Akad. Nauk SSSR, 118(1958), 17–19.
[2] Approximative compactness and Chebyshev sets, Sov. Math. Dokl., 2(1961), 1226–1228.

M. Eidelheit
[1] Zur Theorie der konvexen Mengen in linearen normierten Räumen, Studia Math., 6(1936), 104–111.

G. Faber
[1] Über die interpolatorische Darstellung stetiger Funktionen, Deutsche Mathematiker-Vereinigung Jahresbericht, 23(1914), 192–210.

J. Favard
[1] Sur l'interpolation, J. Math. Pures Appl., 19(1940), 281–306.

L. Fejér
[1] Sur les fonctions bornées et integrables, Comptes Rendus, 131(1900), 984–987.

[2] Über Weierstrassche Approximation, besonderes durch Hermitesche Interpolation, Math. Annal., 102(1930), 707–725.

W. Fenchel
[1] **Convex Cones, Sets, and Functions**, Lecture notes, Princeton University, 1953.

G. B. Folland
[1] **Real Analysis**, John Wiley & Sons, New York, 1984.

C. Franchetti
[1] On the alternating approximation method, Sezione Sci., 7(1973), 169–175.

C. Franchetti and W. Light
[1] The alternating algorithm in uniformly convex spaces, J. London Math. Soc., 29(1984), 545–555.

M. Fréchet
[1] Sur les opérations linéaires I, Trans. Amer. Math. Soc., 5(1904), 493–499.
[2] Sur les opérations linéaires II, Trans. Amer. Math. Soc., 6(1905), 134–140.
[3] Sur quelques points du Calcul fonctionnel, Rend. Circ. mat. Palermo, 22(1906), 1–74.
[4] Sur les opérations linéaires III, Trans. Amer. Math. Soc., 8(1907), 433–446.
[5] Sur les ensembles de fonctions et les opérations linéaires, C. R. Acad. Sci. Paris, 144(1907a), 1414–1416.

I. Fredholm
[1] Sur une nouvelle méthode pour la résolution du problème de Dirichlet, Kong. Vetenskaps-Akademiens Förh. Stockholm, (1900), 39–46.
[2] Sur une classe d'équations fonctionnelles, Acta math., 27(1903), 365–390.

K. Friedrichs
[1] On certain inequalities and characteristic value problems for analytic functions and for functions of two variables, Trans. Amer. Math. Soc., 41(1937), 321–364.

D. Gaier
[1] **Lectures on Complex Approximation**, translated by R. McLaughlin, Birkhäuser, Boston, 1987.

W. B. Gearhart and M. Koshy
[1] Acceleration schemes for the method of alternating projections, J. Comp. Appl. Math., 26(1989), 235–249.

M. Golomb
[1] **Lectures on Theory of Approximation**, Argonne National Laboratory, Applied Mathematics Division, 1962.

B. K. Goodrich and A. Steinhardt
[1] L_2 spectral estimation, SIAM J. Appl. Math., 46(1986), 417–428.

R. Gordon, R. Bender, and G. T. Herman
[1] Algebraic reconstruction techniques (ART) for three-dimensional electron microscopy and X-ray photography, J.Theoretical Biol., 29(1970), 471–481.

M. S. Gowda and M. Teboulle
[1] A comparison of constraint qualifications in infinite-dimensional convex programming, SIAM J. Control Optimiz., 28(1990), 925–935.

J. P. Gram
[1] Om Rackkendvilklinger bestemte ved Hjaelp af de mindste Kvadraters Methode, Copenhagen, 1879. [Translated in: Über die Entwicklung reeler Funktionen in Reihen mittels der Methode der kleinsten Quadrate, J. Reine Angewandte Mathematik, 94 (1883), 41–73.]

I. Grattan–Guiness
[1] **The Development of the Foundations of Mathematical Analysis from Euler to Riemann**, M.I.T. Press, Cambridge, MA, 1970.

C. W. Groetsch
[1] **Generalized Inverses of Linear Operators**, Marcel Dekker, New York, 1977.

L. G. Gubin, B. T. Polyak, and E. V. Raik
[1] The method of projections for finding the common point of convex sets, USSR Computational Mathematics and Mathematical Physics, 7(6) (1967), 1–24.

H. Hahn
[1] Über lineare Gleichungssysteme in linearen Räumen, J. Reine Angew. Math., 157(1927), 214–229.

P. R. Halmos
[1] **Introduction to Hilbert Space and the Theory of Spectral Multiplicity**, Chelsea Publ. Co., New York, 1951.
[2] **A Hilbert Space Problem Book**, Second ed., Springer-Verlag, New York, 1982.

I. Halperin
[1] The product of projection operators, Acta Sci. Math. (Szeged), 23(1962), 96–99.

F. Hausdorff
[1] Zur Theorie der linearen metrischen Räume, J. Reine Angew. Math., 167(1932), 294–311.

H. E. Heine
[1] Über trigonometrische Reihen, Journal für die Reine und Angewandte Mathematik, 71 (1870), 353–365.
[2] Die Elemente der Functionenlehre, Journal für dei Reine und Angewandte Mathematik, 74(1872), 172–188.

M. R. Hestenes
[1] Relative self-adjoint operators in Hilbert space, Pacific J. Math., 11(1961), 1315–1357.

D. Hilbert
[1] Grundzüge einer allgemeinen Theorie der linearen Integralgleichungen, I, Nachrichten Akad. Wiss. Göttingen. Math.-Phys. Kl., 1904, 49–91.
[2] Grundzüge einer allgemeinen Theorie der linearen Integralgleichungen, II, Nachrichten Akad. Wiss. Göttingen. Math.-Phys. Kl., 1905, 213–259.
[3] Grundzüge einer allgemeinen Theorie der linearen Integralgleichungen, III, Nachrichten Akad. Wiss. Göttingen. Math.-Phys. Kl., 1905, 307–338.
[4] Grundzüge einer allgemeinen Theorie der linearen Integralgleichungen, IV, Nachrichten Akad. Wiss. Göttingen. Math.-Phys. Kl., 1906, 157–227.
[5] Grundzüge einer allgemeinen Theorie der linearen Integralgleichungen, V, Nachrichten Akad. Wiss. Göttingen. Math.-Phys. Kl., 1906, 439–480.
[6] Grundzüge einer allgemeinen Theorie der linearen Integralgleichungen, VI, Nachrichten Akad. Wiss. Göttingen. Math.-Phys. Kl., 1910, 355–417.
(These six articles were published in book form by Teubner, Leipzig, 1912. They were reprinted by Chelsea Pub. Co., New York, 1953.)

T. H. Hildebrandt
[1] On uniform limitedness of sets of functional equations, Bull. Amer. Math. Soc., 29(1923), 309–315.

R. A. Hirschfeld
[1] On best approximation in normed vector spaces, II, Nieuw Archief voor Wiskunde (3), VI (1958), 99–107.

R. B. Holmes
[1] **A Course on Optimization and Best Approximation**, Lecture Notes in Mathematics, #257, Springer-Verlag, New York, 1972.

G. N. Hounsfield
[1] Computerized transverse axial scanning (tomography): Part I Description of system, British J. Radiol., 46(1973), 1016–1022.

H. S. Hundal
[1] Generalizations of Dykstra's Algorithm, doctoral dissertation, The Pennsylvania State University, 1995.

H. Hundal and F. Deutsch
[1] Two generalizations of Dykstra's cyclic projections algorithm, Math. Programming, 77 (1997), 335–355.

W. A. Hurwitz
[1] On the pseudo-resolvent to the kernel of an integral equation, Trans. Amer. Math. Soc., 13(1912), 405–418.

G. Illiev and W. Pollul
[1] Convex interpolation with minimal L_∞-norm of the second derivative, Math Z., 186(1984), 49–56.
[2] Convex interpolation by functions with minimal L_p-norm $(1 < p < \infty)$ of the kth derivative, in Proceedings of the Thirteenth Spring Conference of the Union of Bulgarian Mathematicians, Sunny Beach, April 6–9, 1984a.

L. Irvine
[1] Constrained Interpolation, master's thesis, Old Dominion University, 1985.

L. D. Irvine, S. P. Marin, and P. W. Smith
[1] Constrained interpolation and smoothing, Constr. Approx., 2(1986), 129–151.

D. Jackson
[1] Über die Genauigkeit der Annäherung Stetiger Funktionen Durch Ganze Rationale Funktionen Gegebenen Grades und Trigonometrische Summen Gegebener Ordnung, dissertation, Göttingen, 1911.
[2] **The Theory of Approximation**, Amer. Math. Soc. Colloq. Publications, Vol. XI, New York, 1930.

B. Jessen
[1] To Saetninger om konvekse Punktmaengder, Mat. Tidsskrift B, 1940, 66–70.

M. Jiang
[1] On Johnson's example of a nonconvex Chebyshev set, J. Approx. Theory, 74(1993), 152–158.

G. G. Johnson
[1] A nonconvex set which has the unique nearest point property, J. Approx. Theory, 51(1987), 289–332.
[2] Chebyshev foam, Topology and its Applications, 20(1998), 1–9.

P. Jordan and J. von Neumann
[1] On inner products in linear metric spaces, Ann. Math., 36(1935), 719–723.

L. V. Kantorovich and V. I. Krylov
[1] **Approximate Methods of Higher Analysis**, translated by C. D. Benster, Interscience Publ., New York, 1958.

T. Kato
[1] **Perturbation Theory for Linear Operators**, second edition, Springer, New York, 1984.

S. Kayalar and H. Weinert
[1] Error bounds for the method of alternating projections, Math. Control Signal Systems, 1 (1988), 43–59.

B. F. Kelly, III
[1] The convexity of Chebyshev sets in finite dimensional normed linear spaces, master's paper, The Pennsylvania State University, 1978.

V. Klee
[1] Convex bodies and periodic homeomorphisms in Hilbert space, Trans. Amer. Math. Soc., 74 (1953), 10–43.
[2] Convexity of Chebyshev sets, Math. Annalen, 142(1961), 292–304.
[3] On a problem of Hirschfeld, Nieuw Archief voor Wiskunde (3), XI (1963), 22–26.
[4] Remarks on nearest point in normed linear spaces, Proc. Colloq. Convexity, Univ. of Copenhagen, 1966, 168–176.

M. Kline
[1] **Mathematical Thought from Ancient to Modern Times**, Oxford Univ. Press, New York, 1972.

A. N. Kolmogorov and S. V. Fomin
[1] **Elements of the Theory of Functions and Functional Analysis, Volume 1 Metric and Normed Spaces**, translated by L. F. Boron, Graylock Press, Rochester, NY, 1957.

P. P. Korovkin
[1] On the convergence of linear positive operators in the space of continuous functions, Doklady Akademii Nauk SSSR, 90(1953), 961–964 (Russian).
[2] **Linear Operators and Approximation Theory**, Hindustan Publ. Corp., Delhi, 1960.

M. Krein and D. Milman
[1] On extreme points of regular convex sets, Studia Math., 9(1940), 133–138.

M. Kritikos
[1] Sur quelques propriétés des ensembles convexes, Bull. Math. de la Soc. Roumaine des Sciences, 40(1938), 87–92.

H. Kuhn
[1] Ein elementarer Beweis des Weierstrasschen Approximationssatzes, Archiv der Mathematik, 15(1964), 316–317.

C. Kuratowski
[1] La propriété de Baire dans les espaces métriques, Fund. Math., 16(1930), 390–394.
[2] **Topologie**, Vol. I, Warzawa, 1958.
[3] **Topologie**, Vol. II, Warzawa, 1961.

J. L. Lagrange
[1] Oeuvres, t. 3, Gauthier-Villars, Paris, 1869.

E. Landau
[1] Über die Approximation einer stetigen Funktionen durch eine ganze rationale Funktion, Rendiconti del Circolo Matematico di Palermo, 25(1908), 337–345.

R. E. Langer (editor)
[1] **On Numerical Approximation**, Univ. Wisconsin Press, Madison, 1959.

P.-J. Laurent
[1] **Approximation et Optimisation**, Hermann, Paris, 1972.

C. L. Lawson and R. J. Hanson
[1] **Soving Least Squares Problems**, Prentice-Hall, Inc., Englewood Cliffs, NJ, 1974.

H. Lebesgue
[1] Sur l'approximation des fonctions, Bull. Sci. Mathématique, 22(1898), 278–287.

[2] Leçons sur l'intégration et la recherche des fonctions primitives, Gauthier-Villars, Paris, 1904 (second edition 1928).

M. Lerch
[1] On the main theorem of the theory of generating functions (in Czech), Rozpravy České Akademie, no. 33, Volume I, (1892) 681–685.
[2] Sur un point de la théorie des fonctions génératrices d'Abel, Acta Mathematica, 27(1903), 339–352.

B. Levi
[1] Sul principio de Dirichlet, Rend. Circ. Math. Palermo, 22(1906), 293–360.

W. Li
[1] Private communication, 1998.

W. Li, P. M. Pardalos, and C. G. Han
[1] Gauss–Seidel method for least-distance problems, J. Optim. Theory Appl., 75(1992), 487–500.

W. Li and J. Swetits
[1] A new algorithm for solving strictly convex quadratic programs, SIAM J. Optim., 7(1997), 595–619.

W. A. Light and E. W. Cheney
[1] **Approximation Theory in Tensor Product Spaces**, Lecture Notes in Mathematics, # 1169, Springer-Verlag, New York, 1985.

G. G. Lorentz, M. von Golitschek, and Y. Makovoz
[1] **Constructive Approximation Advanced Problems**, Springer, New York, 1996.

H. Löwig
[1] Komplexe euklidische Räume von beliebiger endlicher oder unendlicher Dimensionszahl, Acta Sci. Math. Szeged, 7(1934), 1–33.

D. G. Luenberger
[1] **Optimization by Vector Space Methods**, John Wiley & Sons, New York, 1969.

H. Massam
[1] Optimality conditions for a cone-convex programming problem, J. Australian Math. Soc., Ser. A, 27(1979), 141–162.

S. Mazur
[1] Über konvexe Mengen in linearen normierten Räumen, Studia Math., 4(1933), 70–84.

G. Meinardus
[1] **Approximation of Functions: Theory and Numerical Methods**, translated by L. L. Schumaker, Springer-Verlag, New York, 1967.

C. Micchelli, P. Smith, J. Swetits, and J. Ward
[1] Constrained L_p-approximation, Constr. Approx., 1(1985), 93–102.

C. Micchelli and F. Utreras
[1] Smoothing and interpolation in a convex subset of Hilbert space, SIAM J. Sci. Statist. Computing, 9(1988), 728–746.

G. Mittag-Leffler
[1] Sur la représentation analytique des fonctions d'une variable réelle, Rend. Circ. Math. Palermo, 14(1900), 217–224.

E. H. Moore
[1] On the reciprocal of the general algebraic matrix, Bull. Amer. Math. Soc., 26(1920), 394–395.
[2] General Analysis, Memoirs Amer. Philos. Soc., 1(1935), 147–209.

J. J. Moreau
   [1] Décomposition orthogonale dans un espace hilbertien selon deux cônes mutuellement
       polaires, C. R. Acad. Sci., Paris, 255(1962), 238–240.

T. S. Motzkin
   [1] Sur quelques propriétés caractéristiques des ensembles convexes, Rend. Acad. dei Lincei
       (Roma), 21(1935), 562–567.
   [2] Sur quelques propriétés caractéristiques des ensembles bornés non convexes, Rend. Acad.
       dei Lincei (Roma), 21(1935a), 773–779.

B. Mulansky and M. Neamtu
   [1] Interpolation and approximation from convex sets, J. Approx. Theory, 92(1998), 82–100.

C. Müntz
   [1] Über den Approximationssatz von Weierstrass, H. A. Schwarz Festschrift,
       Mathematische Abhandlungen, Berlin, 1914, 303–312.

F. J. Murray and J. von Neumann
   [1] On rings of operators. I, Ann. Math., 37(1936), 116–229.

H. Nakano
   [1] Spectral theory in the Hilbert space, Japan Soc. Promotion Sc., 1953, Tokyo.

T. D. Narang
   [1] Convexity of Chebyshev sets, Nieuw Archief Voor Wiskunde, 25(1977), 377–402.

M. Z. Nashed
   [1] A decomposition relative to convex sets, Proc. Amer. Math. Soc., 19(1968), 782–786.
   [2] **Generalized Inverses and Applications**, Acad. Press, New York, 1976.

M. Z. Nashed and G. F. Votruba
   [1] A unified approach to generalized inverses of linear operators: I. Algebraic, topological
       and projectional properties, Bull. Amer. Math. Soc., 80(1974), 825–830.
   [2] A unified approach to generalized inverses of linear operators: II. Extremal and proximal
       properties, Bull. Amer. Math. Soc., 80(1974), 831–835.

J. von Neumann
   [1] Allegemeine Eigenwerttheorie Hermitescher Funktionaloperatoren, Math. Annalen, 102
       (1929a), 49–131. (In collected works: Vol. II, pp. 3–85.)
   [2] Sur Algebra der Funktionaloperatoren und Theorie der normalen Operatoren, Mat.
       Annalen, 102(1929b), 370–427. (In collected works: Vol. II, pp. 86–143.)
   [3] Über adjungierte Funktionaloperatoren, Ann. Math., 33(1932), 294–310. (In collected
       works: Vol. II, pp. 242–258.)
   [4] **Functional Operators-Vol. II. The Geometry of Orthogonal Spaces**, Annals
       of Math. Studies #22, Princeton University Press, Princeton, NJ, 1950. [This is a reprint
       of mimeographed lecture notes first distributed in 1933.]
   [5] Collected Works, Vol. II, Operators, Ergodic Theory and Almost Periodic Functions in
       a Group, Permagon Press, New York, 1961.

L. Nirenerg
   [1] **Functional Analysis**, lecture notes by L. Sibner, New York University, 1961.

A. Ocneanu
   [1] personal correspondence, March, 1982.

W. Osgood
   [1] Non uniform convergence and the integration of series term by term, Amer. Jour. Math.,
       19(1897), 155–190.

G. Pantelidis
   [1] Konvergente Iterationsverfahren für flach konvexe Banachräume, Rhein.-Westf. Institut
       für Instrumentelle Mathematik Bonn (IIM), Serie A, 23(1968), 4–8.

S. Paszkowski
[1] On approximation with nodes, Rozprawy Mathematyczne XIV, Warszawa, 1957.

R. Penrose
[1] A generalized inverse for matrices, Proc. Cambridge Philos. Soc., 51(1955), 406–413.

R. R. Phelps
[1] Convex sets and nearest points, Proc. Amer. Math. Soc., 8(1957), 790–797.

E. Picard
[1] Sur la représentation approchée des fonctions, Comptes Rendus, 112(1891), 183–186.

A. Pinkus
[1] Weierstrass and approximation theory, J. Approx. Theory (to appear).

M. J. D. Powell
[1] A new algorithm for unconstrained optimization, in **Nonlinear Programming** (edited by J. B. Rosen, O. L. Mangasarian, and K. Ritter), Academic Press, New York, 1970.
[2] **Approximation Theory and Methods**, Cambridge Univ. Press, Cambridge, 1981.

J. B. Prolla
[1] Weierstrass–Stone, the Theorem, Vol. 5 in the series "Approximation & Optimization," Peter Lang, Frankfurt, 1993.

R. Rado
[1] Note on generalized inverses of matrices, Proc. Cambridge Philos. Soc., 52(1956), 600–601.

T. J. Ransford
[1] A short elementary proof of the Bishop–Stone–Weierstrass theorem, Math. Proc. Cambridge Philos. Soc., 96(1984), 309–311.

C. R. Rao
[1] A note on a generalized inverse of a matrix with applications to problems in mathematical statistics, J. Royal Statist. Soc. Ser. B., 24(1962), 152–158.

C. R. Rao and S. K. Mitra
[1] **Generalized Inverse of Matrices and Its Applications**, John Wiley & Sons, New York, 1971.

W. T. Reid
[1] A simple optimal control problem involving approximation by monotone functions, J. Optim. Theory Appl., 2(1968), 365–377.
[2] Generalized inverses of differential and intergral operators, in **Proceedings of the Symposium on Theory and Applications of Generalized Inverses of Matrices** (edited by T. L. Boullion and P. .L. Odell), Texas Tech Press, Lubbock, Texas, 1968, pp. 1–25.

F. Rellich
[1] Spektraltheorie in nichtseparabeln Räumen, Math. Ann., 110(1935), 342–356.

J. R. Rice
[1] **The Approximation of Functions, Volume 1 Linear Theory**, Addison-Wesley, Reading, MA, 1964.
[2] **The Approximation of Functions, Volume 2 Nonlinear and Multivariate Theory**, Addison-Wesley, Reading, MA, 1969.

F. Riesz
[1] Sur une espèce de géométrie analytiques des systèmes de fonctions sommables, C. R. Acad. Sci. Paris, 144(1907), 1409–1411.
[2] Untersuchungen über Systeme integrierbarer Funktionen, Math. Annalen 69(1910), 449–497.

[3] **Les Systèmes d'Équations Linéaires à une Infinité d'Inconnues**, Gauthier-Villars, Paris, 1913.

[4] Über lineare Funktionalgleichungen, Acta Math., 41(1918), 71–98.

[5] Zur Theorie des Hilbertschen Raumes, Acta Sci. Math. Szeged, 7(1934), 34–38.

F. Riesz and B. Sz.-Nagy

[1] **Functional Analysis**, translated by L. F. Boron, Ungar Publ. Co., New York, 1955.

T. J. Rivlin

[1] **An Introduction to the Approximation of Functions**, Blaisdell, Waltham, MA, 1969.

R. T. Rockafellar

[1] Duality and stability in extremum problems involving convex functions, Pacific J. Math., 21(1967), 167–187.

[2] **Convex Analysis**, Princeton Univ. Press, Princeton, NJ, 1970.

C. Runge

[1] Zur Theorie der eindeutigen analytischen Funktionen, Acta Mathematica, 6(1885), 229–245.

[2] Über die Darstellung willkürlicher Functionen, Acta Mathematica, 7(1885a), 387–392.

A. Sard

[1] **Linear Approximation**, Math. Surveys, No. 9, Amer. Math. Soc., Providence, 1963.

J. Schauder

[1] Über die Umkehrung linearer, stetiger Funktionaloperationen, Studia Math., 2(1930), 1–6.

E. Schmidt

[1] Über die Auflösung linearer Gleichungen mit unendlich vielen Unbekannten, Rend. Circ. mat. Palermo, 25(1908), 53–77.

A. Schönhage

[1] **Approximationstheorie**, W. de Gruyter & Co., New York, 1971.

H. A. Schwarz

[1] **Gesammelte mathematische Abhandlungen**, 2 vols., Springer, Berlin, 1890.

H. S. Shapiro

[1] **Topics in Approximation Theory**, Lecture Notes in Math., # 187, Springer-Verlag, New York, 1971.

C. L. Siegel

[1] Über die analytische Theorie der quadratischen Formen III, Ann. Math., 38(1937), 212–291.

I. Singer

[1] Remarque sur un théorème d'approximation de H. Yamabe, Atti Accad. Naz. Lincei Rend. Cl. Sci. Fis. Mat. Nat., 26(1959), 33–34.

[2]] **Best Approximation in Normed Linear Spaces by Elements of Linear Subspaces**, Springer-Verlag, New York, 1970.

[3] **The Theory of Best Approximation and Functional Analysis**, CBMS #13, SIAM, Philadelphia, 1974.

R. Smarzewski

[1] Iterative recovering of orthogonal projections, preprint, 1986.

K. T. Smith, D. C. Solmon, and S. L. Wagner

[1] Practical and mathematical aspects of the problem of reconstructing objects from radiographs, Bull. Amer. Math. Soc., 83(1977), 1227–1270.

P. W. Smith and J. D. Ward
  [1] Distinguished solutions to an L_∞ minimization problem, Approx. Theory Appl., 4(1988), 29–40.

P. W. Smith and H. Wolkowicz
  [1] A nonlinear equation for linear programming, Math. Programming, 34(1986), 235–238.

G. A. Soukhomlinoff
  [1] Über Fortsetzung von linearen Funktionalen in linearen komplexen Räumen und linearen Quaternionräumen, Mat. Sbornik N.S., 45(1938), 353–358.

J. F. Steffensen
  [1] Interpolation, Second ed., Chelsea Publ. Co., New York, 1950.

E. Steinitz
  [1] Bedingt konvergente Reihen und konvexe Systeme I, J. Reine Angew. Math., 143(1913), 128–175.
  [2] Bedingt konvergente Reihen und konvexe Systeme II, J. Reine Angew. Math., 144(1914), 1–40.
  [3] Bedingt konvergente Reihen und konvexe Systeme III, J. Reine Angew. Math., 146(1916), 1–52.

W. J. Stiles
  [1] Closest point maps and their products, Nieuw Archief voor Wiskunde (3), XIII (1965a), 19–29.
  [2] Closest point maps and their products II, Nieuw Archief voor Wiskunde (3), XIII (1965b), 212–225.
  [3] A solution to Hirschfeld's problem, Nieuw Archief voor Wiskunde (3), XIII (1965c), 116–119.

M. H. Stone
  [1] Applications of the theory of boolean rings to general topology, Trans. Amer. Math. Soc., 41(1937), 375–481.
  [2] The generalized Weierstrass approximation theorem, Mathematics Magazine, 21(1948), 167–183, 237–254.

J. J. Swetits, S. E. Weinstein, and Y. Xu
  [1] Approximation in $L_p[0, 1]$ by n-convex functions, Numer. Funct. Anal. Optimiz., 11(1990), 167–179.

J. Szabados and P. Vértesi
  [1] Interpolation of Functions, World Scientific Publ., New Jersey, 1990.

O. Szász
  [1] Über die Approximation stetiger Funktionen durch lineare Aggregate von Potenzen, Math. Annal., 77(1916), 482–496.

I. G. Tsar'kov
  [1] Bounded Chebyshev sets in finite-dimensional Banach spaces, Math. Notes, 36(1)(1984), 530–537.

Y. Y. Tseng
  [1] The Characteristic Value Problem of Hermitian Functional Operators in a Non-Hilbert Space, doctoral dissertation, Univ. Chicago, 1933.
  [2] Sur les solutions des équations operatrices fonctionnelles entre les espaces unitaires. Solutions extremales. Solutions virtuelles., C. R. Acad. Sci. Paris, 228(1949a), 640–641.
  [3] Generalized inverses of unbounded operators between two unitary spaces, Dokl. Akad. Nauk SSSR (N.S.), 67(1949b), 431–434.
  [4] Properties and classification of generalized inverses of closed operators, Dokl. Akad. Nauk SSSR (N.S.), 67(1949c), 607–610.
  [5] Virtual solutions and general inversions, Uspehi. Mat. Nauk (N.S.), 11(1956), 213–215.

F. A. Valentine
[1] **Convex Sets**, McGraw-Hill, New York, 1964.

C. J. de la Vallée Poussin
[1] Sur l'approximation des fonctions d'une variable réelle et leurs dérivées par de polynomes et des suites limitées de Fourier, Bulletin de l'Académie Royal de Belgique, 3(1908), 193–254.

L. P. Vlasov
[1] Čebyšev sets in Banach spaces, Sov. Math. Dokl., 2(1961), 1373–1374.
[2] Chebyshev sets and approximatively convex sets, Math. Notes, 2(1967), 600–605.
[3] On Čebyšev sets, Sov. Math. Dokl., 8(1967a), 401–404.
[4] Approximative properties of sets in normed linear spaces, Russian Math. Surveys, 28(1973), 1–66.

J. L. Walsh
[1] **Interpolation and Approximation by Rational Functions in the Complex Domain**, Amer. Math. Soc. Colloquium Publications, vol. 20, Providence, RI, 1956.

J. Ward
[1] Some constrained approximation problems, in **Approximation Theory V** (edited by C. K. Chui, L. L. Schumaker, and J. D. Ward), Academic Press, New York, 1986, 211–229.

G. A. Watson
[1] **Approximation Theory and Numerical Methods**, Wiley, New York, 1980.

H. Weyl
[1] **Raum · Zeit · Materie**, Springer-Verlag, Berlin, 1923.

K. Weierstrass
[1] Über die analytische darstellbarkeit sogenannter willkürlicher Funktionen einer reelen Veraenderlichen, Sitzungsberichte der Akademie zu Berlin, 1885, 633–639 and 789–805.

N. Wiener
[1] On the factorization of matrices, Comment. Math. Helv., 29(1955), 97–111.

S. Xu
[1] Estimation of the convergence rate of Dykstra's cyclic projections algorithm in polyhedral case, Acta Math. Appl. Sinica, 16 (2000), 217–220.

W. Wolibner
[1] Sur un polynôme d'interpolation, Colloq. Math., 2(1951), 136–137.

H. Yamabe
[1] On an extension of the Helly's theorem, Osaka Math. J., 2(1950), 15–17.

N. Young
[1] **An Introduction to Hilbert Space**, Cambridge Univ. Press, Cambridge, 1988.

E. H. Zarantonello
[1] Projections on convex sets in Hilbert space and spectral theory. Part I. Projections on convex sets. Part II. Spectral theory, in **Contributions to Nonlinear Functional Analysis** (edited by E. H. Zarantonello), Academic Press, New York, 1971, pp. 237–424.

INDEX